TUTORIALS ON EMERGING METHODOLOGIES AND APPLICATIONS IN OPERATIONS RESEARCH

Recent titles in the

INTERNATIONAL SERIES IN OPERATIONS RESEARCH & MANAGEMENT SCIENCE

Frederick S. Hillier, Series Editor, *Stanford University*

Zhu, J. / *QUANTITATIVE MODELS FOR PERFORMANCE EVALUATION AND BENCHMARKING*
Ehrgott, M. & Gandibleux, X. / *MULTIPLE CRITERIA OPTIMIZATION: State of the Art Annotated Bibliographical Surveys*
Bienstock, D. / *Potential Function Methods for Approx. Solving Linear Programming Problems*
Matsatsinis, N.F. & Siskos, Y. / *INTELLIGENT SUPPORT SYSTEMS FOR MARKETING DECISIONS*
Alpern, S. & Gal, S. / *THE THEORY OF SEARCH GAMES AND RENDEZVOUS*
Hall, R.W./*HANDBOOK OF TRANSPORTATION SCIENCE - 2nd Ed.*
Glover, F. & Kochenberger, G.A. / *HANDBOOK OF METAHEURISTICS*
Graves, S.B. & Ringuest, J.L. / *MODELS AND METHODS FOR PROJECT SELECTION: Concepts from Management Science, Finance and Information Technology*
Hassin, R. & Haviv, M./ *TO QUEUE OR NOT TO QUEUE: Equilibrium Behavior in Queueing Systems*
Gershwin, S.B. et al/ *ANALYSIS & MODELING OF MANUFACTURING SYSTEMS*
Maros, I./ *COMPUTATIONAL TECHNIQUES OF THE SIMPLEX METHOD*
Harrison, T., Lee, H. & Neale, J./ *THE PRACTICE OF SUPPLY CHAIN MANAGEMENT: Where Theory And Application Converge*
Shanthikumar, J.G., Yao, D. & Zijm, W.H./ *STOCHASTIC MODELING AND OPTIMIZATION OF MANUFACTURING SYSTEMS AND SUPPLY CHAINS*
Nabrzyski, J., Schopf, J.M., Węglarz, J./ *GRID RESOURCE MANAGEMENT: State of the Art and Future Trends*
Thissen, W.A.H. & Herder, P.M./ *CRITICAL INFRASTRUCTURES: State of the Art in Research and Application*
Carlsson, C., Fedrizzi, M., & Fullér, R./ *FUZZY LOGIC IN MANAGEMENT*
Soyer, R., Mazzuchi, T.A., & Singpurwalla, N.D./ *MATHEMATICAL RELIABILITY: An Expository Perspective*
Chakravarty, A.K. & Eliashberg, J./ *MANAGING BUSINESS INTERFACES: Marketing, Engineering, and Manufacturing Perspectives*
Talluri, K. & van Ryzin, G./ *THE THEORY AND PRACTICE OF REVENUE MANAGEMENT*
Kavadias, S. & Loch, C.H./*PROJECT SELECTION UNDER UNCERTAINTY: Dynamically Allocating Resources to Maximize Value*
Brandeau, M.L., Sainfort, F., Pierskalla, W.P./ *OPERATIONS RESEARCH AND HEALTH CARE: A Handbook of Methods and Applications*
Cooper, W.W., Seiford, L.M., Zhu, J./ *HANDBOOK OF DATA ENVELOPMENT ANALYSIS: Models and Methods*
Luenberger, D.G./ *LINEAR AND NONLINEAR PROGRAMMING, 2nd Ed.*
Sherbrooke, C.C./ *OPTIMAL INVENTORY MODELING OF SYSTEMS: Multi-Echelon Techniques, Second Edition*
Chu, S.-C., Leung, L.C., Hui, Y. V., Cheung, W./ *4th PARTY CYBER LOGISTICS FOR AIR CARGO*
Simchi-Levi, Wu, Shen/ *HANDBOOK OF QUANTITATIVE SUPPLY CHAIN ANALYSIS: Modeling in the E-Business Era*
Gass, S.I. & Assad, A.A./ *AN ANNOTATED TIMELINE OF OPERATIONS RESEARCH: An Informal History*

***** A list of the early publications in the series is at the end of the book *****

TUTORIALS ON EMERGING METHODOLOGIES AND APPLICATIONS IN OPERATIONS RESEARCH

Presented at INFORMS 2004, Denver, CO

Edited by

HARVEY J. GREENBERG

Mathematics Department

University of Colorado at Denver

Harvey J. Greenberg
Dept. of Mathematics
University of Colorado
Denver, CO 80217-3364
U.S.A

Mathematics Subject Classification (2004): 26024, 1400X, 23067

Library of Congress Cataloging-in-Publication Data

A C.I.P Catalogue record for this book is available from the
Library of Congress.

ISBN (HB) 0-387-22826-8 / e-ISBN 0-387-22827-6 Printed on acid-free paper.

Printed in the United States of America.

9 8 7 6 5 4 3 2 1 SPIN 11053019

springeronline.com

This book is dedicated to the memory of Carl M. Harris, a pioneer in operations research with great vision and perseverance.

Contents

Contributing Authors

David A. Bader is Associate Professor and Regents' Lecturer of the Departments of Electrical & Computer Engineering and Computer Science, University of New Mexico, since 1996. He received the NSF CAREER award for *High-Performance Algorithms for Scientific Applications*, for 2001–06. He is Chair, IEEE Computer Society Technical Committee on Parallel Processing. David is a member of the Founding Editorial Board of *The International Journal of High Performance Computing and Networking (IJHPCN)* and Associate Editor for *The ACM Journal of Experimental Algorithmics*. `dbader@eece.unm.edu`

Robert D. Carr is a Senior Member of the Technical Staff in the Algorithms and Discrete Mathematics Department of Sandia National Laboratories, Albuquerque since 1997. After receiving a Ph.D. in Mathematics from Carnegie Mellon in 1995, he held a post-doctoral position at the University of Ottawa and the ACO Post-doctoral fellowship at Carnegie Mellon. As Adjunct Faculty of the Computer Science Department at the University of New Mexico in 2001, he taught a course in polyhedral combinatorics. The notes from that course became part of this book's chapter. `bobcarr@cs.sandia.gov`

John W. Chinneck is the Associate Dean (Research) for the Faculty of Engineering and Design, and Associate Chair of the Department of Systems and Computer Engineering, at Carleton University. He has developed computer tools that assist in the formulation and analysis of very large and complex optimization models. His algorithms for the analysis of infeasible linear programs have been incorporated in many commercial linear programming solvers, including LINDO, CPLEX, XPRESS-MP, Frontline Systems Excel solvers, and OSL. `chinneck@sce.carleton.ca`

Ioannis Gamvros is a doctoral student at the University of Maryland. His research interests are in Telecommunications Network Design. Currently, he is working on solving network design and revenue management problems that arise in satellite communication systems. `igamvros@rhsmith.umd.edu`

Bruce Golden is the France-Merrick Chair in Management Science in the Robert H. Smith School of Business at the University of Maryland. His research interests include heuristic search, combinatorial optimization, and applied OR. He has received numerous awards, including the Thomas L. Saaty Prize (1994), the University of Maryland Distinguished Scholar-Teacher Award (2000), and the INFORMS Award for the Teaching of OR/MS Practice (2003). He is currently Editor-in-Chief of *Networks* and was previously Editor-in-Chief of the *INFORMS Journal on Computing*. `bgolden@rhsmith.umd.edu`

William E. Hart is Principle Member of the Algorithms and Discrete Mathematics Department, Sandia National Laboratories, Albuquerque. After earning his Ph.D. in Computer Science from the University of California at San Diego in 1994, Bill joined SNL as an AMS Fellow and received their Award for Excellence in 1995, 1999, and 2001. In 1997, Bill received the *Best Paper by a Young Scientist* from RECOMB. He has performed algorithmic research on a variety of global and combinatorial optimization methods, and he has been a primary architect of the SNL Parallel Integer and Combinatorial Optimization system (PICO). `wehart@sandia.gov`

Pascal Van Hentenryck is Professor of Computer Science at Brown University since 1990. He is the main designer and implementor of the CHIP system (the foundation of all modern constraint programming systems), the OPL optimization programming language for constraint and mathematical programming, and the Numerica system for global optimization. These three systems are all described in books published by the MIT Press. `pvh@cs.brown.edu`

Allen Holder is Assistant Professor of Mathematics, Trinity University; Adjunct Professor of Radiation Oncology, University of Texas Health Science Center at San Antonio; and, Research Assistant, Hearin Center for Enterprise Science, The University of Mississippi. He received his Ph.D. in Applied Mathematics from The University of Colorado at Denver, 1998. In 2000, Allen received the INFORMS *William Pierskalla Award in Health Applications*. He currently supervises undergraduate research under an NSF REU. `aholder@trinity.edu`

Goran Konjevod is Assistant Professor of Computer Science and Engineering at Arizona State University since 2000. He is also a consultant for Los Alamos National Laboratory. He has designed and analyzed algorithms for discrete problems in operations research. `goran@asu.edu`

Laurent Michel is Assistant Professor of Computer Science & Engineering at University of Connecticut since 2002. His research interests focus on the design and implementation of domain-specific languages for combinatorial optimization, but they extend to programming languages, constraint programming, artificial intelligence, non-linear programming, and reasoning under uncertainty. He co-developed Newton, Numerica, The Optimization Programming Language (OPL), the constrained-based library Modeler++, and the local search tools Localizer, Localizer++ and Comet. `ldm@engr.uconn.edu`

Michel S. Nakhla is Chancellor's Professor in the Department of Electronics at Carleton University. He arrived in 1988 as holder of the Computer-Aided Engineering Senior Industrial Chair, established by Bell-Northern Research and the Natural Sciences and Engineering Research Council of Canada, and he founded Carleton's high-speed CAD research group. He is Associate Editor of the *IEEE Transactions on Circuits and Systems — Fundamental Theory and Applications* and of the *Circuits, Systems and Signal Processing Journal.* msn@doe.carleton.ca

Cynthia A. Phillips is a Distinguished Member of the Technical Staff in the Algorithms and Discrete Math Department at Sandia National Laboratories. She joined the Labs right after receiving her Ph.D. in computer science from MIT in 1990. She received a Lockheed Martin/Sandia National Laboratories Employee Recognition Award for individual technical excellence in 2000, SNL Awards for Excellence in 1993, 1998, 2002, and 2003, and was the YWCA (Central Rio Grande) *Woman on the Move in Science and Technology* in 2000. Cindy is one of three main developers of the SNL PICO massively-parallel integer programming code. caphill@sandia.gov

S. Raghavan is Assistant Professor of Management Science at the Robert H. Smith School of Business, University of Maryland. In 1996 he received the *George B. Dantzig Dissertation Award* from INFORMS for his thesis, "Formulations and Algorithms for Network Design Problems with Connectivity Requirements." He serves as Associate Editor of *Networks*, and he has recently co-edited a book, *Telecommunications Network Design and Management.* raghavan@umd.edu

Bill Salter is Assistant Professor of Radiation Oncology and Associate Director of Medical Physics at the Cancer Therapy and Research Center (CTRC) at The University of Texas Health Science Center, San Antonio. He is the inventor of technology used to enhance the delivery capabilities of the Nomos Peacock IMRT system for delivery of Intensity Modulated RadioSurgical (IMRS) treatments. bsalter@ctrc.net

Daliborka Stanojević is pursuing a Ph.D. in Management Science at the University of Maryland. Her primary research interests are in Telecommunications Network Design. Most recently she has been working on developing optimization techniques for designing optical communication networks.
dstanoje@rhsmith.umd.edu

Robert J. Vanderbei is Professor of Operations Research and Financial Engineering at Princeton University. He is also an Associated Faculty Member, Departments of Mathematics and Computer Science, and the Program in Applied and Computational Mathematics. He is the Director of the Program in Engineering and Management Systems. His 1997 book, *Linear Programming:*

Foundations and Extensions has been used extensively, supplemented by the wealth of materials on his web site. rvdb@princeton.edu

Stefan Voß is Professor and Director of the Institute of Information Systems at the University of Hamburg. Previous positions include Professor and Head of the Department of Business Administration, Information Systems and Information Management at the University of Technology Braunschweig. Stefan is Associate Editor of *INFORMS Journal on Computing* and Area Editor of *Journal of Heuristics*. He is co-author of the 2003 book, *Introduction to Computational Optimization Models for Production Planning in a Supply Chain*. stefan.voss@uni-hamburg.de

David L. Woodruff is Professor of Management at University of California, Davis. He was Chair of the INFORMS Computing Society, and serves on several editorial boards: *INFORMS Journal On Computing*, *Journal Of Heuristics* (Methodology Area Editor), *Production and Operations Management*, and *International Journal of Production Research*. He is co-author of the 2003 book, *Introduction to Computational Optimization Models for Production Planning in a Supply Chain*. dlwoodruff@ucdavis.edu

Q.J. Zhang is a Professor in the Electronics Department, Carleton University. His research interests are neural network and optimization methods for high-speed/high-frequency electronic design, and has over 160 papers in the area. He is co-author of *Neural Networks for RF and Microwave Design*, co-editor of *Modeling and Simulation of High-Speed VLSI Interconnects*, a contributor to *Analog Methods for Computer-Aided Analysis and Diagnosis*, and twice a Guest Editor for the Special Issues on Applications of Artificial neural networks to RF and Microwave Design for the *International Journal of RF and Microwave Computer-Aided Engineering*. qjz@doe.carleton.ca

Preface

This volume reflects the theme of the INFORMS 2004 Meeting in Denver: *Back to OR Roots.* Emerging as a quantitative approach to problem-solving in World War II, our founders were physicists, mathematicians, and engineers who quickly found peace-time uses. It is fair to say that Operations Research (OR) was born in the same incubator as computer science, and it has spawned many new disciplines, such as systems engineering, health care management, and transportation science. Although people from many disciplines routinely use OR methods, many scientific researchers, engineers, and others do not understand basic OR tools and how they can help them.

Disciplines ranging from finance to bioengineering are the beneficiaries of what we do — we take an interdisciplinary approach to problem-solving. Our strengths are modeling, analysis, and algorithm design. We provide a quantitative foundation for a broad spectrum of problems, from economics to medicine, from environmental control to sports, from e-commerce to computational geometry. We are both producers and consumers because the mainstream of OR is in the interfaces.

As part of this effort to recognize and extend OR roots in future problem-solving, we organized a set of tutorials designed for people who heard of the topic and want to decide whether to learn it. The 90 minutes was spent addressing the questions:

- What is this about, in a nutshell?
- Why is it important?
- Where can I learn more?

In total, we had 14 tutorials, and eight of them are published here.

Chapter 1, *Heuristic Search for Network Design*, by Ioannis Gamvros, Bruce Golden, S. Raghavan, and Daliborka Stanojević, is a great meeting of OR roots with modern solution techniques. Network design is an early OR problem, but still too complex for exact solution methods, like integer programming. The success of heuristic search, as advanced through many OR applications, makes it a natural approach. This chapter considers simple heuristics first (a piece of good advice to problem solvers), then takes a hard look at local search and a variety of metaheuristics, such as GRASP, simulated annealing, tabu search, and genetic algorithms. They proceed to describe a series of highly effective genetic algorithms, giving insight into how the algorithm controls are developed for various network design problems.

Chapter 2, *Polyhedral Combinatorics*, by Robert D. Carr and Goran Konjevod, draws from some early mathematical foundations, but with a much deeper importance for solving hard combinatorial optimization problems. They carefully describe problem formulations and impart an intuition about how to recognize better ones, and why they are better. This chapter also brings together concepts of modeling with a view of algorithm design, such as understanding precisely the separation problem and its relation with duality.

Chapter 3, *Constraint Languages for Combinatorial Optimization*, by Pascal Van Hentenryck and Laurent Michel, brings together OR and AI vantages of combinatorial optimization. Traditionally, OR has focused on methods and hence produced very good algorithms, whereas AI has focused on models and hence produced some languages of immediate value to OR problem-solving. Constraint-based languages comprise a class, and these authors have been major contributors to this marriage. This chapter illustrates how constraint-based languages provide a powerful and natural way to represent traditional OR problems, like scheduling and sequencing.

Chapter 4, *Radiation Oncology and Optimization*, by Allen Holder and Bill Salter, reflects on how directly OR applies to science. This particular problem in radiation therapy was modeled with linear programming in 1968, but this was not extended until 1990. It is now a core problem in medical physics being addressed by modern OR techniques, aimed at giving physicians accurate information and helping to automate radiation therapy. Further, the authors' interdisciplinary collaboration personifies an element of OR roots.

Chapter 5, *Parallel Algorithm Design for Branch and Bound*, by David A. Bader, William E. Hart, and Cynthia A. Phillips, draws from the authors' vast experiences, and includes a section on how to debug an algorithm implemented on a parallel architecture. As computers become more powerful, we expand our scope of problem domains to keep the challenges ongoing. Parallelism is a natural way to meet this challenge. OR has always been at the forefront of computer science, not by coincidence, but by design. In the 1950's, linear programming was used to debug hardware and software of new computers because it challenged every aspect of symbolic and numerical computing. This chapter reflects a modern version of this, with combinatorial optimization posing computational challenges, and the OR/CS interfaces continue to flourish.

Chapter 6, *Computer-Aided Design for Electrical and Computer Engineering*, by John W. Chinneck, Michel S. Nakhla, and Q.J. Zhang, exemplifies the application of OR techniques to an area of engineering whose problems challenge our state-of-the-art due to size and complexity. This chapter achieves the daunting task of addressing the two, disjoint communities, so that not only will OR people learn about these CAD problems, but also the electrical and computer engineering community will learn more about the applicability of OR techniques. After carefully developing a mixed-integer, nonlinear programming model of physical design, this tutorial goes further into circuit optimization and routing.

Chapter 7, *Nonlinear Programming and Engineering Applications*, by Robert J. Vanderbei, is a perfect example of drawing from OR roots to solve engineering problems. After describing LOQO, the author's interior point code for nonlinear programming, this chapter describes its application to Finite Impulse Response (FIR) filter design, telescope design — optics, telescope design — truss structure, and computing stable orbits for the n-body problem. Many of the early engineering applications used nonlinear programming, and this chapter shows how his interior point algorithm can push that envelope.

Chapter 8, *Connecting mrp, MRP II and ERP Supply Chain Planning via Optimization Models*, by Stefan Voß and David L. Woodruff, is concerned with a problem that is dubbed in modern parlance "supply chain planning." They point out that this stems from the OR roots in *materials requirements planning* and proceed to develop a family of mathematical programming models. This chapter presents the model development step-by-step, moving toward greater flexibility and scope, such as incorporating substitutions at various levels of the supply chain and identifying bottlenecks.

Collectively, these chapters show how OR continues to apply to business, engineering, and science. The techniques are mainstream OR, updated by recent advances.

HARVEY J. GREENBERG

Foreword

To keep a lamp burning we have to keep putting oil in it.
– Mother Teresa

The field of operations research has a glorious history, distinguished by impressive foundations steeped in deep theory and modeling and an equally impressive range of successful applications. However, to remain healthy and vigorous, any field must continually renew itself, embracing new methodologies and new applications.

With the Denver meeting, INFORMS is both bringing attention to its OR roots and also looking toward the future. The tutorials on emerging technologies and applications in this collection are a part of that effort, not only in reflecting a wide range of theory and applications, but also in making operations research known to a much broader population who might not know about the profession. As an engineer, I am delighted that this book is designed to bring attention to the profession's roots and to an array of analytic methods and intellectual ideas that have had, and will continue to have, considerable impact on practice. The editor is to be congratulated for assembling a first-rate cast of authors with impressive experience in developing OR methods and applying OR to engineering, science, and technology.

Through the impressive range of topics covered in this volume, readers will learn directly about the breadth of operations research, and also find outstanding guides to the literature. In particular, one cannot help but see the many ties between operations research and engineering.

In fact, I see something of a renewal these days in the application of operations research to engineering and science, with many notable and visible applications in such fields as medicine, communications, supply chain management, and systems design, as reflected in the pages to follow. But I also see less visible, but equally impressive, imbedded applications, a sort of 'OR Inside,' in contexts such as the design of computer compilers, machine learning, and human memory and learning.

I am delighted that this volume is dedicated to Carl Harris. Carl had an unwavering exuberance for operations research and especially the interplay between theory and practice. He would very much have enjoyed both the content and intended objective of these tutorial papers.

The editor, Harvey Greenberg, who himself is noted for the breadth of his experiences in research and applications, is to be congratulated for assembling such a fine volume and for his efforts to keep the lamp of OR burning by highlighting important developments and, hopefully, fueling new fields of inquiry and application within operations research.

Thomas L. Magnanti
Dean, School of Engineering
Massachusetts Institute of Technology
Cambridge, MA - June 2004

Acknowledgments

First, and foremost, I thank the authors, without whom this book could not have been written. Amidst busy schedules, they worked hard to enable this book to be published before the INFORMS 2004 meeting, and the results are evidently outstanding. I also thank Manuel Laguna, our Meeting Chair, who encouraged this to happen and helped with sorting out some of the early details. Gary Folven, from Kluwer Academic Press, also supported this from the beginning, and he helped with the production process. Last, but not least, it is the memory of visionaries like Carl Harris that inspire books like this one to be produced.

Chapter 1

HEURISTIC SEARCH FOR NETWORK DESIGN

Ioannis Gamvros, Bruce Golden, S. Raghavan, and Daliborka Stanojević
The Robert H. Smith School of Business, University of Maryland
{igamvros,bgolden,rraghava,dstanoje}@rhsmith.umd.edu

Abstract In this chapter, we focus on heuristics for network design problems. Network design problems have many important applications and have been studied in the operations research literature for almost 40 years. Our goal here is to present useful guidelines for the design of intelligent heuristic search methods for this class of problems. Simple heuristics, local search, simulated annealing, GRASP, tabu search, and genetic algorithms are all discussed. We demonstrate the effective application of heuristic search techniques, and in particular genetic algorithms, to four specific network design problems. In addition, we present a selected annotated bibliography of recent applications of heuristic search to network design.

Keywords: Heuristics, Local Search, Network Design, Genetic Algorithms

Introduction

Network design problems arise in a wide variety of application domains. Some examples are telecommunications, logistics and transportation, and supply chain management. While the core network design problems such as the minimum spanning tree problem, and the shortest path problem are well-solved, adding additional (or different) restrictions on the network design frequently results in an $\mathcal{NP}$-complete problem. Coincidentally, most network design problems that arise in practice are $\mathcal{NP}$-complete. When it is difficult to solve these network design problems using an exact approach, we are interested in finding good heuristic solutions to them.

In this chapter, we provide practitioners guidelines for designing heuristic search techniques for network design problems. Our guidelines encompass a wide variety of search techniques starting from simple heuristics, to local search and large-scale neighborhood search, and finally to genetic algorithms. We illustrate these guidelines by sharing our experience in developing and applying these heuristic search techniques to four different network design problems: the minimum labeling spanning tree problem, the Euclidean non-uniform Steiner tree problem, the prize collecting generalized minimum spanning tree problem, and the multi-level capacitated spanning tree problem. Finally, we provide an annotated bibliography to selectively describe some recent research work in this area.

We hope to convey through this chapter some of the key ideas in developing good heuristic search techniques for network design problems. In addition, we illustrate the level of success that one can expect when applying heuristic search to specific problems involving network design.

The rest of this chapter is organized as follows. In Section 1.1 we describe a set of general guidelines for developing heuristic search procedures for network design problems. We then illustrate, in Sections 1.2–1.5, the application of these guidelines on four different network design problems. In Section 1.6 we provide a brief annotated bibliography on some recent and successful applications of heuristic search techniques to network design problems. Finally, in Section 1.7 we provide concluding remarks.

1.1 Guidelines for Network Design Heuristics

In this section we present a set of guidelines to develop heuristic solution approaches for network design problems. Our goal is to inform the reader about popular approaches that have been favored in the literature, discuss their strengths and weaknesses, and provide some notes on how they can be used for related problems. The reader should keep in mind that in almost every case procedures that achieve a high level of performance take advantage of problem-specific structures. However, we believe, that most network design problems share many common characteristics and often the search for an efficient algorithm can follow the same steps and adhere to the same principles regardless of the problem. We will begin by presenting simple heuristic procedures that are usually fairly shortsighted and obtain moderate to poor solutions. We then look at more advanced local search (LS) procedures that rely on the definition of a neighborhood and specify the way in which this neighborhood is explored. In the context of local search we also discuss metaheuristic procedures, such as simulated annealing, GRASP, and tabu search. Finally, we discuss genetic algorithms (GA), which usually require a significant amount of computation time but can find very high quality solutions.

1.1.1 Simple Heuristics

It is always a good idea to experiment with simple heuristics first, before exploring more complex procedures. They usually provide initial feasible solutions for more advanced procedures, act as a benchmark against which other algorithms are compared, and, if nothing else, help better understand the structure of the problem. Moreover, these simple heuristics are usually employed by more complex procedures for solving smaller subproblems efficiently. Conceptually, we can split the different types of heuristics into three different categories depending on how they find solutions.

Construction. The most popular approach is to attempt to find a solution by starting with the problem graph, without any edges, and to construct the solution by adding edges one at a time. Usually this is done by selecting the *best* edge, according to a specified rule, and adding it to the solution only if it does not violate feasibility. The simplest example of this is Kruskal's algorithm [36] for the Minimum Spanning Tree (MST) problem in which we select the cheapest edge in the graph and add it to the solution if it does not create a cycle. Another equally simple and well-known example is Prim's algorithm [46] for the same problem where the rule is to choose the cheapest edge that connects the nodes in the partially constructed tree T with $T \setminus V$ (V is the node set of the graph, and initially T is a random vertex).

Deletion. A less favored approach works in the opposite way. That is, to start with all the edges in the graph and delete the *worst* edge at each iteration if its deletion results in a network that contains a feasible solution. The procedure stops when no edges in the solution can be deleted. Typically, for minimization problems, the edges are selected for deletion by decreasing order of costs.

Savings. A somewhat different approach starts with a very simple solution and attempts to improve that solution by calculating the savings generated when replacing existing edges with new edges while maintaining feasibility. For tree network design problems these approaches usually start with a MST or star solution (generated by connecting all nodes to a given node). For example, the Esau-Williams heuristic [18] for the capacitated minimum spanning tree (CMST) problem and the Clarke and Wright heuristic [9] for the vehicle routing problem (VRP) are based on the savings approach.

The first two approaches are easy to understand and implement and usually have very good running times (depending on the feasibility check) but typically result in very poor solutions, especially for more complex problems. The savings approach can give slightly better solutions and has been used extensively in practice. It is advantageous to look into all three types of simple heuristics

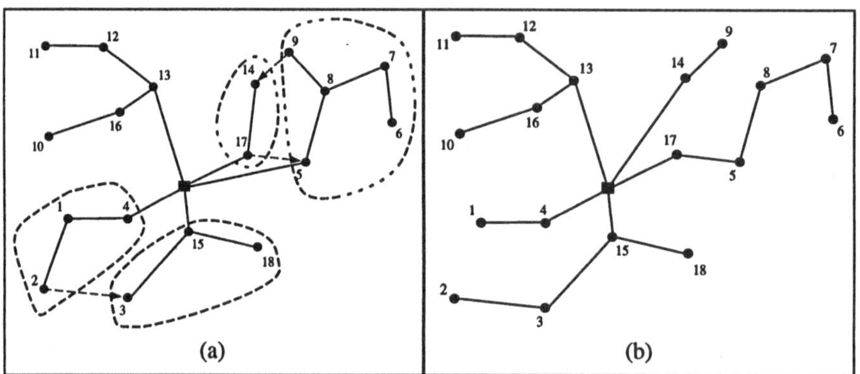

Figure 1.1. Two-exchange example for the CMST problem. (a) Original solution with the proposed exchanges. (b) Final solution after the exchanges.

since their performance is problem dependent (i.e., some of them may work better in some cases than others).

1.1.2 Local Search

All of the above heuristics, although fast and easy to implement, have the significant drawback of sometimes generating poor quality solutions (especially for large problems). One approach that is widely used to improve on the simple heuristic solutions uses the notion of a neighborhood of feasible solutions and defines a search procedure to find a more promising solution from the neighborhood. In each iteration of a LS algorithm, the neighborhood of the current solution is explored for a suitable alternative solution. If one is found, then it replaces the current solution and the procedure starts over. If none is found, then the procedure terminates. In the case of LS, both the neighborhood structure and the search algorithm play a significant role in the success of the procedure and have to be designed carefully.

Neighborhood. Simple neighborhood structures are usually defined through exchanges between nodes or edges and are evaluated by the savings generated much like the ones used in the savings heuristic we discussed earlier. A popular way of generating neighboring solutions is to first partition the node set in the graph into disjoint sets of nodes and then define exchanges between the different sets. This approach relies on the fact that solving the problem on the partition sets is usually much easier than solving it over the entire set of nodes. An example of a neighborhood used in this context is the so-called node-based, two-exchange neighborhood (e.g., see Amberg et al. [4]) in which we perform a shift move for a single node (i.e., the node moves from its current set to a new one) or an exchange move between two nodes (i.e., the two nodes

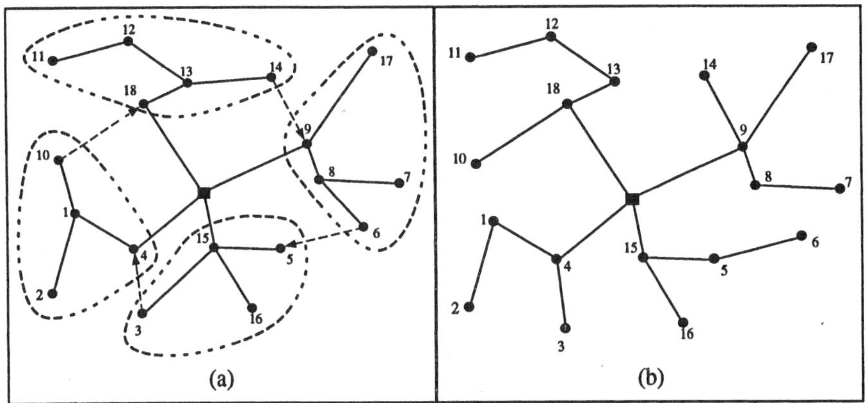

Figure 1.2. Multi-exchange example for the CMST. (a) Original solution with proposed cyclic exchange. (b) Final solution after exchanges.

exchange sets). In Figure 1.1 we give an example of a shift and an exchange move and the resulting tree. In the figure, the sets that participate in the move are highlighted with dashed ellipses while a dashed arc (i, j) indicates that node i moves from its set to node j's set. A similar neighborhood is the tree-based, two-exchange neighborhood (see Shariaha et al.[49]) in which multiple nodes (as opposed to single nodes) shift from one set to another. More involved, multi-exchange neighborhoods can be defined in the same way (see Ahuja et al. [2]). In these neighborhoods, many sets (not just two) participate in the exchange either by contributing a single node or multiple nodes. In Figure 1.2, we present a simple multi-exchange move and the resulting tree. These approaches (i.e., two-exchange, multi-exchange) have been extensively used for problems like the CMST where the nodes in the different subtrees are considered to belong to the different sets of the partition. We note that, in the CMST, after determining the assignment of nodes into sets, the problem of interconnecting the nodes in the same set is a simple MST problem.

Another way of generating neighborhoods is by changing the placement, number, or type of the edges for a given solution. A simple example of this is the 2-opt neighborhood for the traveling salesman problem (TSP) in which we delete edges (x, y) and (z, w) from the current solution and add edges (x, z) and (y, w). The 3-opt exchange that has also been used for the TSP works in a similar way. Three edges are deleted from the current solution and their endpoints are then connected in a different way that guarantees a feasible tour. Figure 1.3 presents an example of a 2-opt and a 3-opt exchange for a TSP instance. A few indicative examples of the successful application of these ideas are the 2-opt neighborhood that was used on a survivable network design problem by Monma and Shallcross [39] and Park and Han [42], and a k-opt neighborhood for a network loading problem by Gendron et al. [26].

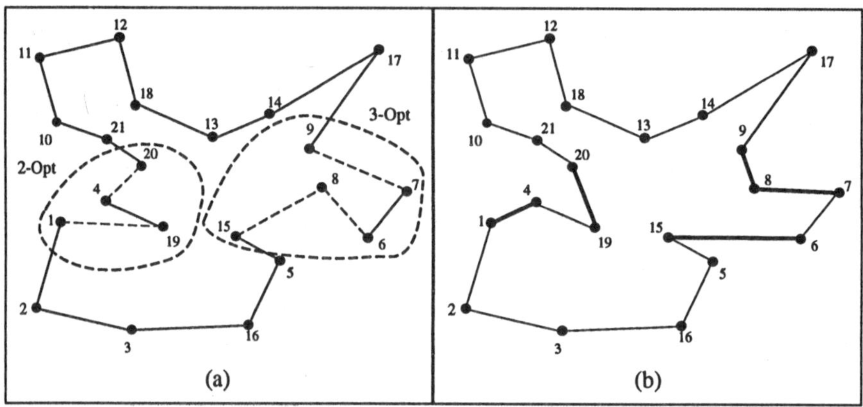

Figure 1.3. 2-opt and 3-opt example for the TSP. (a) Original solution (dashed edges will participate in the exchanges). (b) Final solution after the exchanges (the edges that have been added after the exchanges are noted in bold).

In general, the most efficient neighborhood structures are going to be problem specific. However, the ideas of two-exchange, multi-exchange, 2-opt, and 3-opt can serve as valid starting points for a wide variety of problem contexts and lead to more suitable neighborhoods.

Search. It is fairly easy to see that depending on the definition of the neighborhood the number of possible neighbors ranges from exactly one (in the 2-opt case for the TSP) to $O(nK)$ (for the shift move of the node-based two-exchange neighborhood), to $O(n^2)$ (in the tree-based two-exchange case) to a significantly larger $O(n^K)$ (for the multi-exchange case), where n is the number of nodes in the graph and K is the number of sets participating in the exchanges. Because of the large number of possible neighbors, we need to determine a procedure that searches the neighboring solutions efficiently and finds one with which to replace the current solution. There are a few well-documented and extensively tested procedures that concentrate on this step of the procedure. The most prominent ones are steepest descent, simulated annealing (SA), and the greedy randomized adaptive search procedure (GRASP). The main idea behind most of these procedures is essentially the same. It specifies that the neighboring solution could be chosen at random, characterized as improving or non-improving, and accordingly accepted with a certain probability specific for each case (i.e., improving or not). Compared to steepest descent, SA and GRASP are more sophisticated local search procedures. Due to their advanced search strategies, they are often classified as metaheuristics in the literature.

Steepest descent is the simplest and the most short-sighted of the search procedures. At each iteration, we compute all neighbors for the existing solution and choose the one which reduces (in the case of a minimization problem) the cost of the solution the most. A potential disadvantage of this procedure is that it never accepts *uphill moves* (i.e., neighbors with higher cost than the existing solution) and as a result can potentially terminate after a few iterations at a poor quality local optimum. Moreover, in some cases it is impractical, because of the computational effort required, to compute all neighbors. However, if the neighborhood structure used is well suited to the problem and there is an efficient way to find the best neighbor (i.e., other than enumeration) then a Steepest descent procedure can still achieve a high level of performance and will be computationally efficient. For applications of this approach in the context of the CMST problem, see [2, 3].

Simulated annealing (SA) is a very popular local search procedure. In Figure 1.4 we present the steps of a generic SA algorithm. The main feature of this procedure is that it can accept *uphill moves* with a certain probability that decreases at each iteration. The way in which this acceptance probability decreases (i.e., the so-called cooling schedule) plays an important role in the efficiency of the procedure. If the decrease is too steep then the procedure will terminate early at a poor solution. On the other hand, if it is very gentle we will have to wait a long time before we reach the final solution. However, the slower the cooling schedule, the better the solution obtained by the procedure. In particular, for theoretical convergence to the optimal solution the temperature needs to be reduced infinitesimally in each step. While this result is reassuring, it also implies that there is no finite-time implementation that guarantees the optimal solution. In practice, a balance is to be obtained between the running time of the procedure, quality of solutions obtained, and the cooling schedule. Determining an appropriate cooling schedule is not an easy task, and most of the time it takes a lot of computational testing to adequately justify a particular selection. Aarts et al. provide a nice synopsis of the simulated annealing procedure in [1].

The GRASP procedure (see [43]) effectively searches the solution space rapidly by implementing local search many times while starting from different points in the solution space. In Figure 1.5 we present the steps of a generic GRASP procedure. The different starting solutions, f_s, that GRASP uses are generated by a greedy construction heuristic that in each iteration selects any edge that is within $100 * \alpha$ percent of the cheapest edge. In this way starting solutions have considerable variability, but are also fairly reasonable. GRASP then proceeds by repeatedly updating the current solution with the best known neighbor, f_n, until there are no more improvements. In the end, the procedure returns the best solution it has encountered, f_b. The total number of iterations or the number of successive iterations without an improvement can both be used as a termination condition. Most of the criticism of GRASP has to do with the fact that successive iterations of the procedure are independent and as a result the algorithm does not learn from solutions found in previous iterations. More advanced versions of GRASP attempt to resolve this issue. Some of them include

```
Begin
   randomly generate an initial solution f_s
   compute the cost of f_s, C(f_s)
   set the effective temperature, T, to the initial value, T_0
   while T ≠ T_N do
      randomly select a neighbor, f_n, of the current solution, f_s
      compute the cost C(f_n) and Δ = C(f_n) − C(f_s)
      if Δ ≤ 0 then
         f_s ⟵ f_n
      else
         with probability e^{−Δ/T}, f_s ⟵ f_n
      end if
      update the temperature T
   end while
end
```

Figure 1.4. Steps of a Simulated Annealing procedure.

path-relinking approaches (see [37]) that allow for intensification and diversification strategies, and adaptive mechanisms, called Reactive GRASP [45], that change the value of α depending on the results of the search. Another issue with GRASP has to do with the fact that starting points might in fact be identical solutions. An easy way to resolve this issue is to use hash tables (see [10]) to guarantee independent starting points. For an extensive annotated bibliography of GRASP literature (including many applications in network design) see Festa and Resende [21].

All these search procedures have been used successfully in different contexts but, in general, greatest attention should be given to defining the neighborhood structure. In our opinion, in most cases where local search procedures have performed well in the literature, it has been due to the properties of the neighborhood. Often, with an effective neighborhood simple search strategies such as steepest descent can provide results competitive with more sophisticated search strategies such as GRASP and simulated annealing. Additionally, specifying the parameters used in the search procedures—like the rate for decreasing the temperature in SA, or the value of α for GRASP—is critical to their success and a lot of testing is required to find values that work well for a wide range of problem instances.

```
Begin
    t ⟵ 0
    f_b = inf
    while (not termination-condition) do
        generate a random greedy solution, f_s
        using local search find f_n from f_s
        if C(f_n) < C(f_b) then
        f_b ⟵ f_n
        end if
    end while
    return f_b
end
```

Figure 1.5. Steps of a Greedy Randomized Adaptive Search Procedure.

1.1.3 Tabu Search

An approach that is similar to LS in some ways but deserves to be noted separately is tabu search (TS). Tabu search, like LS procedures, requires the definition of a neighborhood for each solution in the search space. However, TS differs in the way it searches the neighborhood. The main idea of TS is to incorporate memory in the search process. The notion is that by effectively using memory, one may be better able to guide the search to regions in the search space containing the global optimum or near-optimal solutions to the problem.

In TS, memory concerning the search path is created by the use of a tabu list. This list keeps a record of solutions, or solution attributes that should not be considered (i.e., are tabu) in the next step of the search process. By doing so, TS guides the search process away from local optima and visits new areas in the solution space. Obviously, selecting the rules under which a solution or a specific attribute becomes tabu and the number of iterations for which it will remain tabu (i.e., the tabu tenure) are of vital importance when applying this procedure.

To illustrate the use tabu moves, consider the k-th best spanning tree problem. In this problem we wish to find the k-th best spanning tree in a graph. A feasible solution is a spanning tree, and the neighborhood of a solution can be defined as the set of spanning trees obtained by dropping and adding an edge to the given spanning tree. Here are three different ways in which a move can be classified as tabu. 1) The candidate move is tabu if either the edge added or the edge dropped is tabu. 2) The candidate move is tabu if a specific edge in the solution is tabu. 3) The candidate move is tabu if both the edge added and the edge dropped are tabu. In general, when evaluating choices for tabu moves, the tabu rules should be simple and ensure that the search can escape from local optima and visit new areas in the solution space.

Apart from the definition of the tabu rules another way to control the search is by determining the tabu tenure associated with the different rules. Defining these tenures plays a very important role since it greatly affects the outcome of the search. Small tenures will result in cycling (i.e., revisiting the same solutions) while large tenures will result in a gradual degradation of the search results. It is possible to define tenures as constant, in which case the tenure never changes during the search, or as dynamic with either a deterministic or a random way in which the value of the tenure can change. In any case it is important to experiment until one finds good values for the tabu tenures.

The issues discussed so far although important usually come under the label of *short-term memory* and do not cover the more powerful features of TS that have to do with *long-term memory.* We will not attempt to give a complete account of all these possibilities nor provide guidelines for their use. However, we would like to bring to the attention of the reader some of the more important features of long-term memory like frequency based measures, strategic oscillation, and path-relinking. The use of long-term memory can significantly increase the efficiency of a TS algorithm. Frequency based measures are built by counting the number of iterations in which a desirable attribute has been present in the current solution. These measures can then be used to guide subsequent explorations of the solution space to regions that are likely to contain near-optimal solutions. Strategic oscillation allows the search to continue past a predetermined boundary which usually coincides with the space of feasible solutions. Once the search process passes the boundary we allow it to proceed to a specified depth beyond the oscillation boundary before forcing it to turn around. This time the boundary is crossed in the other direction (from infeasibility to feasibility) and the search continues normally. Path-relinking generates new solutions by exploring trajectories that connect elite solutions and has been found to be very promising. TS procedures with long-term memory features can significantly outperform simpler heuristic search algorithms, but can be quite complex and intricate to design. For an in depth discussion on TS and all of its features see the text by Glover and Laguna [29].

1.1.4 Genetic Algorithms

In this section, we describe a popular heuristic search procedure known as an evolutionary algorithm or a genetic algorithm (GA). When carefully applied

```
Begin
  t ⟵ 0
  initialize P(t)
  evaluate P(t)
  while (not termination-condition) do
    t ⟵ t + 1
    select r solutions from P(t − 1)
    apply genetic operators on the r solutions
    insert the new r solutions to P(t)
    evaluate P(t)
  end while
end
```

Figure 1.6. Steps of a Genetic Algorithm.

to a problem, GAs can generate high-quality solutions that are often better than LS procedures.

Genetic algorithms are powerful search procedures motivated by ideas from the theory of evolution. They have been successfully used for a variety of problems. Although many different variations exist, the main steps that GAs follow are essentially the same (see Figure 1.6). At first, an initial population of solutions is generated and the *fitness* of all solutions is evaluated. The fitness essentially describes how good each solution is, and is usually a function of the cost of the solution. A specific set of solutions is then *selected* and on that set different genetic operators are applied. The genetic operators are usually simple heuristic procedures that combine two solutions to get a new one (crossover operator) or transform an existing solution (mutation operator). The new solutions generated by the genetic operators are then included in a new population and the above steps are repeated until a terminating condition is satisfied. The selection of solutions from a population is done with respect to their fitness value. Specifically, fitter solutions have a higher chance of being selected than less-fit solutions.

Representation. The most challenging and critical aspect of using a GA is the determination of the way in which solutions are represented. A good representation can lead to near-optimal solutions while a bad one will cause the GA to perform very poorly. Selecting a representation is by no means a trivial task. The main thing to keep in mind when selecting a representation is that it should have meaningful blocks of information that can be related to elements of a desirable solution and that these blocks should be preserved from one generation to the next. These are essentially the main ideas described in the *Building Block Hypothesis* [38] and the guidelines presented by Kershenbaum in [34]. If we are interested in representing trees, then we can look into some of the approaches already developed and tested for various tree problems. Some of the more traditional approaches include the so-called characteristic vectors, predecessor encoding, and Prüfer numbers. Characteristic vectors represent trees by a vector $v \in \{0,1\}^{|E|}$ where $|E|$ is the set of edges in the graph. In predecessor encoding each entry in the encoding corresponds to a particular node and signifies the predecessor of that node, with respect to a predetermined root node in the tree. Finally, Prüfer numbers indirectly encode trees and require specific algorithms to convert a number into a tree and vice versa. All three of these approaches have been criticized (see Palmer and Kershenbaum [41]) because they violate some of the principles we mentioned above.

Palmer and Kershenbaum introduce a link-and-node-bias (LNB) encoding in [41]. In order to encode one tree with the LNB encoding we need to specify one bias, b_i, for each node i in the graph ($|N|$ biases) and one bias, b_{ij} for each edge, $\{i, j\}$, in the graph ($|E|$ biases). The cost matrix is then biased by using: $C'_{ij} = C_{ij} + P_1 b_{ij} C_{max} + P_2(b_i + b_j)C_{max}$, where C_{ij} is the original cost of edge $\{i, j\}$, C_{max} is the maximum edge cost in the graph and P_1, P_2 are two multipliers that are constant for each GA run. The tree can then be found by running a MST algorithm with the new costs. We observe that this representation can encode any tree by setting $b_i = 0$, for all nodes i, and $b_{ij} = 0$ for the edges $\{i, j\}$ that belong to the tree we wish to represent and $b_{ij} = M$, otherwise (where $M > C_{max}$). This encoding satisfies most of the guidelines discussed earlier and has been found to work well for the optimal communication spanning tree problem (OCSTP).

Sometimes, even when we are interested in constructing trees, it is more efficient to look at other representations that do not specifically deal with tree structures. One example is the CMST problem. As in the case of the LS procedures for the CMST, it may be desirable to find a partition of the nodes in the graph. Once a partition is found it is much easier to construct a feasible solution by solving the subproblems on the partition (for the CMST we find a MST on the union of the central node and the nodes in the partition). In these cases, using a representation that is geared towards groupings/partitions can prove to be far more valuable. In Figure 1.7 we present an example of a capacitated tree encoded using Falkenauer's [19] grouping representation. In this representation there is an item part which defines the assignment of nodes into groups and a group part that specifies the groups in the solution.

Another example of a representation originally developed for problems that are not strictly associated with trees is the network random key encoding

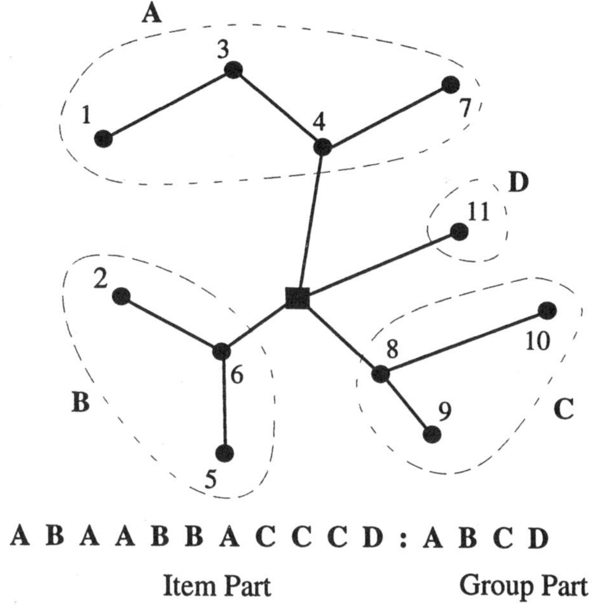

A B A A B B A C C C D : A B C D

Item Part Group Part

Figure 1.7. Example of a group assignment and the respective representation for a CMST instance.

(NetKey) originally introduced by Bean [5] for problems where the relative order of tasks is important (e.g., scheduling problems). Rothlauf et al. [47] present a way in which NetKey can be used for tree representations in network design problems, compare it with characteristic vector encoding, and experimentally show that NetKey is better. The idea behind NetKey when used for tree representations is to introduce a relative order in which the edges in the graph will be considered. Based on the order represented in the encoding, edges are added to an empty graph. When the addition of an edge introduces a cycle, we skip it and move on to the next edge until we have $|N| - 1$ edges in the graph and, therefore, a tree.

Operators. Genetic operators are usually distinguished as crossover operators or mutation operators. Typical crossover operators like single-point or two-point crossover are usually a good choice and can achieve good results depending on the strengths and weaknesses of the representation. On the other hand, mutation operators should be designed carefully to complement what cannot be achieved by the crossover operators. Historically, mutation operators started out as random changes to individual solutions that were applied with very low probability. This approach, however, makes the mutation operators very weak and hardly ever adds true value to the search procedure. It would be much more beneficial to design a mutation operator that is likely to improve the solution and can be applied with a much higher probability. When we think along these lines, a natural choice for a mutation operation is a simple local search procedure. In most cases, the operators should be defined with the specific characteristics of the representation in mind. For example the strength of the grouping representation described earlier lies in the fact that operators are applied on the group part of the representation and, therefore, allow for meaningful parts of the solution (i.e., the subtrees) to be carried on from one generation to the next.

In general, GAs can be extremely involved procedures and as a result the aim should be to keep the number of parameters as small as possible. In any case, the selected values for the parameters should be able to achieve high quality results regardless of the problem type.

1.1.5 Remarks

Obviously, the above discussion on heuristics, local search methods, and metaheuristics is by no means exhaustive. However, we believe that it showcases a set of directions that one can follow when approaching a challenging problem in network design. In our opinion, a fruitful attempt at problem solving should begin by the exploration of naive heuristics that are easy to implement and understand. These first attempts can then lead to a neighborhood definition. The development of the neighborhood could possibly start with known and tested ideas and expand to more sophisticated structures that are specifically tailored to the problem. Depending on the efficiency of the neighborhood, different algorithms for the search procedure can then be tested and compared. Generally, strong and complex neighborhoods require very simple search methods (i.e., steepest descent) while simpler neighborhoods could benefit from more advanced search methods (e.g., simulated annealing, GRASP, tabu search). For genetic algorithms we have given a few examples of representations that have been used in network design problems, commented on aspects such as parameter setting and genetic operators, and recommended the use of local search procedures as mutation operators. In the next few sections, based on our experience applying heuristic search, we discuss several $\mathcal{NP}$-complete network design problems, and illustrate the successful application of heuristic search techniques to them.

1.2 The Minimum Labeling Spanning Tree Problem

A problem in communications network design that has attracted attention recently is the minimum labeling spanning tree (MLST) problem. In this problem, we are given a graph with labeled edges as input. Each label represents a type of edge and we can think of a label as a unique color. For example, in

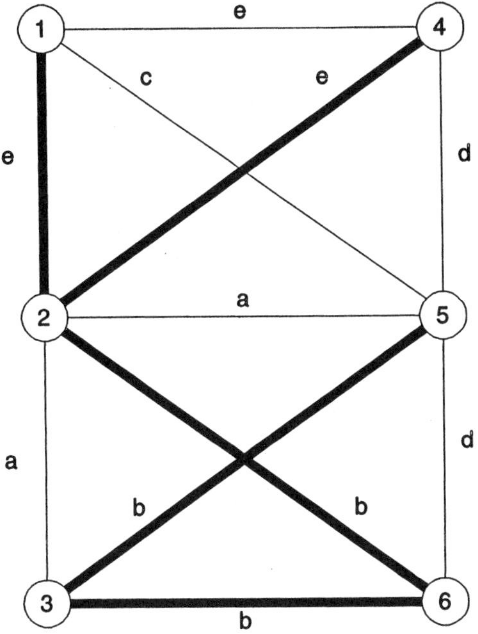

Figure 1.8. An MLST example and solution.

Figure 1.8, there are six nodes, 11 edges, and five labels (denoted by letters). We would like to design a spanning tree using the minimum number of labels. For the example in Figure 1.8 an optimal MLST is indicated in bold.

Chang and Leu [8] introduced the problem, proved that it is NP-hard, and proposed two heuristics. One of these, called MVCA, clearly outperformed the other. Krumke and Wirth [35] proposed an improvement to MVCA and proved that the ratio of the (improved) MVCA solution value to the optimal solution value is no greater than $2\ln n + 1$. Wan et al. [51] improved upon this bound and were able to show that the MVCA solution value divided by the optimal solution value is no greater than $\ln(n-1)+1$. Brüggeman et al. [7] introduced local search techniques for the MLST and proved a number of complexity results. For example, they were able to show that if no label appears more than twice, the problem is solvable in polynomial time. Building on some of these ideas, Xiong et al. [53] were able to further improve the worst-case bound. In particular, they proved that, in a graph where the label that appears most frequently appears b times, the ratio of the MVCA solution value to the optimal solution value is no greater than the bth harmonic number

$$H_b = 1 + \frac{1}{2} + \ldots + \frac{1}{b} < 1 + \ln b.$$

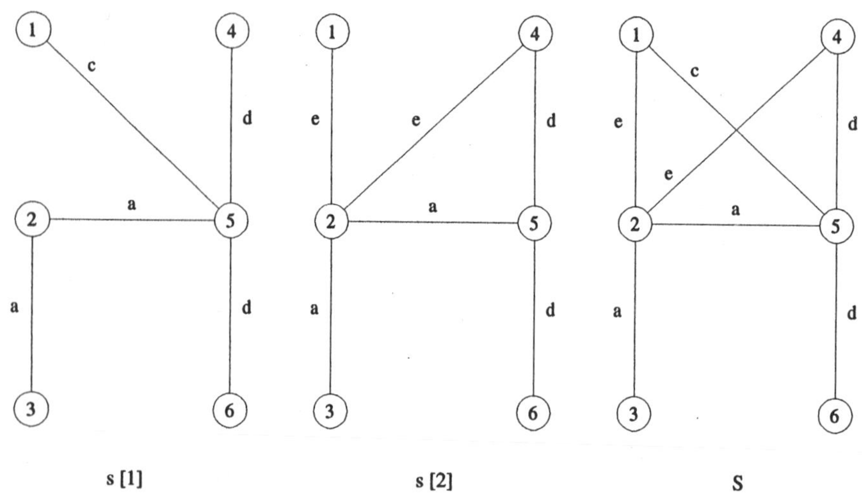

Figure 1.9. The crossover operator in the GA for the MLST problem.

In addition, Xiong et al. present a family of graphs for which this worst-case ratio can be attained. Therefore, the worst-case ratio cannot be improved.

In a second paper, Xiong et al. [52] present a simple genetic algorithm for the MLST problem and they compare it with MVCA over a wide variety of problem instances. The GA consists of crossover and mutation operations.

Before we describe these operations, we must define what we mean by a solution. Suppose we are given a graph $G = (V, E, L)$, where V is the set of nodes, E is the set of edges, and L is the set of labels. A solution $s[i]$ is a subset C of the labels in L such that all the edges with labels in C construct a connected subgraph of G and span all the nodes in G. If we consider the MLST instance in Figure 1.8, we can see that $\{a, c, d\}$ and $\{a, d, e\}$ are two solutions. Thus, any spanning tree of $s[i]$ will be a feasible solution to the MLST problem.

Given an initial population of solutions, the crossover operator works as follows. Two solutions $s[1]$ and $s[2]$ are selected for crossover. First, we combine the labels from the two solutions. For example, if $s[1] = \{a, c, d\}$ and $s[2] = \{a, d, e\}$ in Figure 1.9, then $S = \{a, c, d, e\}$ is the union of these two. Next, we sort S in decreasing order of label frequency, yielding $\{a, d, e, c\}$. Finally, we add labels of S in sorted order to an empty set, T, until T represents a solution. In this case, we output $\{a, d, e\}$ which is the same as $s[2]$.

The mutation operator is also quite simple. Suppose we begin with solution S. First, we randomly select a label c not in S and let $T = S \cup c$. Next, we sort T in decreasing order of label frequency. Finally, starting at the end of the list, we delete one label at a time from T provided that the result is itself a solution.

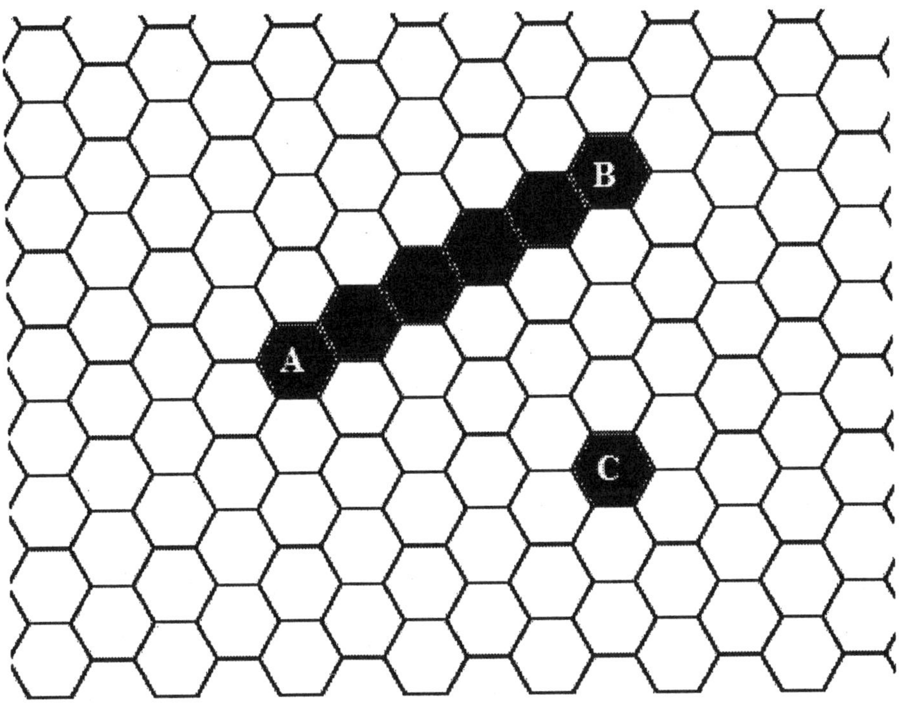

Figure 1.10. Hexagonal tiling of a 2-dimensional space.

From the above description it should be evident that the GA is fairly simple. To compare the performance of the GA to existing heuristics for the MLST problem, the GA and MVCA were tested over 81 problem instances. The largest of these involved 200 nodes and 250 labels. Both procedures are extremely fast. The GA outperformed MVCA in 53 cases (sometimes by a substantial margin), MVCA outperformed the GA in 12 cases (by a very small amount in each case), and there were 16 ties.

1.3 The Euclidean Non-uniform Steiner Tree Problem

In Steiner tree problems, we are given a set of terminal nodes which we seek to connect at minimum cost. Additional nodes, known as Steiner points, may be added to the terminal nodes in order to reduce the overall construction cost. The problem requires that specific Steiner points and connecting edges be determined.

Many variants of the Steiner tree problem have been studied in the literature including the Euclidean Steiner problem, the Steiner problem in graphs, and the rectilinear Steiner problem (see Du et al. [15] for recent work on different

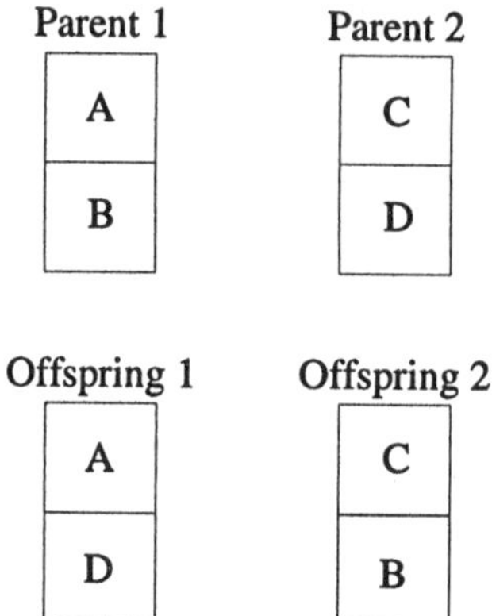

Figure 1.11. Horizontal crossover operator in the GA for the Euclidean non-uniform Steiner tree problem.

variants of the Steiner tree problem). In most variants, the cost of an edge typically depends on its distance only. That is, costs are uniform with respect to the location of the edge.

Coulston [11] recently introduced an interesting variant. In this problem, the cost of an edge depends on its location on the plane as well as its distance. Suppose the problem involves laying cable, constructing pipeline, or routing transmission lines. In such a setting, certain streets (routes) are more expensive to rip apart and re-build than others. The Euclidean non-uniform Steiner tree problem reflects this realism. To begin with, the 2-dimensional region is covered with a hexagonal tiling. Each tile has an associated cost and terminal nodes (located at specific tiles), which must be connected, are given as input. Other nodes (equivalently, tiles) may be selected for use as Steiner nodes. Two nodes can be directly connected if and only if a straight line of tiles can be drawn between the nodes (see Figure 1.10). The total cost of the Euclidean non-uniform Steiner tree is the sum of the cost of tiles in the tree. Coulston uses heuristic search techniques such as genetic algorithms and ant colony optimization in his computational work.

In Frommer et al. [22], a simple genetic algorithm is introduced that exploits the geometric nature of the problem (i.e., it makes use of (x, y) coordinates). In addition, Frommer et al. allow an additional charge for each Steiner node.

At the heart of their genetic algorithm is a spatial crossover operator. In its simplest implementation, the grid is split in half horizontally. Let A, B, C, and D be sets of Steiner nodes. A and C can have some common nodes, as can B and D. The parents and offspring are shown in Figure 1.11. In each of the four cases, there are a set of terminal nodes also (not shown). If we combine Steiner nodes and terminal nodes and solve a minimal spanning tree problem over these

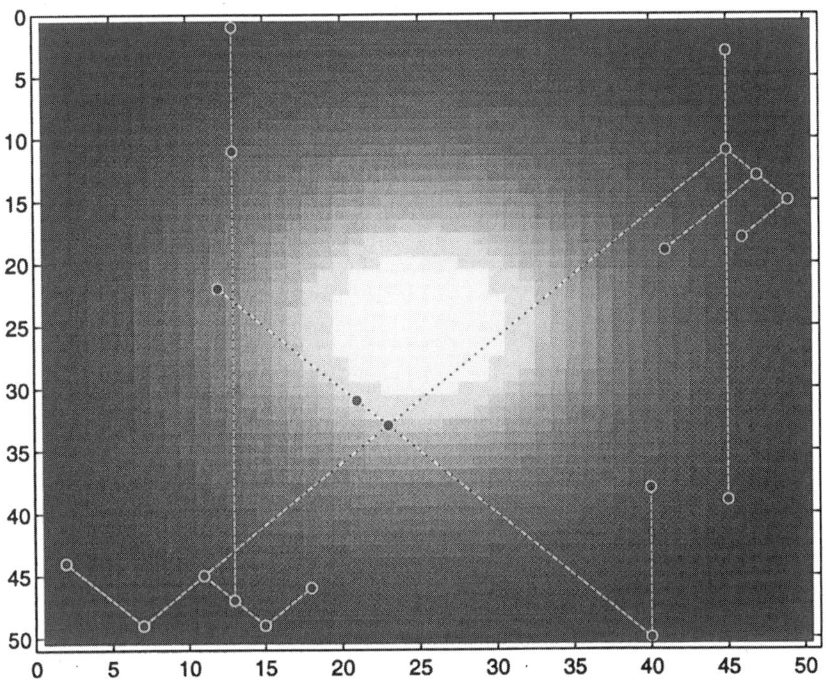

Figure 1.12. MST solution (cost = 56.24) for the Euclidean non-uniform Steiner tree problem.

nodes, we can obtain a Steiner tree. (Note, we must eliminate degree-1 Steiner nodes first.)

The genetic algorithm is tested over a wide variety of problem instances and is compared with a heuristic based on the progressive addition of Steiner nodes. The overall performance is comparable, but progressive addition is significantly slower as problem size grows.

The GA behaves quite reasonably as indicated by Figures 1.12 and 1.13. In Figure 1.12, a minimal spanning tree solution is presented. The lightest regions have the highest costs. The terminal nodes are connected in a least cost way, without using any Steiner nodes. In Figure 1.13, the GA finds a better solution, by using Steiner nodes in order to avoid the lightest (most expensive) regions.

1.4 The Prize-Collecting Generalized Minimum Spanning Tree Problem

The prize-collecting generalized minimum spanning tree (PCGMST) problem occurs in the regional connection of local area networks (LAN), where several LANs in a region need to be connected with one another. For this purpose, one gateway node needs to be identified within each LAN, and the gateway nodes are to be connected via a minimum spanning tree. Additionally, nodes within the same cluster are competing to be selected as gateway nodes, and each node offers certain compensation (a prize) if selected.

Formally, the PCGMST problem is defined as follows: Given an undirected graph G = (V, E), with node set V, edge set E, cost vector $c \in R_+^{|E|}$ on the edges E, prize vector $p \in R_+^{|V|}$ on the nodes V, and a set of K mutually exclusive and exhaustive node sets V_1, ..., V_K (i.e., $V_i \bigcap V_j = \emptyset$, if $i \neq j$, and $\bigcup_{k=1}^{K} V_k = V$), find a minimum cost tree spanning exactly one node from each cluster. This problem represents a generalization of the $\mathcal{NP}$-hard generalized minimum spanning tree (GMST) problem, where all nodes are equally important, that is, each node offers equal compensation if selected for the regional network design.

Several exact and heuristic procedures have been developed for the GMST problem so far. Myung et al. [40] developed a dual-ascent procedure based on a multicommodity flow formulation for the problem. Feremans [20] developed a branch-and-cut algorithm that uses a tabu search algorithm for the initial upper bound, and a local search algorithm that improves upper bounds found during the course of the branch-and-cut procedure. Pop [44] developed a relaxation procedure based on a polynomial-size MIP formulation for the GMST problem. Golden et al. [31] developed several upper bounding procedures including simple local search and genetic algorithm heuristics.

Building upon our previous work in Golden et al. [31] for the GMST problem, we describe the application of a local search procedure and a genetic algorithm to the PCGMST problem.

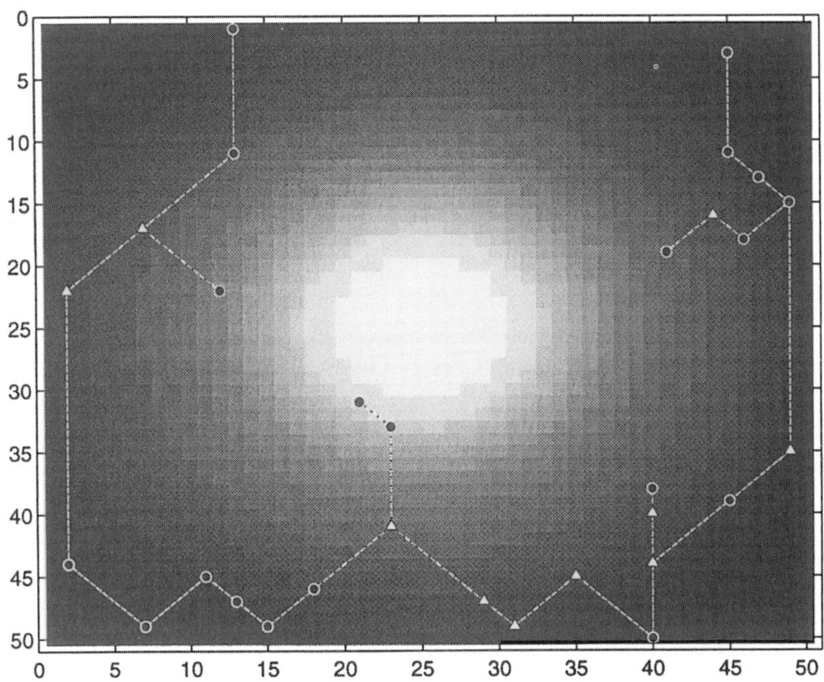

Figure 1.13. GA solution (cost = 29.01) for the Euclidean non-uniform Steiner tree problem.

1.4.1 Local Search Procedure

The local search (LS) procedure we developed in [31] for the GMST problem may be directly applied to the PCGMST problem. It differs only in that an additional objective function term for node prizes needs to be taken into account here.

The LS procedure is an iterative 1-opt procedure. It visits clusters in a wraparound fashion following a randomly defined order. In each cluster visit, the neighborhood of a feasible generalized spanning tree is explored by examining all feasible trees obtained by replacing the node (in the tree) from the current cluster. In other words, a GST of least cost (the cost of the tree is the sum of the edge costs minus the rewards on the nodes) is found by trying to use every node from that cluster, while fixing nodes in other clusters. The local search procedure continues with visiting clusters until no further improvement is possible. The procedure is applied to a pre-specified number of starting solutions (denoted by t). The steps of the procedure are outlined in Figure 1.14.

1.4.2 Genetic Algorithm

We present a genetic algorithm for the PCGMST problem that is similar to the one we developed in [31], with a few differences in the initial population and genetic operators applied. Figure 1.15 shows an outline of our genetic algorithm. The initial population is created by randomly generating a pre-specified number of feasible GSTs. Before adding a new chromosome to the population $P(0)$, we apply local search and add the resulting chromosome as a new population member. Within each generation t, new chromosomes are created from population $P(t-1)$ using two genetic operators: local search enhanced crossover and random mutation. The total number of offspring created using these operators is equal to number of chromosomes in the population $P(t-1)$, with $\alpha P(t-1)$ offspring created using crossover, and $\beta P(t-1)$ offspring created using mutation (fractions α and β are experimentally determined). Once the pre-specified number of offspring is generated, a subset of chromosomes is selected to be carried over to the next generation. The algorithm terminates when the termination condition, a pre-specified number of generations, is met.

We now give a more detailed description of the genetic algorithm.

Representation. We represent a chromosome by an array of size K, so that the gene values correspond to the nodes selected for the generalized spanning tree. Figure 1.16 provides an example of a generalized spanning tree and its corresponding chromosome representation.

Initial Population. The initial population is generated by random selection of nodes for each cluster. If possible, a minimum spanning tree is built over the

Local Search Heuristic (LS):

Step 0. Specify the number of feasible solutions to be generated - t. Repeat Steps 1 through 3 t times.

Step 1. Generate a feasible solution as follows. Randomly select a single node from each cluster, and build an MST on the selected nodes (using any MST algorithm). If no MST exists on the selected nodes, repeat Step 1. Otherwise, continue to Step 2.

Step 2. Randomly define an order in which clusters will be searched.

Step 3. Follow the order defined in Step 2 in visiting clusters. Repeat the following steps until K sequential cluster visits result in no improvement.

- While visiting a cluster, consider each node in the cluster as a replacement for the current node in the cluster contained in the GST. Compute the cost of the solution for each replacement.
- Among the solutions computed in the previous step, identify the one giving the greatest improvement in the objective function. If there is an improvement, implement it by replacing the node representing the cluster, and update the current tree.

Figure 1.14. Steps in the Local Search procedure for the PCGMST problem.

Begin
  $t \longleftarrow 0$
  initialize $P(t)$
  while (**not** termination-condition) **do**
    $t \longleftarrow t+1$
    Generate $\alpha P(t-1)$ offspring using local search enhanced
      crossover operator
    Generate $\beta P(t-1)$ offspring using random mutation
      operator
    Select new generation $P(t)$, with population size equal
      to $\theta(1+\alpha+\beta)P(t-1)$
  end
end

Figure 1.15. Steps of our Genetic Algorithm for the PCGMST problem.

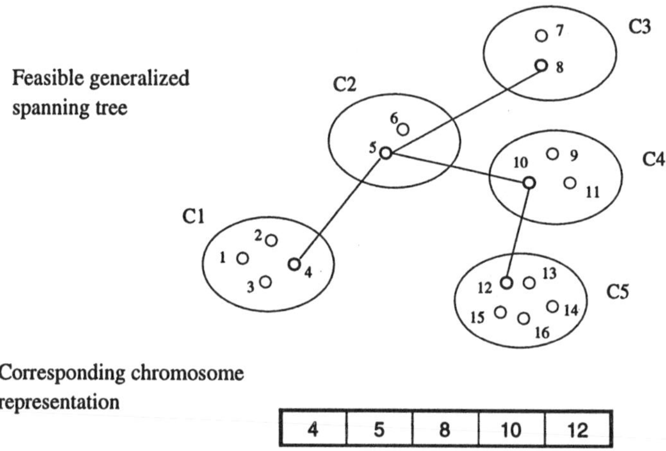

Figure 1.16. An example of a feasible GST and the corresponding chromosome representation

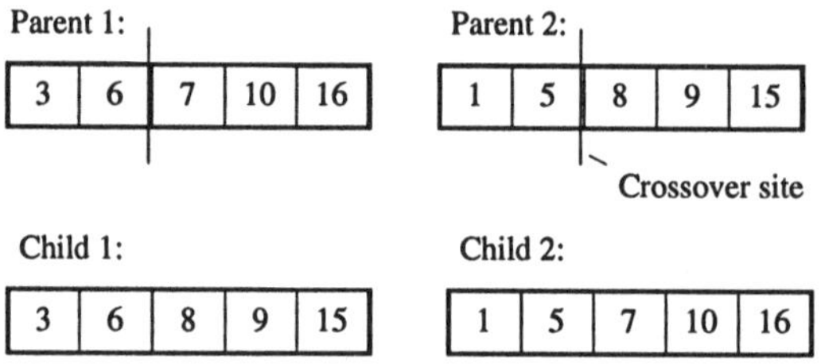

Figure 1.17. One-point crossover operator example.

selected nodes. Otherwise, the chromosome is discarded, since it represents an infeasible solution. Each feasible minimum spanning tree built in this way is then used as input for the local search procedure. The resulting solution is then added to the initial population as a new chromosome with a fitness value defined as the difference between the cost of the MST and the prizes associated with the nodes in the chromosome.

Crossover. We apply a standard one-point crossover operation as shown in Figure 1.17. As in the initial population, only the feasible solutions are accepted. Each child chromosome created using this operator is used as input to the local search procedure, and the resulting chromosome is added to the population.

Random Mutation. A random mutation operator randomly selects a cluster to be modified and replaces its current node by another, randomly selected, node from the same cluster. The new chromosome is accepted if it results in a feasible

GST. In order to maintain diversity of the population, we do not apply local search to new chromosomes created by random mutation.

Selection/Replacement. At the end of each generation, a fraction, θ, of the current population is selected to be carried to the next generation, while the remaining chromosomes are discarded. This selection is a combination of elitism and rank-based selection, where the top 10% of the current population is selected using elitism and the remaining 90% is selected using rank-based selection (see Michalewicz [38]).

1.4.3 Computational Experiments

In this subsection, we provide a summary of some computational experiments with the local search (LS) heuristic and the genetic algorithm (GA) procedure. Both LS and GA were coded in Microsoft Visual C++ on a workstation with 2.66 GHz Xeon processor and 2GB RAM. The heuristics were tested on two classes of instances—those from TSPLIB (identical to those in [20], with edge costs satisfying the triangle inequality) and those with random edge costs (identical to those in [31]). For these experiments we have added integer node prizes generated randomly in the range [0, 10].

The TSPLIB instances consist of 5 sets of problems differing in terms of a user-defined parameter μ that may be thought of as the average number of nodes per cluster. The size of these instances varies from 7 to 84 clusters, with 47 to 226 nodes, and up to 25,118 edges. The size of random instances varies from 15 to 40 clusters, with 120 to 200 nodes, and up to 15,000 edges.

Parameter settings for LS and GA. Based on the initial testing over a separate set of large instances with random edge costs, we selected the GA parameters as follows. The size of the initial population $P(0)$ was set to 100. α and β were both set to 0.5. θ, the fraction of the population to survive, was set to 0.5. The stopping criterion was 15 generations. For LS the number of starting solutions was set to 500.

Computational results. For the TSPLIB instances, we were able to find optimal solutions for 133 of 169 instances using a branch-and-cut algorithm for the PCGMST problem that we proposed in [30]. The branch-and-cut procedure was able to find optimal solutions for all random instances. The summary of results for these instances is shown in Table 1.1.

The first two columns in the table represent the type of problem set and number of instances within each set. The third column in the table indicates the number of instances where the optimal solution is known (from the branch-and-cut procedure). The first column under LS and GA indicates the number of instances where these procedures found the optimal solution. In all TSPLIB

Table 1.1. Summary of computational results for the PCGMST problem.

Problem Set	*Inst*	*Opt Known*	*LS*		*GA*	
			Opt	*Time (sec)*	*Opt*	*Time (sec)*
TSPLIB, geog	41	35	35	14.44	35	7.48
TSPLIB, $\mu = 3$	32	22	22	70.95	22	24.34
TSPLIB, $\mu = 5$	32	22	22	28.82	22	11.13
TSPLIB, $\mu = 7$	32	27	27	8.61	27	5.35
TSPLIB, $\mu = 10$	32	27	27	2.60	27	3.09
TSPLIB	169	133	133	24.51	133	10.25
Random	42	42	40	6.33	40	5.96
Summary	211	175	173	20.89	173	9.39

instances where the optimal solution is known (there are 133 such instances), our GA and LS found the optimum. In the remaining 36 TSPLIB instances, GA always provided solutions that were at least as good as the ones obtained by LS. In 8 of these instances, GA provided a solution better than the one obtained by LS. In the case of random instances, where the optimum solution is known for all 42 instances, both LS and GA did not find the optimum in 2 of 42 instances. In one of these instances, GA provided a better solution than LS, and in the other instance LS was better than GA. The branch-and-cut algorithm took an average of 726 seconds per instance to solve the 175 instances to optimality, while LS and GA take an average of 20.89 and 9.39 seconds per instance respectively for all 211 problem instances.

1.4.4 Remarks

The computational results with the two heuristic search techniques, local search and the genetic algorithm, presented in this section have shown that relatively simple heuristic search techniques can provide high quality results for the PCGMST problem, while requiring only a small fraction of time compared to an exact procedure like branch-and-cut.

It is noteworthy that we obtained similar compelling results with local search and a genetic algorithm for the GMST problem in [31]. The LS procedure provided results that were on average 0.09% from optimality (where the optimal solution was known), while the genetic algorithm (with a different structure than the one described here) provided results that were on average only 0.01% from optimality.

Finally, we note that there is a slightly different version of the GMST problem that has been studied in the literature. In this version the clusters are collectively exhaustive, but not necessarily mutually exclusive, and *at least* one node from each cluster needs to be selected for the GMST. Several heuristic search procedures have been applied to this at least version of the GMST problem. Dror et al. [14] developed a genetic algorithm for the *at least* version of the GMST problem. Shyu et al. [50] developed an ant colony optimization procedure for this version of the GMST problem, with faster but similar results (in terms of solution quality) to the GA developed by Dror et al. More recently, Duin and Voß [17] use a transformation of the *at least* version of the GMST problem to the Steiner problem in graphs, and show that the problem can be efficiently solved using heuristic pilot methods (the pilot method is a technique proposed by Duin and Voß [16] that is designed to improve solutions obtained by other heuristics, referred to as sub-heuristics or pilots, through tailored repetitions of these heuristics).

1.5 The Multi-Level Capacitated Minimum Spanning Tree Problem

In this section, we describe the multi-level capacitated minimum spanning tree (MLCMST) problem, a generalization of the well-known capacitated minimum spanning tree (CMST) problem, and discuss two heuristics for it. In the CMST problem, we are given a set of terminal nodes, each with a specific traffic requirement that we wish to transport to a given *central* node. Furthermore, a single type of facility with a fixed capacity is available for installation between any two nodes. We wish to design a feasible, minimum cost, *tree* network to carry the traffic. The CMST is a fundamental problem in network design and has been extensively studied by many researchers over the years (see [25] for a nice survey).

In many practical applications in telecommunications, utility networks, etc., there is more than a single facility type available to the network planner. The MLCMST addresses this practical issue and generalizes the CMST by allowing for multiple facilities with different costs and capacities in the design of the network. Other than our own work [23, 24], little has been done on the MLCMST problem.

Formally, the MLCMST problem can be defined as follows. Given a graph $G = (V, E)$, with node set $V = \{0, 1, 2, \ldots, N\}$ and edge set E. Node 0 represents the central node (which we will also denote by c) and the rest are terminal nodes. W_i is the traffic requirement (or weight) of node i to be transported to the central node c. We are also given a set of facility types $\Lambda = \{0, 1, \ldots, L\}$ with capacities $Z_0 < Z_1 < \ldots < Z_L$ and cost functions C_{ij}^l denoting the cost of a facility of type l installed between nodes i and j. We wish to find a minimum cost *tree* network on G to carry the traffic from the terminal nodes to the central node.

We will restrict our attention to (realistic) cost functions that exhibit economies of scale and are typical in communication networks. In other words, the cost of

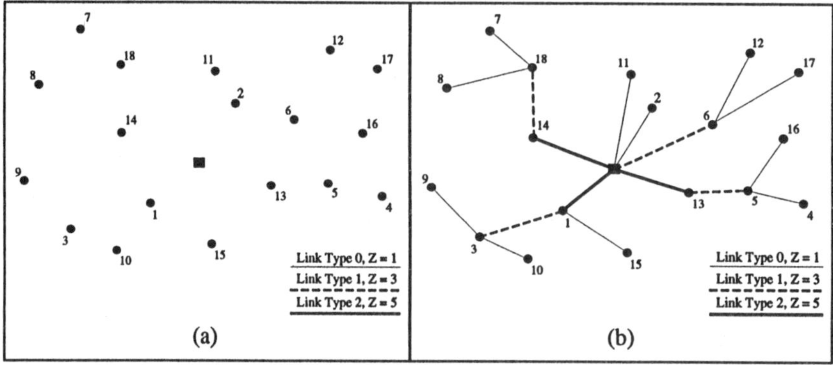

Figure 1.18. Multi-Level Capacitated Minimum Spanning Tree. (a) Nodes in the network. (b) Feasible Multi-Level Capacitated Spanning Tree.

each facility satisfies the relationship $C_{ij}^{y} \leq \frac{Z_y}{Z_x} C_{ij}^{x}$ for every edge $\{i, j\} \in E$, and $x < y$. We also impose the condition that only a single facility type is permitted to be installed on an edge. This condition is actually not restrictive as a problem without this restriction can be transformed to one with this restriction (see [48]). Finally, in the following discussion, we only deal with the homogenous demand case in which $W_i = \kappa$, $\forall i \in V$. Like most of the capacitated network design problems, the MLCMST is very difficult to solve and since the CMST is a special case for $|\Lambda| = 1$, it is NP-hard.

1.5.1 Savings Heuristic

We first briefly describe a savings heuristic for the MLCMST problem (complete details of this heuristic procedure are in a working paper [23]). This heuristic starts with a feasible star solution in which all nodes in V are connected to the central node c, with the lowest capacity link (i.e., type 0). The central idea behind the heuristic is to upgrade the connection of a node i that is currently connected with a type 0 link, to a higher capacity link, while connecting other nodes to i. Obviously, we would like to reduce the overall cost of the network after the upgrade and the reconnections. In order to ensure this for each node i, we compute the savings, D_i^l introduced by upgrading to a type l link,

$$D_i^l = C_{i\text{pred}(i)}^{0} - C_{i\text{pred}(i)}^{l} + \sum_{j \in H} d_{ij}, \tag{1.1}$$

where $\text{pred}(i)$ denotes the first node on the path from node i to the central node. The first two terms in (1.1) represent the change in cost involved in upgrading the current connection of i to a type l link. The third term is the sum of the savings introduced when reconnecting the nodes j in H, to i. Note that D_i^l is only defined for nodes that are currently connected with a type 0 link (for all other nodes, we assume $D_i^l = 0$). For a given node j, $d_{ij} = C^0_{j\mathrm{pred}(j)} - C^0_{ji}$ is the savings obtained by reconnecting node j from its current connection to node i. Again, notice that the reconnection is only possible if node j is currently connected with a type 0 link. We would like to define the set H in a way that maximizes the sum of the savings in (1.1). This can be done by finding all j for which $d_{ij} > 0$ and sorting them in decreasing order. Next, consider all j in sorted order and attempt to connect them to i. If j can be connected to i without violating any of the capacity constraints of the links that are on the path from i to the central node, we add j to H. Otherwise, we consider the next node.

The heuristic starts with $l = L$. Once the calculation of all savings D_i^l, for $i = V\backslash\{c\}$, is complete, we select the largest one and implement the upgrade and the reconnections. We then repeat the computations and implement the changes as long as $D_i^l > 0$. When we can no longer find positive savings, we set $l := l - 1$ and repeat the same steps while $l \geq 0$.

1.5.2 Genetic Algorithm

The solutions provided by the savings heuristic, although reasonable, can be significantly improved. We now discuss the main characteristics of a genetic algorithm (GA) developed for the MLCMST that improves upon the savings heuristic. Conceptually, our genetic algorithm approach splits the MLCMST problem into two parts: a grouping problem and a network design problem. The grouping problem aims at finding the lowest cost, feasible partition, $S_1, S_2, \ldots, S_m$ of the set $V\backslash\{c\}$. The partition is feasible if $\sum_{i \in S_\mu} W_i \leq Z_L$, for all $\mu = 1, 2, \ldots, m$. The cost of a partition is determined by finding a minimum cost tree T_μ on all of the sets $S_\mu \cup \{c\}$ and then summing the costs of the individual trees. Although the network design problem on the partition (finding T_μ) is also an MLCMST problem, in practice, we will have to solve it on a much smaller graph than the original. Our GA algorithm searches through possible partitions and evaluates their cost by calling on the savings heuristic we presented earlier.

The steps of the GA procedure are described in Figure 1.19. First, an initial population of feasible partitions (or individuals) $P(t)$ is created. Next, the fitness of each individual in that population is evaluated. In this case, the fitness of a partition is equal to $\sum_{\mu=1\ldots m} c(T_\mu)$ ($c(T_\mu)$ denotes the cost of T_μ). The evaluation of the fitness of the population is required to check the termination condition of the algorithm and during the selection of individuals. At first, we *select* r individuals (parents) from the old population $P(t-1)$ for reproduction (crossover). The crossover operation results in r new individuals (children) which are assigned to the new population, $P(t)$. Next, ($pop\_size - r - m$) individuals are *selected* from $P(t-1)$ and are copied to $P(t)$. Additionally,

```
Begin
  t ⟵ 0
  initialize P(t)
  evaluate P(t)
  while (not termination-condition) do
    t ⟵ t+1
    select r parents from P(t − 1)
    let the r parents crossover and generate r offspring
    insert the r offspring to P(t)
    select (pop_size − r − m) individuals from P(t − 1) and
       copy them to P(t)
    take the best m individuals of P(t − 1) and mutate them
    insert the mutated individuals to P(t)
    evaluate P(t)
  end while
end
```

Figure 1.19. Steps of the Genetic Algorithm for the MLCMST problem.

the best m individuals are chosen from $P(t-1)$ to be mutated. The mutated individuals then become part of $P(t)$. Finally, the new population is evaluated and the procedure starts over. Note that all selections throughout the algorithm are done with regard to the fitness values of the individuals (i.e., fitter individuals have a higher chance of being selected than less-fit individuals). We now give a brief overview of the most important aspects of the GA. Additional details can be found in our working paper [23].

Initial Population. The initial population is generated by running our savings heuristic and the Esau-Williams [18] heuristic (with link capacity Z_L) on perturbed versions of the problem graph.

Selection. All selections are done with the typical roulette wheel mechanism on fitness values that were transformed through a *Sigma Truncation* function [38]. This function, $f'(i) = f(i) + (\overline{f} - \gamma\sigma)$, computes the transformed fitness value $f'(i)$ of an individual i, from the original $f(i)$, the population average $\overline{f}$, and the population standard deviation σ. γ is a constant that can be used to control selective pressure.

Representation. Our genetic algorithm uses a representation proposed by Falkenauer [19] for grouping problems in general. This representation consists of an item part and a group part. The item part consists of the group labels assigned to each node and the group part contains a list of the group labels. The representation is shown in Figure 1.7 in the context of the CMST problem. In that case, nodes that belong in the same group are connected with a MST. In our problem, nodes in the same group have to still be connected by a multi-level capacitated spanning tree.

Crossover. Crossovers are handled as suggested by Falkenauer [19]. The crossover operator is a typical two-point crossover but acts only on the group part of the representation. The item part of the representation is updated once the crossover is complete based on the changes done in the group part.

Mutation. Mutations can be handled in many different ways. We use an elaborate local search procedure, which we briefly discuss in Section 1.5.4.

1.5.3 Computational Results

Extensive computational testing on both the savings heuristic and the genetic algorithm was done on problem instances with different problem sizes ($N = 20$, 30, 50, and 100) and different central location positions (center, edge, and random). Each problem set (combination of size and central location position) contained 50 randomly generated problem instances. The results of the heuristics were compared to optimal solutions for smaller problems (for which $N = 20$ and some of the instances for which $N = 30$) and to lower bounds. The lower bounds were generated by linear programming relaxations of mathematical formulations for the MLCMST problem. Both procedures were coded in C++ and all runs were done on a Dual Processor Pentium III PC running Windows 2000 with 1GHz clock speed and 512MB of RAM.

When compared to the optimal solutions, the savings heuristic and the GA found solutions that were on average, within 4.29% and 0.25% of the optimal solutions, respectively. Specifically, the GA solutions were never more than 2.12% away from optimality. For the savings heuristic, however, in the worst case the results were almost 12% away from optimality. When compared to the lower bounds, the savings heuristic was, on average, within 9.91% of the bound with 20.00% being the worst case. The GA was within 6.09% of the lower bound, on average, and never exceeded 11.18%. It is important to note here that for smaller problems the lower bounds were found to be within 6.14% of the optimal solution, on average. In terms of running times, the savings heuristic runs within 0.05 of a second for all problems and the GA runs on average within 45 minutes for larger problems (i.e., $N = 100$).

1.5.4 Remarks

The MLCMST problem is a challenging network design problem that arises in practice. The results presented here reveal a fairly strong performance from the GA. The savings heuristic provides weaker solutions than the GA, which might be expected from a greedy heuristic for a challenging problem like this. One possible avenue for research is to develop local search (LS) procedures that use neighborhood structures similar to those originally developed for the CMST problem. As we mentioned earlier Ahuja et al. [2, 3] present a multi-exchange neighborhood structure that defines a very extensive neighborhood for CMST problems, and present efficient search procedures for this neighborhood structure. Based on this neighborhood, and in conjunction with our savings heuristic, we have developed a LS algorithm for the MLCMST problem that performs quite well. Complete details on this procedure are in our working paper [23].

There are many potential directions for future research. Different possibilities include improving the savings heuristics, genetic algorithms based on tree (instead of group) representations, and alternative local search procedures. Finally, the heterogenous demand case is even more challenging and many promising directions for a solution procedure exist.

1.6 Selected Annotated Bibliography

In this section, we provide an annotated bibliography of recent and interesting applications of heuristic search to network design. While the list is certainly not complete, it includes representative work in network design. We have also limited the list to the less extensively studied problems, so we did not include problems such as the Steiner tree problem. The following selection of papers is classified into five groups depending upon the heuristic search technique used: local search, tabu search, genetic algorithms, simulated annealing, and finally approaches that use multiple heuristic search techniques in one place. Under each classification we describe a few papers (usually containing a novel idea or with very good computational results) that are representative of the recent work in that particular heuristic search area.

Interested readers may find more on the general use of these heuristic search techniques in the following references. Surveys of different strategies related to local search, such as iterated local search, variable neighborhood, and guided local search, along with applications of these procedures can be found in the recently edited handbook on metaheuristics by Glover and Kochenberger [28]. More recent applications of tabu search can be found in the survey paper by Hindsberger and Vidal [33], while additional selected applications of heuristic approaches in telecommunication networks can be found in the special issue of the *Journal of Heuristics* edited by Doverspike and Saniee [13].

Applications of Local Search

R. K. Ahuja, J. B. Orlin, and D. Sharma, 2001. "A composite neighborhood search algorithm for the capacitated minimum spanning tree problem," Operations Research Letters, Vol. 31, pp. 185–194.
Ahuja et al. present a neighborhood structure for the CMST problem that allows for multiple exchanges between the subtrees of a feasible tree. Each subtree can be part of the exchange by contributing one of its nodes or one of its subtrees. A new, directed graph is constructed on which all the exchanges and their savings are represented by cycles. The authors also present an exact algorithm that finds the best multi-exchange by searching this new graph. Their computational results indicate that a local search procedure based on this large-scale multi-exchange neighborhood is able to find the best-known solutions for a set of benchmark problems and improve upon the best-known solution in 36% of the cases.

B. Gendron, J.-Y. Potvin, and P. Soriano, 2002. "Diversification strategies in local search for a nonbifurcated network loading problem," European Journal of Operational Research, Vol. 142, pp. 231–241.
Gendron et al. studied the impact of different diversification strategies on the performance of the local search procedure for a nonbifurcated network loading problem. This problem occurs in telecommunications networks, where we are given a central supply node and a set of demand nodes. We are also given multiple facilities of different capacity and cost that can be installed on links between the nodes. The goal is to lay out facilities in the network so that all demand requirements are met and the total network cost is minimized. Figure 1.20 shows an example of this problem. The proposed solution procedure alternates between construction and local search phase. Gendron et al. presented four diversification strategies: adaptive memory [6], greedy multistart, 2-opt neighborhood, and greedy k-opt. Computational experiments on networks with up to 500 nodes indicate that the best diversification strategy is greedy k-opt, which significantly outperforms other diversification strategies.

M. G. C. Resende, and C. C. Ribeiro, 2003. "A GRASP with path-relinking for private virtual circuit routing," Networks, Vol. 41, No. 3, pp. 104–114.
Resende and Ribeiro have developed a version of GRASP for the private virtual circuit (PVC) routing problem, where the goal is to route off-line a set of PVC demands over a backbone network, so that a linear combination of propagation and congestion-related delays is minimized. The PVC problem occurs in a frame relay service, which provides private virtual circuits for customers connected through a large backbone network. Each customer's request can be seen as a commodity specified by its origin and destination, and required bandwidth,

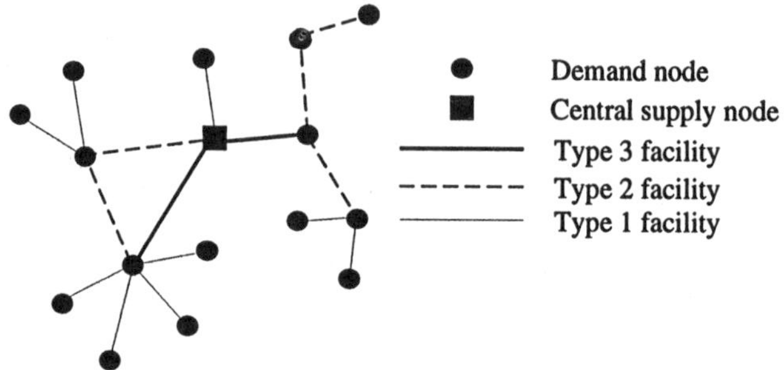

Figure 1.20. An example of a nonbifurcated network loading problem.

and must be routed over a single path, that is, without bifurcation. The backbone network may consist of multiple links between pairs of nodes and any given link can support a limited bandwidth and a limited number of PVCs. Resende and Ribeiro discussed different path-relinking strategies (a technique described in Glover [27]) that can be used along with their GRASP procedure, and tested their performance on networks with up to 100 nodes and 9900 commodities. These tests showed that, among the tested variants, the most effective variant of GRASP is backward path-relinking from an elite solution to a locally optimal solution. These experiments also showed that the proposed version of GRASP for the PVC routing significantly outperforms other, simpler, heuristics used by network planners in practice.

D. Fontes, E. Hadjiconstantinou, and N. Christofides, 2003. "Upper bounds for single-source uncapacitated concave minimum-cost network flow problems," Networks, Vol. 41, No. 4, pp. 221–228.
The single-source uncapacitated concave min-cost network flow problem (SSU concave MCNFP) is a network flow problem, with a single source node and concave costs on the arcs. This models many real-world applications. Fontes et. al. proposed a local search algorithm for the SSU concave MCNFP. It starts by finding a lower bound obtained by a state-space relaxation of a dynamic programming formulation of the problem. This lower bound is improved by using a state-space ascent procedure (that considers penalties for constraint violations and state-space modifications). The solutions obtained by the lower bound procedure are first corrected for infeasiblities by an upper-bounding procedure and then iteratively improved by applying local search over extreme flows (extreme flows correspond to spanning tree solutions rooted at the source node and spanning all demand vertices). The extensive tests for two types of cost functions (linear fixed-charge and second-order concave polynomial) over a large set of instances with up to 50 nodes and 200 arcs have shown that the new procedure provides optimal solutions in most instances, while for the remaining few, the average percent gap is less than 0.05%.

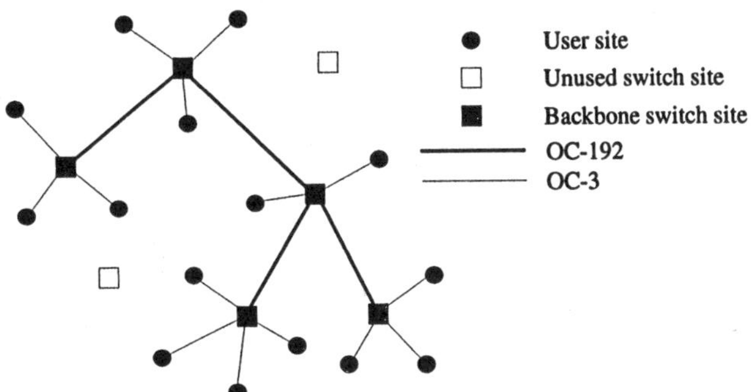

Figure 1.21. An example of a two-level telecommunication network with a tree-backbone network and a star-access network.

Applications of Tabu Search

S. Chamberland, B. Sanso, and O. Marcotte, 2000. "Topological design of two-level telecommunication networks with modular switches," Operations Research, Vol. 48, No. 5, pp. 745–760.

Chamberland et al. have developed a tabu search algorithm that improves upon a greedy heuristic for the design of two-level telecommunication networks. In this problem we are given a set of possible switch sites that need to be connected by high capacity links (OC-192 links) into a tree or ring backbone network. We are also given a set of user nodes that need to be connected to a backbone network using low capacity links (OC-3 links) and a star network structure. Figure 1.21 shows an example of a two-level telecommunication network with a tree backbone network. The low capacity (OC-3) links from users are connected to the switch either directly or via a multiplexer. In either case a "port" of the appropriate capacity must also be installed on the switch (in a "slot" on the switch). There are multiplexers of different capacities and costs, and each switch port can connect to at most one multiplexer. By using a multiplexer, multiple users can be connected to a single port (for example an OC-12 multiplexer can handle 4 users). A backbone link between two switches occupies a port on each switch. Each port is installed into a slot on the switch. There is a constraint on the number of slots at each switch site. The network design must be determined so that the total traffic requirements are met and the total network cost is minimized. Chamberland et al. created initial solutions using a greedy heuristic, which were then improved by a tabu search algorithm where the neighborhood structure is defined by the states of slots at switch sites (these states indicate whether a slot is occupied, and if so by what type of port). The tabu search algorithm was tested on networks with up to 500 users and 30 potential switch sites. The proposed algorithm provided results that were, on average, 0.64% and 1.07% from the optimum for the ring and tree topologies, respectively.

D. Berger, B. Gendron, J. Potvin, S. Raghavan, and P. Soriano, 2000. "Tabu search for a network loading problem with multiple facilities," Journal of Heuristics, Vol. 6, No. 2, pp. 253–267.
Berger et al. presented the first tabu search algorithm for a network loading problem with multiple facilities. (This is the same problem as in [26], and an example can be seen in Figure 1.20). The neighborhood of the search space is explored by using the kth shortest path algorithm to find alternative paths for demand nodes in a given solution. Computational experiments on networks with up to 200 nodes and 100 demand nodes indicated that tabu search provides solutions better than using a steepest-descent local search heuristic based on the 1-opt and 2-opt neighborhoods.

L. A. Cox and J. R. Sanchez, 2000. "Designing least-cost survivable wireless backhaul networks," Journal of Heuristics, Vol. 6, No. 4, pp. 525–540.
Cox and Sanchez developed a short-term memory tabu search heuristic, with embedded knapsack and network flow problems for the design of survivable wireless backhaul networks. In this problem we need to select hub locations and types of hubs along with types of links used in the network so that capacity and survivability constraints are satisfied and network cost is minimized. The tabu search algorithm solved to optimality all problems where the optimum is known in less than 0.1% of time needed to solve the corresponding MIP. On the real-world problems, this algorithm provided 20% improvement over the previously best known designs.

M. Vasquez and J.-K. Hao, 2001. "A heuristic approach for antenna positioning in cellular networks," Journal of Heuristics, Vol. 7, No. 5, pp. 443–472.
Vasquez and Hao developed a three-step procedure utilizing a tabu search algorithm for antenna positioning in cellular networks. In this problem, we need to determine the position, number, and type of antennas at a set of potential locations, while satisfying a set of constraints and optimizing multiple objectives (minimization of the number of the sites used, minimization of the level of the noise within the network, maximization of the total traffic supported by the network, and maximization of the average utilization of each base station). The authors developed a heuristic procedure that was successful in finding feasible solutions for problems where no feasible solution was known previously.

T. G. Crainic and M. Gendreau, 2002. "Cooperative parallel tabu search for capacitated network design," Journal of Heuristics, Vol. 8, pp. 601–627.
Crainic and Gendreau developed a parallel cooperative tabu search met-hod for the fixed charge, capacitated multicommodity flow network design problem. In this problem we are given a set of demand nodes, and we need to route all the traffic through links of limited capacity. However, aside from routing cost, which is proportional to the number of units of each commodity transported over any given link, there is also a fixed charge cost for the use of a link. The proposed method is based on using multiple tabu search threads, which exchange their search information through a common pool of good solutions. Computational experiments indicate that cooperative parallel tabu search outperforms sequential tabu search algorithms, including the version of tabu search developed by the authors previously in [12]. (The previous tabu search algorithm works in combination with column generation, and, in the earlier paper, was found to significantly outperform other solution approaches in terms of the solution quality.)

I. Ghamlouche, T. G. Crainic, and M. Gendreau, 2003. "Cycle-based neighborhoods for fixed-charge capacitated multicommodity network design," Operations Research, Vol. 51, No. 4, pp. 655–667.
Ghamlouche et al. have defined a new neighborhood structure that can be used for heuristics for fixed-charge capacitated multicommodity network design. The new neighborhood structure differs from the previous ones in the sense that moves are implemented on cycles in the network, instead of individual arcs. More specifically, the proposed neighborhood structure is based on identifying two paths for a given node pair (thus identifying a cycle) and deviating flow from one to another path in this cycle so that at least one currently used arc becomes empty. The exploration of the neighborhood is based on the notion of γ-residual networks, which together with a heuristic for finding low-cost cycles in these networks, is designed to identify promising moves. The new neighborhood structure was implemented within a simple tabu search algorithm, and computational experiments on problems with up to 700 arcs and 400 commodities show that the proposed tabu search algorithm outperforms all existing heuristics for fixed-charge capacitated multicommodity networks.

Applications of Genetic Algorithms

L. Davis, D. Orvosh, L. Cox, and Y. Qiu, 1993. "A genetic algorithm for survivable network design," in Proceedings of the Fifth International Conference on Genetic Algorithms, pp. 408–415, Morgan Kaufmann, San Mateo, CA.
Although this paper is more than a decade old, we include it in this bibliography, since it represents a nice example of how a chromosome structure can be

used to simplify evaluation of chromosomes in problems with complex constraints/requirements. Davis et al. developed a GA with a hybrid chromosome structure for the survivable network design problem where a set of links in the network needs to be assigned a bandwidth (capacity) so that all the demand is routed and survivability requirements are met (these requirements may impose only a certain percentage of the traffic be routed in case of a single link failure in the network). The proposed hybrid chromosome structure contains three groups of genes designed to reduce the complexity of chromosome evaluation with respect to routing and survivability requirements. The first group is the list of capacities assigned to each link, while the remaining two groups represent ordered lists (permutations) of traffic requirements used for the approximate evaluation of the chromosome using greedy heuristics. Computational experiments indicate that the GA provides near optimal solutions (obtained by solving an MIP for this problem) on small problems with 10 nodes and up to 21 links. On the larger problems, with 17 nodes and up to 31 links, where the optimum is not known, the GA outperforms greedy heuristics.

C. C. Palmer and A. Kershenbaum, 1995. "An approach to a problem in network design using genetic algorithms," Networks, Vol. 26 pp. 151–163.
In the optimal communication spanning tree problem (OCSTP) we are given a set of nodes, flow requirements between pairs of nodes, and an underlying graph with the cost per unit of flow for each edge. We would like to design a minimum cost tree network to carry the flow. The cost of an edge in this tree network is the product of cost per unit flow of the edge and the total flow (over all demand pairs) carried on that edge in the tree network design. Palmer and Kershenbaum proposed a set of criteria that can be used for selection of adequate genetic encoding for the OCSTP. They discuss applicability of existing encodings (Prüfer number, characteristic vector, and predecessor encoding), and present a new type of chromosome encoding called link and node bias (LNB) that meets desired criteria for encodings to a greater extent. The GA with the LNB encoding was tested on networks with up to 98 nodes. The results indicate that the proposed procedure provides results up to 7.2% better than results obtained by pure random search or a previously developed heuristic for the OCSTP.

Y. Li and Y. Bouchebaba, 1999. "A new genetic algorithm for the optimal communication spanning tree problem," in Artificial Evolution, 4th European Conference, Dunkerque, France, November 3-5, 1999, Lecture Notes in Computer Science, 1829, Springer, 2000.
Li and Bouchebaba developed a new genetic algorithm for the optimal communication spanning tree problem, where a chromosome is represented by a tree structure. The genetic operators are specific for the proposed tree chromosome

representation, and work directly on the tree structure through exchange of paths and subtrees (crossover operators), or insertion of randomly selected paths or subtrees (mutation operators). Experiments over a small set of instances on complete graphs with up to 35 nodes indicate that the proposed GA provides higher quality solutions than previously developed GAs.

H. Chou, G. Premkumar, and C.-H. Chu, 2001. "Genetic algorithms for communications network design–An empirical study of the factors that influence performance," IEEE Transactions on Evolutionary Computation, Vol. 5, No. 3, pp. 236–249.
Chou et. al. looked at the impact of three factors on the performance of the GA developed for the degree-constrained minimum spanning tree problem: two forms of genetic encoding (Prüfer and determinant encoding) and different versions of mutation and crossover operators. Computational experiments indicated that, in terms of solution quality, the best combination of genetic factors is determinant encoding, exchange mutation (randomly selects two positions in a given chromosome and exchange genes), and uniform crossover (a set of positions, called a mask, is chosen for a chromosome, and their alleles are exchanged with each other based on the generated positions). The results also indicate that, among tested factors, genetic encoding has the greatest impact on solution quality.

F. G. Lobo and D. E. Goldberg. "The parameter-less genetic algorithm in practice," to appear in Information Sciences.
Lobo and Goldberg provided an overview of a parameter-less genetic algorithm that was introduced by Harik and Lobo [32]. The notion behind a parameter-less genetic algorithm is to minimize or eliminate the specification of parameters for a genetic algorithm. Typically, a significant amount of testing is necessary to come up with a good choice of values for GA parameters. The parameter-less genetic algorithm solely uses the crossover operator. It uses a fixed value for the selection rate and probability of crossover. These values are shown to comply with the schema theorem and ensure the growth of building blocks (see [38] for more on the schema theorem). The parameter-less genetic algorithm continuously increases the population size until a solution of desired quality is obtained. This is done in a novel fashion, by running the genetic algorithm on multiple populations. Larger number of iterations are run on smaller populations, while a smaller number of iterations are run on larger populations (in particular if population a is 2^k times larger than population b, the algorithm runs 4^k iterations on population b for every iteration on population a). When the average fitness of a small population is lower than the average fitness of a larger population, the smaller population is eliminated. The algorithm is run until a user-specified stopping criteria, is met. The authors illustrate the parameter-

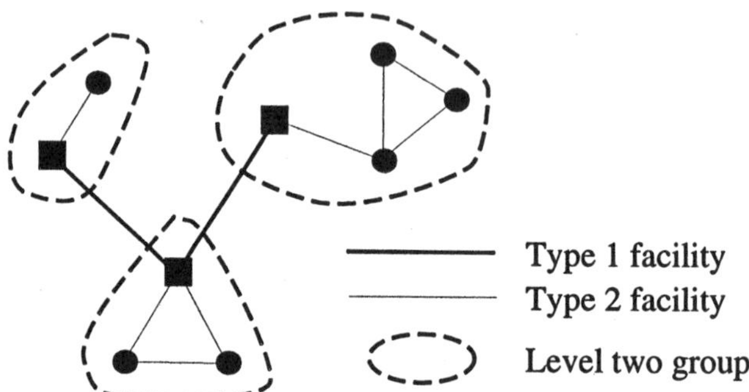

Figure 1.22. An example of a two-level hierarchical network. Type 1 facilities have higher capacity than type 2 facilities.

less genetic algorithm by applying it to an electrical utility network expansion problem.

Applications of Simulated Annealing

T. Thomadsen and J. Clausen, 2002. "Hierarchical network design using simulated annealing," Technical report, Informatics and Mathematical Modelling, Technical University of Denmark, DTU.
Thomadsen and Clausen define a problem in which a minimum cost hierarchical network has to be designed to satisfy demands between origin/destination pairs of nodes. In this problem, we are given a set of different facility types $\Gamma = \{1, 2, \ldots, L\}$ with varying capacities that can be installed between a pair of nodes. We are also given the cost of installing these facilities. We would like to design a minimum cost *hierarchical network* with sufficient capacity to route the traffic. The requirements of a hierarchical network are specified as follows. The network must be connected. Further, each connected subgraph defined by facilities of type l defines a group of level l. A hierarchical network requires that every group, other than those groups defined by the highest capacity facility in the network, contains exactly one node that is connected to a higher capacity facility. In other words, a group of level l contains exactly one node that is connected to a facility with capacity greater than that of facility type l. Figure 1.22 provides an example of a two-level hierarchical network. Thomadsen and Clausen develop a simulated annealing algorithm that is based on a two-exchange neighborhood and report solutions with up to 100 nodes and two levels in the network hierarchy.

K. Holmberg and D. Yuan, 2004, "Optimization of Internet protocol network design and routing," Networks, Vol. 43, No. 1, pp. 39–53.
Holmberg and Yuan developed a simulated annealing (SA) heuristic for the network design and routing of Internet Protocol traffic. In this problem, we need to determine both the topology of the network and the capacity of the links so that the total network cost is minimized. (In this version of the problem, two types of costs are encountered: link costs represented by a fixed charge and linear capacity expansion cost, and a penalty cost for lost demand.) Computational experiments on networks with up to 24 nodes and 76 arcs have shown that, in general, SA provides results as good as a MIP-based sequential procedure developed in the same paper. SA, however, proves to be much better in instances with a large number of commodities.

Other Applications

F. Altiparmak, B. Dengiz, and A. E. Smith, 2003, "Optimal design of reliable computer networks: A comparison of metaheuristics," Journal of Heuristics, Vol. 9, pp. 471–487.
Altiparmak et al. looked at the design of reliable computer networks, where network topology is fixed, i.e., the node locations and links connecting nodes are already specified. However, the type of equipment (differing in reliability and cost) used at each node and link needs to be determined, so that subject to a given budget, the total network reliability is maximized. This paper provides a comparison between three heuristic search techniques—steepest descent, simulated annealing, and genetic algorithm—and hybrid forms of SA and GA (referred to as seeded SA and GA), where both SA and GA use the best solution found by steepest descent at the initial stage of the search. Additionally, a memetic algorithm, a GA with steepest descent applied to all new solutions generated in the reproduction phase, was tested against other proposed heuristic search techniques. Computational experiments performed on graphs with up to 50 nodes and 50 links indicated that SA and GA outperform steepest descent, while seeded SA and GA provide only minor improvements over the original SA and GA. The memetic algorithm, on the other hand, turned out to be the best heuristic search technique for this problem in terms of solution quality.

G. Carello, F. Della Croce, M. Ghirardi, and R. Tadei, "Solving the hub location problem in telecommunication network design: a local search approach," to appear in Networks.
Carello et al. developed a two-step solution procedure for the hub location problem in telecommunication networks. In this problem, we are given a set of access nodes with certain traffic demand, and a set of candidate transit nodes that are used to interconnect the access nodes. The selected transit nodes represent

a fully connected backbone network, and the access nodes along with the links connecting these nodes to the backbone network form the tributary network. The costs encountered in this problem include fixed costs for opening a given transit node and installing the necessary node equipment, and the connection costs for all links deployed in the network. More formally, the network hub location problem can be defined as follows: Given a set of access nodes, a set of candidate transit nodes, and a traffic matrix for each type of traffic, select a set of transit nodes and route all the traffic in the network so that all capacity constraints are satisfied and the total network cost is minimized. In the first step of the solution procedure, where only transit nodes are selected, four heuristic procedures were proposed: local search, tabu search, iterated local search, and a random multistart procedure. For the second step, a local search procedure with a choice of limited or complete neighborhood search was proposed. Computational experiments over 19 test instances indicated that a version of tabu search with quadratic neighborhood and with a complete neighborhood search in the second step produces the best results with average error of 0.2%.

1.7 Conclusion

The number of applications of heuristic search approaches to network design problems in the literature has significantly increased over the last decade. There are several reasons for this. First, these procedures are often simpler to implement compared to more complex exact procedures. Second, researchers have made significant advances in developing neighborhoods and searching effectively through them. Third, there have also been significant advances in the design of genetic algorithms, and in particular in combining genetic algorithms with other heuristic search procedures. Finally, there has been a rapid and steady increase in the average processor speed of desktop/personal computers. Due to all of these factors, when artfully applied, heuristic search techniques enable us to obtain near-optimal (i.e., high-quality) solutions for difficult problems where existing exact procedures fail or require prohibitively long computation times.

In this chapter, we have tried to provide specific guidelines applicable to a wide class of network design problems. First, we have described some general guidelines, where we have tried to cover a broad range of solution approaches, starting from the most basic ones to the more complex procedures such as local search, steepest descent, GRASP, simulated annealing, tabu search, and genetic algorithms. We have paid special attention to guidelines for using genetic algorithms in network design, since in our experience, this heuristic search technique can provide very good results for this class of problems. To illustrate the application of GAs, we have provided four examples of GAs applied to different network design problems. These four applications differ in the complexity of the GA design, which varies with the nature of the problem. Our sample applications show that for certain problems, such as MLST and PCGMST, very simple genetic algorithms can provide excellent results. More complex design problems, such as the Euclidean non-uniform Steiner tree problem and MLCMST, on the other hand, require more complex GA designs tailored to specific problem characteristics. Finally, we have provided an annotated bibliography that includes a selection of recent applications of heuristic search techniques for numerous network design problems found in the literature.

As a final remark, we would like to emphasize once again that although heuristic search techniques can be a powerful tool for network design problems, the degree to which they are successful depends upon a careful selection of procedures used in the algorithm design. In order to develop a good algorithm, one must keep in mind the specific properties of a particular problem as well as general guidelines for the intelligent application of heuristics.

Acknowledgments

We thank Ian Frommer for assistance with the figures in Section 1.3. The first, third, and fourth authors' research was partially supported by the DoD, Laboratory for Telecommunications Sciences, through a contract with the University of Maryland Institute for Advanced Computer Studies.

References

[1] E. H. L. Aarts, J. H. M. Kost, and P. J. M van Laarhoven. Simulated annealing. In E. H. L. Aarts and J. K. Lenstra, editors, *Local Search in Combinatorial Optimization*, pages 91–120. Wiley, New York, 1997.

[2] R. K. Ahuja, J. B. Orlin, and D. Sharma. A composite neighborhood search algorithm for the capacitated minimum spanning tree problem. *Operations Research Letters*, 31:185–194, 2001.

[3] R. K. Ahuja, J. B. Orlin, and D. Sharma. Multi-exchange neighborhood structures for the capacitated minimum spanning tree problem. *Mathematical Programming*, 91:71–97, 2001.

[4] A. Amberg, W. Domschke, and S. Voß. Capacitated minimum spanning trees: Algorithms using intelligent search. *Combinatorial Optimization: Theory and Practice*, 1:9–40, 1996.

[5] J. C. Bean. Genetic algorithms and random keys for sequencing and optimization. *ORSA Journal on Computing*, 6(2):154–160, 1994.

[6] D. Berger, B. Gendron, J. Y. Potvin, S. Raghavan, and P. Soriano. Tabu search for a network loading problem with multiple facilities. *Journal of Heuristics*, 6(2):253–267, 2000.

[7] T. Brüggeman, J. Monnot, and G. J. Woeginger. Local search for the minimum label spanning tree problem with bounded color classes. *Operation Research Letters*, 31:195–201, 2003.

[8] R. S. Chang and S. J. Leu. The minimum labeling spanning trees. *Information Processing Letters*, 63(5):277–282, 1997.

[9] G. Clarke and J. Wright. Scheduling of vehicles from a central depot to a number of delivery points. *Operations Research*, 12(4):568–581, 1964.

[10] T. H. Cormen, C. E. Leiserson, and R. L. Rivest. *Introduction to Algorithms*. MIT Press, Cambridge, MA, 1990.

[11] C. S. Coulston. Steiner minimal trees in a hexagonally partitioned space. *International Journal of Smart Engineering System Design*, 5:1–6, 2003.

[12] T. G. Crainic and M. Gendreau. A simplex-based tabu search method for capacitated network design. *INFORMS Journal on Computing*, 12(3):223–236, 2000.

[13] R. Doverspike and I. Saniee, editors. Heuristic approaches for telecommunications network management, planning and expansion. *Journal of Heuristics*, 6(1):1–188, 2000.

[14] M. Dror, M. Haouari, and J. Chaouachi. Generalized spanning trees. *European Journal of Operational Research*, 120:583–592, 2000.

[15] D. Du, J. H. Rubenstein, and J. M. Smith, editors. *Advances in Steiner Trees*. Kluwer Academic Publishers, 2000.

[16] C. Duin and S. Voß. The pilot method: A strategy for heuristic repetition with application to the Steiner problem in graphs. *Networks*, 34(3):181–191, 1999.

[17] C. Duin and S. Voß. Solving group Steiner problems as Steiner problems. *European Journal of Operational Research*, 154:323–329, 2004.

[18] L. R. Esau and K. C. Williams. On teleprocessing system design. *IBM System Journal*, 5:142–147, 1966.

[19] E. Falkenauer. A hybrid grouping genetic algorithm for bin packing. *Journal of Heuristics*, 2:5–30, 1996.

[20] C. Feremans. *Generalized Spanning Trees and Extensions*. PhD thesis, Universitè Libre dc Bruxelles, 2001.

[21] P. Festa and M. G. C. Resende. GRASP: An annotated bibliography. In P. Hansen and C. C. Ribeiro, editors, *Essays and Surveys on Metaheuristics*. Kluwer Academic Publishers, Norwell, MA, 2001.

[22] I. Frommer, B. Golden, and G. Pundoor. Heuristic methods for solving Euclidean non-uniform Steiner tree problems. Working paper, Smith School of Business, University of Maryland, 2003.

[23] I. Gamvros, B. Golden, and S. Raghavan. The multi-level capacitated minimum spanning tree problem. Working paper, Smith School of Business, University of Maryland, 2003.

[24] I. Gamvros, S. Raghavan, and B. Golden. An evolutionary approach to the multi-level capacitated minimum spanning tree problem. In G. Anandalingam and S. Raghavan, editors, *Telecommunications Network Design and Management*, chapter 6, pages 99–124. Kluwer Academic Publishing, Norwell, MA, 2003.

[25] B. Gavish. Topological design of telecommunications networks - local access design methods. *Annals of Operations Research*, 33:17–71, 1991.

[26] B. Gendron, J. Y. Potvin, and P. Soriano. Diversification strategies in local search for a nonbifurcated network loading problem. *European Journal of Operational Research*, 142:231–241, 2002.

[27] F. Glover. Tabu search and adaptive memory programming - advances, applications and challenges. In R. S. Barr, R. V. Helgason, and J. L. Kennington, editors, *Interfaces in Computer Science and Operations Research*, pages 1–75. Kluwer, Boston, 1996.

[28] F. Glover and G. Kochenberger, editors. *Handbook of Metaheuristics*, volume 57 of *International Series in Operations Research & Management Science*. Kluwer Academic Publishers, 2002.

[29] F. Glover and M. Laguna. *Tabu Search.* Kluwer Academic Publishers, Norwell, MA, 1997.

[30] B. Golden, S. Raghavan, and D. Stanojević. The prize-collecting generalized minimum spanning tree problem. Working paper, Smith School of Business, University of Maryland, 2004.

[31] B. Golden, S. Raghavan, and D. Stanojević. Heuristic search for the generalized minimum spanning tree problem. *INFORMS Journal on Computing*, to appear.

[32] G. R. Harik and F. G. Lobo. A parameter-less genetic algorithm. In W. Banzhaf, J. Daida, A. E. Eiben, M. H. Garzon, V. Honavar, M. Jakiela, and R. E. Smith, editors, *GECCO-99: Proceedings of the Genetic and Evolutionary Computation Conference*, pages 258–265, San Francisco, CA, USA, 1999. Morgan Kaufmann.

[33] M. Hindsberger and R. V. V. Vidal. Tabu search - a guided tour. *Control and Cyberentics*, 29(3):631–661, 2000.

[34] A. Kershenbaum. When genetic algorithms work best. *INFORMS Journal on Computing*, 9(3):254–255, 1997.

[35] S. O. Krumke and H. C. Wirth. On the minimum label spanning tree problem. *Information Processing Letters*, 66(2):81–85, 1998.

[36] J. B. Kruskal. On the shortest spanning subtree of a graph and the traveling salesman problem. In *Proc. American Math. Society*, volume 2, pages 48–50, 1956.

[37] M. Laguna and R. Martí. GRASP and path relinking for 2-layer straight line crossing minimization. *INFORMS Journal on Computing*, 11:44–52, 1999.

[38] Z. Michalewicz. *Genetic Algorithms + Data Structures = Evolution Programs.* Springer, Heidelberg, 1996.

[39] C. L. Monma and D. F. Shallcross. Methods for designing communications networks with certain two-connected survivability constraints. *Operations Research*, 37(4):531–541, 1989.

[40] Y. S. Myung, C. H. Lee, and D. W. Tcha. On the generalized minimum spanning tree problem. *Networks*, 26:231–241, 1995.

[41] C. C. Palmer and A. Kershenbaum. An approach to a problem in network design using genetic algorithms. *Networks*, 26:151–163, 1995.

[42] J. C. Park and C. G. Han. Solving the survivable network design problem with search space smoothing. *in Network Optimization, Lecture Notes in Economics and Mathematical Systems, Springer Verlag*, 450, 1997.

[43] L. S. Pitsoulis and M. G. C. Resende. Greedy randomized adaptive search procedure. In P. M. Pardalos and M. G. C. Resende, editors, *Handbook of Applied Optimization*, chapter 3, pages 168–183. Oxford University Press, 2002.

[44] P. C. Pop. *The Generalized Minimum Spanning Tree Problem.* PhD thesis, University of Twente, 2002.

[45] M. Prais and C. C. Ribeiro. Reactive GRASP: An application to a matrix decomposition problem in TDMA traffic assignment. *INFORMS Journal on Computing*, 12(3):164–176, 2000.

[46] R. C. Prim. Shortest connecting networks and some generalizations. *Bell System Tech. J.*, 36:1389–1401, 1957.

[47] F. Rothlauf, D. E. Goldberg, and A. Heinzl. Network random keys - a tree representation scheme for genetic and evolutionary algorithms. *Evolutionary Computation*, 10(1):75–97, 2002.

[48] F. Salman, R. Ravi, and J. Hooker. Solving the local access network design problem. Working paper, Krannert Graduate School of Management, Purdue University, 2001.

[49] Y. M. Sharaiha, M. Gendreau, G. Laporte, and I. H. Osman. A tabu search algorithm for the capacitated shortest spanning tree problem. *Networks*, 29:161–171, 1997.

[50] S. J. Shyu, P. Y. Yin, B. M. T. Lin, and M. Haouari. Ant-tree: an ant colony optimization approach to the generalized minimum spanning tree problem. *Journal of Experimental and Theoretical Artificial Intelligence*, 15:103–112, 2003.

[51] Y. Wan, G. Chen, and Y. Xu. A note on the minimum label spanning tree. *Information Processing Letters*, 84:99–101, 2002.

[52] Y. Xiong, B. Golden, and E. Wasil. A one-parameter genetic algorithm for the minimum labeling spanning tree problem. Working paper, Smith School of Business, University of Maryland, 2003.

[53] Y. Xiong, B. Golden, and E. Wasil. Worst-case behavior of the MVCA heuristic for the minimum labeling spanning tree problem. *Operations Research Letters*, to appear.

Chapter 2

POLYHEDRAL COMBINATORICS

Robert D. Carr
Discrete Mathematics and Algorithms Department, Sandia National Laboratories
bobcarr@cs.sandia.gov

Goran Konjevod
Computer Science and Engineering Department, Arizona State University
goran@asu.edu

Abstract Polyhedral combinatorics is a rich mathematical subject motivated by integer and linear programming. While not exhaustive, this survey covers a variety of interesting topics, so let's get right to it!

Keywords: combinatorial optimization, integer programming, linear programming, polyhedron, relaxation, separation, duality, compact optimization, projection, lifting, dynamic programming, total dual integrality, integrality gap, approximation algorithms

Introduction

There exist several excellent books [14, 24, 30, 35, 37] and survey articles [15, 33, 36] on combinatorial optimization and polyhedral combinatorics. We do not see our article as a competitor to these fine works. Instead, in the first half of our Chapter, we focus on the very foundations of polyhedral theory, attempting to present important concepts and problems as early as possible. This material is for the most part classical and we do not trace the original sources. In the second half (Sections 5–7), we present more advanced material in the hope of drawing attention to what we believe are some valuable and useful, but less well explored, directions in the theory and applications of polyhedral combinatorics. We also point out a few open questions that we feel have not been given the consideration they deserve.

2.1 Combinatorial Optimization and Polyhedral Combinatorics

Consider an undirected graph $G = (V, E)$ (i.e., where each edge is a set of 2 vertices (endpoints) in no specified order). A *cycle* $C = (V(C), E(C))$ in G is a subgraph of G (i.e., a graph with $V(C) \subseteq V$ and $E(C) \subseteq E$) that is connected, and in which every vertex has degree 2. If $V(C) = V$, then C is said to be a *Hamilton cycle* in G. We define the *traveling salesman problem* (TSP) as follows. Given an undirected graph $G = (V, E)$ and a cost function $c : E \to \mathbf{R}$, find a minimum cost Hamilton cycle in G. The TSP is an example of a *combinatorial optimization problem.* In fact, it is an NP-hard optimization problem [31], although some success has been achieved both in approximating the TSP and in solving it exactly [19].

More generally, a *combinatorial optimization problem* is by definition a set, and we refer to its elements as *instances.* Each instance further specifies a finite set of objects called *feasible solutions*, and with each solution there is an associated *cost* (usually an integer or rational number). Solving an instance of a combinatorial optimization problem means finding a feasible solution of minimum (in the case of a *minimization* problem)—or maximum (in the case of a *maximization* problem)—cost. The solution to a combinatorial optimization problem is an algorithm that solves each instance.

A first step in solving a combinatorial optimization problem instance is the choice of representation for the set of feasible solutions. Practitioners have settled on representing this set as a set of vectors in a finite-dimensional space over $\mathbf{R}$. The dimension depends on the instance, and the set of component indices typically has a combinatorial meaning special to the problem, for example, the edge-set E of a graph $G = (V, E)$ that gives rise to the instance (Section 2.2). However, we sometimes just use the index set $[n] = \{1, 2, \ldots, n\}$.

There often are numerous ways to represent feasible solutions as vectors in the spirit of the previous paragraph. Which is the best one? It is often most desirable to choose a vector representation so that the cost function on the (finite) set of feasible solutions can be extended to a linear function on $\mathbf{R}^I$ (where I is an index set) that agrees with the cost of each feasible solution when evaluated at the vector representing the solution.

In Theorems 2.1 and 2.2, we give just one concrete reason to strive for such a linear cost function. But first, some definitions (for those missing from this paragraph, see Section 2.2.2). Let $S \subset \mathbf{R}^I$ be a set of vectors (such as the set of all vectors that represent the (finitely many) feasible solutions to an instance of a combinatorial optimization problem, in some vector representation of the problem instance). A *convex combination* of a finite subset $\{v^i \mid i \in I\}$ of vectors in S is a linear combination $\sum_{i \in I} \lambda_i v^i$ of these vectors such that the scalar multiplier λ_i is nonnegative for each i ($\lambda_i \geq 0$) and

$$\sum_{i \in I} \lambda_i = 1.$$

Geometrically, the set of all convex combinations of two vectors forms the line segment having these two vectors as endpoints. A *convex set* of vectors is a set (infinite unless trivial) of vectors that is closed with respect to taking convex combinations. The *convex hull* of S (conv(S)) is the smallest (with respect to inclusion) convex set that contains S. For a finite S, it can be shown that conv(S) is a bounded set that can be obtained by intersecting a finite number of closed halfspaces. The latter condition defines the notion of a *polyhedron.* The first condition (that the polyhedron is also bounded) makes it a *polytope* (see Theorem 2.31). Polyhedra can be specified algebraically, by listing their defining halfspaces using inequalities as in (2.1).

Now, the theorems that justify our mathematical structures.

THEOREM 2.1 *If the cost function is linear, each minimizer over S is also a minimizer over* conv(S).

THEOREM 2.2 *(Khachiyan, see the reference [14].) If the cost function is linear, finding a minimizer over a polyhedron is a polynomially solvable problem as a function of the size of the input, namely the number of bits in the algebraic description of the polyhedron (in particular, this will also depend on the number of closed halfspaces defining the polyhedron and the dimension of the space the polyhedron lies in).*

Proof: Follows from an analysis of the *ellipsoid method* or an *interior point method.* □

We refer to any problem of minimizing (maximizing) a linear function over a polyhedron as a *linear programming* problem, or just a *linear program.* Often we just use the abbreviation *LP.*

Consider a combinatorial optimization problem defined by a graph $G = (V, E)$ and a cost function on E. We define the size of this input to be $|V| + |E|$ and denote it by $|G|$. Denote the set of feasible solutions for an instance where the input graph is G by S_G. The following possibility, of much interest to polyhedral combinatorics, can occur. It is usually the case that $|S_G|$ grows as an exponential function of $|G|$. In this case, it may be hard to tell whether this combinatorial optimization problem is polynomially solvable. However, it is possible here that both the number of closed halfspaces defining conv(S_G) and the dimension of the space that the convex hull conv(S_G) lies in grow only as polynomial functions of $|G|$. Hence, if we can find this set of closed halfspaces in polynomial time, the above theorems allow us to immediately conclude that our optimization problem is polynomially solvable! This polyhedral approach may not be the most efficient way of solving our problem (and admittedly, there are annoying technical difficulties to overcome when the minimizer found for $\mathrm{conv}(S_G)$ is not in S_G), but it is a sure-fire backup if more combinatorial approaches are not successful. We will find out later that polynomial solvability is also guaranteed if there exist polynomially many (as a function of $|G|$) sufficiently well-behaved *families* of halfspaces defining $\mathrm{conv}(S_G)$. This involves the concept of *separation*, which we discuss in Section 2.3.2.

2.2 Polytopes in Combinatorial Optimization

Let us recall the traveling salesman problem of the previous section. When the input graph is undirected, we will refer to the *symmetric* traveling salesman problem (STSP), in contrast to the *asymmetric* traveling salesman problem (ATSP) that results when the input graph is directed. (For more on polytopes and polyhedra related to STSP and ATSP see the surveys by Naddef [28] and Balas and Fischetti [1].) It is convenient to assume that the input graph for the STSP is a complete graph $K_n = (V_n, E_n)$ on n vertices for some $n \geq 3$. Along with the complete graph assumption, we usually assume that the cost function satisfies the *triangle inequality* (i.e., defines a *metric* on V_n).

Recall that we wish to have a vector representation of Hamilton cycles so that our cost function is linear. If H is a Hamilton cycle, then the cost $c(H) := \sum_{e \in E(H)} c_e$. So, we cleverly represent each Hamilton cycle H in K_n by its *edge incidence vector* $\chi^{E(H)}$ defined by

$$\chi_e^{E(H)} := \begin{cases} 1 & e \in E(H), \\ 0 & e \in E_n \setminus E(H). \end{cases}$$

Note that now $c(H) = c \cdot \chi^{E(H)}$. Hence, we can extend our cost function from edge incidence vectors of Hamilton cycles to a linear function on the entire space $\mathbf{R}^{E_n}$ by simply defining $c(x) := c \cdot x$ for each $x \in \mathbf{R}^{E_n}$.

The convex hull of all edge incidence vectors of Hamilton cycles in K_n is a polytope called $STSP(n)$. We noted earlier that optimizing a linear cost function over $STSP(n)$ can be done in polynomial time in the number of closed halfspaces defining the polytope, the dimension of the space and the maximum number of bits needed to represent any coefficient in the algebraic description of the halfspaces (the size of the *explicit description*), using linear programming. Unfortunately, every known explicit description of $STSP(n)$ is exponential in size with respect to n. Note that in our representation the vector representing any Hamilton cycle has only integer values, in fact only values in $\{0, 1\}$. This suggests the new problems of *integer programming* (IP) and *0-1 programming*. For each of these problems, analyzing the geometry of polytopes is of value. We introduce some geometric concepts related to polyhedra and polytopes, starting with the notion of a polyhedron's dimension.

2.2.1 Affine Spaces and Dimension

Let $S = \{v^i \mid i \in I\}$ be a set of points in $\mathbf{R}^I$, where I is a finite index set. An *affine combination* on S is a linear combination $\sum_{i \in I} \lambda_i v^i$, where the scalar multipliers λ_i satisfy

$$\sum_{i \in I} \lambda_i = 1.$$

An *affine set (space)* is a set which is closed with respect to taking affine combinations. The *affine hull* $\text{aff}(S)$ of a set S is the smallest (with respect to inclusion) affine set containing S.

We now give two definitions of *affine independence.*

DEFINITION 2.3 *A finite set S is affinely independent iff for any point which can be expressed as an affine combination of points in S, this expression is unique.*

For the second definition, we introduce the notion of an *affine$_0$ combination* on S, which is a linear combination $\sum_{i \in I} \lambda_i v^i$ where the scalar multipliers λ_i satisfy

$$\sum_{i \in I} \lambda_i = 0.$$

DEFINITION 2.4 *A finite set S is affinely independent iff the only affine$_0$ combination of points in S that gives the 0 (the origin) is that where the scalar multipliers satisfy $\lambda_i = 0$ for all $i \in I$.*

DEFINITION 2.5 *A set B is an affine basis of an affine space U if it is affinely independent and* $\text{aff}(B) = U$.

THEOREM 2.6 *The affine hull* $\text{aff}(S)$ *of a set S of points in a finite-dimensional space $\mathbf{R}^I$ has an affine basis of cardinality at most $|I| + 1$. Furthermore, every affine basis of* $\text{aff}(S)$ *has the same cardinality.*

Finally, we can define the *dimension* of a polyhedron P.

DEFINITION 2.7 *The dimension* $\dim(P)$ *of a polyhedron P (or any convex set) is defined to be $|B| - 1$, where B is any affine basis of* $\text{aff}(P)$.

For example, consider the one-dimensional case. Any two distinct points $v^1, v^2 \in \mathbf{R}^I$ define a one-dimensional affine space $\text{aff}\{v^1, v^2\}$ called the *line* containing v^1, v^2.

If we replace "affine" in Definition 2.4 by "linear," and "points" by "vectors," we get the definition of *linear independence*:

DEFINITION 2.8 *A set of vectors $V = \{v^1, \ldots, v^k\}$ is linearly independent, if the only linear combination of vectors in V that gives the 0 vector is that where the scalar multipliers satisfy $\lambda_i = 0$ for all $i \in I$.*

Note that linear independence could also be defined using a property analogous to that of Definition 2.3. Either affine or linear independence can be used for a development of the notion of dimension:

THEOREM 2.9 *A set* $\{x^1, \ldots, x^k, x^{k+1}\}$ *is affinely independent iff the set* $\{x^1 - x^{k+1}, \ldots, x^k - x^{k+1}\}$ *is linearly independent.*

As in the definition of an affine space, a *linear (vector) space* is closed with respect to linear combinations. Let $\text{span}(B)$ denote the smallest set containing all linear combinations of elements of B. There is also a notion of basis for linear spaces:

DEFINITION 2.10 *A set B is a linear basis of a linear space U if it is linearly independent and* $\text{span}(B) = U$.

THEOREM 2.11 *The linear space* $\text{span}(S)$, *for a set S of vectors in a finite-dimensional space* $\mathbf{R}^I$ *has a linear basis of cardinality at most* $|I|$. *Furthermore, every linear basis of* $\text{span}(S)$ *has the same cardinality.*

Affine independence turns out to be more useful in polyhedral combinatorics, because it is invariant with respect to translation. This is not so with linear independence. Another way to see this distinction is to understand that linear (in)dependence is defined for vectors, and affine is for points. The distinction between points and vectors, however, is often blurred because we use the same representation ($|I|$-tuples of real numbers) for both.

2.2.2 Halfspaces and Polyhedra

A simple but important family of convex sets is that of *halfspaces*. We say that a set H is a halfspace in $\mathbf{R}^I$, if both H and its complement H^c are convex and nonempty. An equivalent definition can be given algebraically: a (*closed*) halfspace is a set of the form

$$\{x \in \mathbf{R}^I \mid a \cdot x \geq a_0\}, \tag{2.1}$$

where $a \in \mathbf{R}^I$ and $a_0 \in \mathbf{R}$. Every halfspace is either closed or open and every open halfspace is defined analogously but with a strict inequality instead. (Closed and open halfspaces are respectively closed and open with respect to the standard Euclidean topology on $\mathbf{R}^I$.) This algebraic representation is unique up to a positive multiple of a and a_0. The boundary of a (closed or open) halfspace is called a *hyperplane*, and can be described uniquely up to a non-zero multiple of (a, a_0) as the set

$$\{x \in \mathbf{R}^I \mid a \cdot x = a_0\}.$$

We denote the (topological) *boundary* of a halfspace H by ∂H. It is obvious which hyperplane forms the boundary of a given halfspace.

A *polyhedron* is defined as any intersection of a finite number of closed halfspaces, and thus is a closed convex set. The space $\mathbf{R}^I$ (intersection of 0 closed halfspaces), a closed halfspace (intersection of 1 closed halfspace, namely itself), and a hyperplane (intersection of the 2 distinct closed halfspaces that share a common boundary), are the simplest polyhedra. A bounded polyhedron is called a *polytope*. (The convex hull of a finite set of points is always a polytope, and conversely any polytope can be generated in this way: see Theorem 2.31.)

EXERCISE 2.12 *An alternative definition of a closed halfspace: a set H is a closed halfspace if and only if H is a closed convex set strictly contained in $\mathbf{R}^I$, the dimension of H is $|I|$, and its boundary intersects every line it does not contain in at most one point.*

EXERCISE 2.13 *The intersection of two polyhedra is a polyhedron.*

2.2.3 Faces and Facets

Consider a polyhedron $P \subseteq \mathbf{R}^I$. We say that a closed halfspace (inequality) is *valid* for P if P is contained in that closed halfspace. Similarly, a hyperplane (equation) can be *valid* for P. A halfspace H is *slack* for a set P if H is valid for P and the intersection of P with ∂H is empty.

Every closed halfspace H valid but not slack for a polyhedron P defines a (nonempty) *face* F of P, by

$$F := \partial H \cap P,$$

where ∂H is the boundary of H. Of course, F is also a polyhedron. A face of P that is neither empty nor P is said to be *proper*. A face F of P whose dimension satisfies

$$\dim(F) = \dim(P) - 1,$$

in other words, a maximal proper face, is said to be a *facet* of P. Finally, when a face of P consists of a single point x, this point is said to be an *extreme point* of P.

Corresponding naturally to each facet (face) are one or more *facet-defining* (*face-defining*) inequalities. The only case where this correspondence is unique is for facet-defining inequalities of a *full-dimensional* polyhedron P (i.e., $\dim(P) = |I|$). A polyhedron P that is not full-dimensional is said to be *flat*. The following theorem gives an interesting *certificate* that a polyhedron P is full-dimensional.

THEOREM 2.14 *A polyhedron $P \neq \emptyset$ is full-dimensional iff one can find $x \in P$ such that every closed halfspace defining P is slack for x (i.e., the interior of P is non-empty, with x being in P's interior).*

A set of closed halfspaces whose intersection is $P \neq \emptyset$ forms a *description* of P. It turns out that there is a facet-defining inequality for every facet of P

in any description of P. A *minimal description* consists only of facet-defining inequalities and a minimal set of inequalities (or equations if permitted) that define the affine hull $\text{aff}(P)$ of P. Given a minimal description of P, we refer to the set of hyperplanes (equations) that are valid for P as the *equality system* $P^=$ of P. That is,

$$P^= := \{(a, a_0) \in \mathbf{R}^I \times \mathbf{R} \mid a \cdot x = a_0 \text{ is valid for } P\}.$$

Consider the normal vectors for the hyperplanes of $P^=$, namely

$$N(P^=) := \{a \in \mathbf{R}^I \mid \exists a_0 \in \mathbf{R} \text{ s.t. } (a, a_0) \in P^=\}.$$

Let B be a linear basis for $N(P^=)$. The minimum number of equations needed to complete the description of P will then be $|B|$. We define $dim(P^=) := |B|$. An important theorem concerning the dimension of a polyhedron P follows.

THEOREM 2.15 *If the polyhedron $P \neq \emptyset$ lies in the space $\mathbf{R}^I$, then*

$$\dim(P) = |I| - \dim(P^=).$$

A useful technique for producing a set of affinely independent points $\{x^1, \ldots, x^k\}$ of P is to find a set of valid inequalities (a^i, a_0^i) for $i \in [k-1]$ such that for each i,

$$\begin{aligned} a^i \cdot x^i &> a_0^i, \\ a^i \cdot x^j &= a_0^i, \quad \forall j > i. \end{aligned}$$

THEOREM 2.16 *$\{x^1, \ldots, x^k\}$ is affinely independent iff for each $i \in [k-1]$, one can find (a^i, a_0^i) such that*

$$\begin{aligned} a^i \cdot x^i &\neq a_0^i, \\ a^i \cdot x^j &= a_0^i \quad \forall j > i. \end{aligned}$$

One can determine that a face F of P is a facet by producing $\dim(P)$ affinely independent points on F (showing F is at least a facet), and one point in $P \setminus F$ (showing F is at most a facet).

For a valid inequality $a \cdot x \geq a_0$, we say that it is *less than facet-defining* for P, if the hyperplane defined by the inequality intersects P in a nonmaximal proper face of P. We can use the following theorem to determine that an inequality $a \cdot x \geq a_0$ is less than facet-defining.

THEOREM 2.17 *The inequality $a \cdot x \geq a_0$ is less than facet-defining iff it can be derived from a (strictly) positive combination of valid inequalities $a^i \cdot x \geq a_0^i$, for $i = 1, 2$ and one can find $x^1, x^2 \in P$ such that*

$$\begin{aligned} a^2 \cdot x^1 &> a_0^2, \\ a^2 \cdot x^2 &= a_0^2, \\ a \cdot x^2 &> a_0. \end{aligned}$$

To understand the above, note that the first condition implies that the hyperplane $\{x \mid a^2 \cdot x = a_0^2\}$ does not contain P (that is, the second inequality is not valid as an equation). Together, the second and third condition imply that $a^1 \cdot x \geq a_0^1$ is not valid as an equation and the two valid inequalities $a^i \cdot x \geq a_0^i$, $i = 1, 2$, define distinct faces of P.

Finally, we give a certificate important for determining the dimension of an equality system for a polyhedron.

THEOREM 2.18 *A system of equations $a^i \cdot x = a_0^i$ for $i \in [k]$ is consistent and linearly independent iff one can find x^0 satisfying all k equations and x^i satisfying precisely all equations except equation i for all $i \in [k]$.*

2.2.4 Extreme Points

Extreme points, the proper faces of dimension 0, play a critical role when minimizing (maximizing) a linear cost function. Denote the set of extreme points of a polyhedron P by $\mathrm{ext}(P)$. For any polyhedron P, $\mathrm{ext}(P)$ is a finite set. For the following four theorems, assume that P is a polytope.

THEOREM 2.19 *Let $x \in P$. Then $x \in \mathrm{ext}(P)$ iff it cannot be obtained as a convex combination of other points of P iff it cannot be obtained as a convex combination of other points of $\mathrm{ext}(P)$ iff $\forall x^1, x^2 \in P$ s.t.*

$$\frac{1}{2}x^1 + \frac{1}{2}x^2 = x,$$

we have $x^1 = x^2 = x$.

THEOREM 2.20 *If there is a minimizer in P of a linear cost function, at least one of these minimizers is also in $\mathrm{ext}(P)$. Moreover, any minimizer in P is a convex combination of the minimizers also in $\mathrm{ext}(P)$.*

THEOREM 2.21 *Any point $x \in P$ can be expressed as a convex combination of at most $\dim(P) + 1$ points in $\mathrm{ext}(P)$.*

THEOREM 2.22 *Any extreme point can be expressed (not necessarily in a unique fashion) as an intersection of $\dim(P^=)$ valid hyperplanes with linearly independent normals and another $\dim(P)$ hyperplanes coming from facet-defining inequalities. Conversely, any such non-empty intersection yields a set consisting precisely of an extreme point.*

A set S is *in convex position* (or *convexly independent*) if no element in S can be obtained as a convex combination of the other elements in S. Suppose S is a finite set of solutions to a combinatorial optimization problem instance. We saw that forming the convex hull $\mathrm{conv}(S)$ is useful for optimizing a linear cost function over S.

We have the following theorem about the extreme points of $\text{conv}(S)$.

THEOREM 2.23 $\text{ext}(\text{conv}(S)) \subseteq S$, *and* $\text{ext}(\text{conv}(S)) = S$ *if and only if* S *is in convex position.*

Hence, when minimizing (maximizing) over $\text{conv}(S)$ as a means of minimizing over S, one both wants and is (in principle) able to choose a minimizer that is an extreme point (and hence in S).

The extreme point picture will be completed in the next section, but for now we add these elementary theorems.

THEOREM 2.24 *Every extreme point of a polyhedron P is a unique minimizer over P of some linear cost function.*

Proof: Construct the cost function from the closed halfspace that defines the face consisting only of the given extreme point. □

Conversely, only an extreme point can be such a unique minimizer of a linear cost function. Also, we have the following deeper theorem, which should be compared to Theorem 2.22.

THEOREM 2.25 *If an extreme point x is a minimizer for P, it is also a minimizer for a cone Q (defined in Section 2.2.5) pointed at x and defined by a set of precisely* $\dim(P^=)$ *valid hyperplanes and a particular set of* $\dim(P)$ *valid closed halfspaces whose boundaries contain x.*

Define the *relative interior* of a (flat or full-dimensional) polyhedron P to be the set of points $x \in P$ such that there exists a sphere $S \subset \mathbf{R}^I$ centered at x such that $S \cap \text{aff}(P) \subset P$. A point in the relative interior is topologically as far away from being an extreme point as possible. We then have

THEOREM 2.26 *If x is a minimizer for P and x is in the relative interior of P, then every point in P is a minimizer.*

The famous *simplex method* for solving a linear program has not yet been implemented so as to guarantee polynomial time solution performance [40], although it is fast in practice. However, a big advantage of the simplex method is that it only examines extreme point solutions, going from an extreme point to a neighboring extreme point without increasing the cost. Hence, the simplex method outputs an extreme point minimizer every time. In contrast, its polynomial time performance competitors, interior point methods that go through the interior of a polyhedron, and the ellipsoid method that zeros in on a polyhedron from the outside, have a somewhat harder time obtaining an extreme point minimizer.

2.2.5 Extreme Rays

A *conical combination* (or *non-negative combination*) is a linear combination where the scalar multipliers are all non-negative. A finite set S of vectors is *conically independent* if no vector in S can be expressed as a conical combination of the other vectors in S. Denote the set of all non-negative combinations of vectors in S by $\text{cone}(S)$. Note that $0 \in \text{cone}(S)$.

A (convex) *cone* is a set of vectors closed with respect to non-negative combinations. A *polyhedral cone* is a cone that is also a polyhedron. Note that the facet-defining inequalities for a (full-dimensional or flat) polyhedral cone are all of the form $a^i \cdot x \leq 0$.

THEOREM 2.27 *A cone C is polyhedral iff there is a conically independent finite set S such that* $\text{cone}(S) = C$.

Such a finite set S is said to *generate* the cone C and (if minimal) is sometimes referred to as a *basis* for C. Suppose $x \in C$ has the property that when

$$\frac{1}{2}x^1 + \frac{1}{2}x^2 = x$$

for some $x^1, x^2 \in C$, it follows that x^1 and x^2 differ merely by a nonnegative scalar multiple. When this property holds, x is called an *extreme ray* of C. When C is a polyhedral cone that doesn't contain a line, then the set of extreme rays of C is its unique basis.

Given two convex sets $A, B \subseteq \mathbf{R}^I$, let $A + B = \{u + v \mid u \in A,\ v \in B\}$. The set $A + B$ is usually called the *Minkowski sum* of A and B and is also convex. This notation makes it easy to state some important representation theorems.

A *pointed cone* is any set of the form $a + C$, where a is a point in $\mathbf{R}^I$ and C a polyhedral cone that does not contain a line. Here, a is the unique extreme point of this cone. Similarly, a *pointed polyhedron* is a polyhedron that has at least one extreme point.

THEOREM 2.28 *A non-empty polyhedron P is pointed iff P does not contain a line.*

THEOREM 2.29 *A set $P \subset \mathbf{R}^I$ is a polyhedron iff one can find a polytope Q (unique if P is pointed) and a polyhedral cone C such that $P = Q + C$.*

An equivalent representation is described by the following theorem.

THEOREM 2.30 *Any (pointed) polyhedron P can be decomposed using a finite set S of (extreme) points and a finite set S^0 of (extreme) rays, i.e., written in the form*

$$P = \text{conv}(S) + \text{cone}(S^0).$$

This decomposition is unique up to positive multiples of the extreme rays only for polyhedra P that are pointed. On the other hand, the decomposition of

Theorem 2.30 for the polyhedron formed by a single closed halfspace, which has no extreme points, is far from unique.

A consequence of these theorems, anticipated in the earlier sections is

THEOREM 2.31 *A set $S \subseteq R^I$ is a polytope iff it is the convex hull of a finite set of points in R^I.*

2.3 Traveling Salesman and Perfect Matching Polytopes

A combinatorial optimization problem related to the symmetric traveling salesman problem is the *perfect matching* problem. Consider an undirected graph $G = (V, E)$. For each $v \in V$, denote the set of edges in G incident to v by $\delta_G(v)$, or just $\delta(v)$. Similarly, for $\emptyset \subset S \subset V_n$, define $\delta(S) \subset E_n$ to be the set of those edges with exactly one endpoint in S. For $F \subset E_n$, we also use the notation $x(F)$ to denote $\sum_{e \in F} x_e$. A *matching* in G is a subgraph $M = (V(M), E(M))$ of G such that for each $v \in V(M)$,

$$|E(M) \cap \delta_G(v)| = 1.$$

A *perfect matching* in G is a matching M such that $V(M) = V$. The *perfect matching problem* is: given a graph $G = (V, E)$ and a cost function $c : E \to \mathbf{R}$ as input, find a minimum cost perfect matching. Unlike the STSP (under the assumption $P \neq NP$), the perfect matching problem is polynomially solvable (i.e. in P). However, as we will see later, there are reasons to suspect that in a certain sense, it is one of the hardest such problems.

As in the STSP, we use the edge incidence vector of the underlying graph to represent perfect matchings, so that we obtain a linear cost function. We consider the special case where the input graph is a complete graph $K_n = (V_n, E_n)$ on n vertices. We denote the perfect matching polytope that results from this input by $PM(n)$. (Of course, just as is the case with the traveling salesman problem, there is one perfect matching polytope for each value of the parameter n. Most of the time, we speak simply of *the perfect matching polytope*, or *the traveling salesman polytope*, etc.)

2.3.1 Relaxations of Polyhedra

We would like a complete description of the facet-defining inequalities and valid equations for both $STSP(n)$ and $PM(n)$ (we assume in this section that n is even so that $PM(n)$ is non-trivial). For $STSP(n)$, this is all but hopeless unless $P = NP$. But, there is just such a complete description for $PM(n)$. However, these descriptions can get quite large and hard to work with. Hence, it serves our interest to obtain shorter and more convenient approximate descriptions for these polytopes. Such an approximate description can actually be a description of a polyhedron that approximates but contain the more difficult original polyhedron Z_n, and this new polyhedron P_n is then called a *relaxation* of Z_n. A polyhedron P_n is a *relaxation* of Z_n (e.g. $STSP(n)$ or $PM(n)$), if $P_n \supseteq Z_n$. Often, an advantage of a relaxation P_n is that a linear function can be optimized over P_n in polynomial time in terms of n, either because it has a polynomial number of facets and a polynomial-size dimension in terms of n, or because its facets have a nice combinatorial structure. Performing this optimization then gives one a lower bound to the original problem of minimizing over the more intractable polytope, say $STSP(n)$.

The description of a relaxation will often be a subset of a minimal description for the original polyhedron. An obvious set of valid equations for $STSP(n)$ and $PM(n)$ results from considering the number of edges incident to each node $v \in V_n$ in a Hamilton cycle or a perfect matching. Hence, for each $v \in V_n$, we have the *degree constraints*:

$$\begin{aligned} x(\delta(v)) &= 2 \quad (STSP(n)), \\ x(\delta(v)) &= 1 \quad (PM(n)). \end{aligned}$$

We also have inequalities valid for both $STSP(n)$ and $PM(n)$ that result directly from our choice of a vector representation for Hamilton cycles and perfect matchings. These are $0 \leq x_e \leq 1$ for each $e \in E_n$. Thus,

$$\begin{aligned} x(\delta(v)) &= 2 \quad \forall v \in V_n, \\ 0 \leq x_e &\leq 1 \quad \forall e \in E_n, \end{aligned} \tag{2.2}$$

describes a relaxation of $STSP(n)$. Similarly,

$$\begin{aligned} x(\delta(v)) &= 1 \quad \forall v \in V_n, \\ 0 \leq x_e &\leq 1 \quad \forall e \in E_n, \end{aligned} \tag{2.3}$$

describes a relaxation of $PM(n)$. The polytopes $STSP(n)$ and $PM(n)$ both have the property that the set of integer points they contain, namely $\mathbf{Z}^{E_n} \cap STSP(n)$ ($\mathbf{Z}^{E_n} \cap PM(n)$) is in one-to-one correspondence with the set of extreme points of these polytopes, which are in fact also the vector representations of all the Hamilton cycles in K_n (perfect matchings in K_n). If all the extreme points are integral, and the vector representations for all the combinatorial objects are precisely these extreme points, we call such a polytope an *integer polytope*. When its integer solutions occur only at its extreme points, we call this polytope a *fully integer polytope*.

For the integer polytope Z_n of a combinatorial optimization problem, it may happen that a relaxation P_n is an *integer programming (IP) formulation*. We give two equivalent definitions of an IP formulation that are appropriate even when Z_n is not a fully integer polytope.

DEFINITION 2.32 *The relaxation P_n together with integrality constraints on all variables forms an integer programming formulation for the polytope Z_n iff*

$$\mathbf{Z}^I \cap P_n = \mathbf{Z}^I \cap Z_n.$$

The second definition is:

DEFINITION 2.33 *The relaxation P_n (together with the integrality constraints) forms an IP formulation for the polytope Z_n, iff*

$$P_n \supseteq Z_n \supset P_n \cap \mathbf{Z}^I.$$

These definitions can be modified for the *mixed integer* case where only some of the *variables* are required to be integers. One can see that the constraint set (2.3) results in an IP formulation for $PM(n)$. However, the constraint set (2.2) does not result in an IP formulation for $STSP(n)$ because it does not prevent an integer solution consisting of an edge-incidence vector for 2 or more simultaneously occurring vertex-disjoint non-Hamiltonian cycles, called *subtours*. To prevent these subtours, we need a fairly large number (exponential with respect to n) of *subtour elimination constraints*. That is, for each $\emptyset \subset S \subset V_n$, we need a constraint

$$x(\delta(S)) \geq 2 \tag{2.4}$$

to prevent a subtour on the vertices in S from being a feasible solution. Although (2.3) describes an IP formulation for $PM(n)$, this relaxation can also be strengthened by preventing a subtour over every odd-cardinality set S consisting of edges of value $1/2$. That is, for each $\emptyset \subset S \subset V_n$, $|S|$ odd, we add a constraint

$$x(\delta(S)) \geq 1 \tag{2.5}$$

to prevent the above mentioned fractional subtour on S. We call these *odd cut-set constraints*. It is notable that for even $|S|$ the corresponding fractional subtours (of $1/2$'s) actually are in the $PM(n)$ polytope.

We saw that adding (2.4) to further restrict the polytope P_n of (2.2), we obtain a new relaxation P'_n of $STSP(n)$ that is *tighter* than P_n, that is

$$P_n \supset P'_n \supseteq STSP(n).$$

In fact, P'_n is an IP formulation of $STSP(n)$, and is called the *subtour relaxation* or *subtour polytope* ($SEP(n)$) for $STSP(n)$. Similarly, we can obtain a tighter relaxation P_n^{PM} for $PM(n)$ by restricting the polytope of (2.3) by the constraints of (2.5). In fact, adding these constraints yields the complete description of $PM(n)$, i.e., $P_n^{PM} = PM(n)$ [34]. However, we cannot reasonably hope that $SEP(n) = STSP(n)$ unless $P = NP$.

2.3.2 Separation Algorithms

Informally, we refer to the subtour elimination inequalities of $STSP(n)$ and the odd cut-set inequalities of $PM(n)$ as *families* or *classes* of inequalities. We will now discuss the idea that a relaxation P_n consisting of a family of an exponential number of inequalities with respect to n can still be solved in polynomial time with respect to n if the family has sufficiently nice combinatorial properties. The key property that guarantees polynomial solvability is the existence of a so-called *separation algorithm* that runs in polynomial time with respect to n and its other input (a fractional point x^*).

DEFINITION 2.34 *A separation algorithm for a class $\mathcal{A}$ of inequalities defining the polytope $P_n^{\mathcal{A}}$ takes as input a point $x^* \in \mathbf{R}^I$, and either outputs an inequality in $\mathcal{A}$ that is violated by x^* (showing $x^* \notin P_n^{\mathcal{A}}$) or an assurance that there are no such violated inequalities (showing $x^* \in P_n^{\mathcal{A}}$).*

Without giving a very formal treatment, the major result in the field is the following theorem that shows the equivalence of separation and optimization.

THEOREM 2.35 *(Grötschel, Lovász and Schrijver, see the reference [14].) One can minimize (maximize) over the polyhedron Z_n using any linear cost function c in time polynomial in terms of n and size(c) if and only if there is a polynomial-time separation algorithm for the family of inequalities in a description of Z_n.*

(The notion of *size* used in the statement of this theorem is defined as the number of bits in a machine representation of the object in question.) The algorithm used to optimize over Z_n given a separation algorithm involves the so-called *ellipsoid algorithm* for solving linear programs (Section 2.4.3).

It turns out that there are polynomial time separation algorithms for both the subtour elimination constraints of $SEP(n)$ and the odd cut-set constraints of $PM(n)$. Hence, we can optimize over both $SEP(n)$ and $PM(n)$ in polynomial time, in spite of the exponential size of their explicit polyhedral descriptions. This means in particular that, purely through polyhedral considerations, we know that the perfect matching problem can be solved in polynomial time.

By analogy to the ideas of complexity theory [31], one can define the notion of *polyhedral reduction.* Intuitively, a polyhedron A can be reduced to the polyhedron B, if by adding polynomially many new variables and inequalities, A can be restricted to equal B. A combinatorial optimization problem L can then be polyhedrally reduced to the problem M, if any polyhedral representation A of L can be reduced to a polyhedral representation B of M.

EXERCISE 2.36 *Given a graph $G = (V, E)$ and $T \subseteq V$, a subset $F \subseteq E$ is a T-join if T the set of odd-degree vertices in the graph induced by F. Show that the V-join and perfect matching problems are equivalent via polyhedral reductions. (One of the two reductions is easy.)*

The idea that the perfect matching problem is perhaps the hardest problem in P is can be expressed using the following informal conjectures.

CONJECTURE 2.37 *[43] Any polyhedral proof that $PM(n)$ is polynomially solvable requires the idea of separation.*

CONJECTURE 2.38 *Every polynomially solvable polyhedral problem Z has a polyhedral proof that it is polynomially solvable, and the idea of separation is needed in such a proof only if Z can be polyhedrally reduced to perfect matching.*

2.3.3 Polytope Representations and Algorithms

According to Theorem 2.31, every polytope can be defined either as the intersection of finitely many halfspaces (an *$\mathcal{H}$-polytope*) or as the convex hull of a finite set of points (a *$\mathcal{V}$-polytope*). It turns out that these two representations have quite different computational properties.

Consider the $\mathcal{H}$-polytope P defined by $\{x \mid Ax \geq b\}$, for $A \in \mathbf{R}^{I \times J}$, $b \in \mathbf{R}^I$. Given a point $x^0 \in \mathbf{R}^J$, deciding whether $x^0 \in P$ takes $O(|I||J|)$ elementary arithmetic operations. However, if P is a $\mathcal{V}$-polytope, there is no obvious simple separation algorithm for P.

The separation problem for the $\mathcal{V}$-polytope $P = \text{conv}\{v^i \mid i \in I\}$ can be solved using linear programming:

$$\begin{array}{lrcl} \text{minimize} & \lambda_{i'} & & \\ \text{subject to} & & & \\ & \sum_{i \in I} \lambda_i v^i & = & x^0 \\ & \sum_{i \in I} \lambda_i & = & 1 \\ & \lambda & \geq & 0, \end{array}$$

where i' is an arbitrary element of I (the objective function doesn't matter in this case, only whether the instance is feasible). The LP (2.3.3) has a feasible solution iff x^0 can be written as a convex combination of extreme points of P iff $x^0 \in P$. Therefore any linear programming algorithm provides a separation algorithm for $\mathcal{V}$-polytopes. Even for special cases, such as that of *cyclic polytopes* [44], it is not known how to solve the membership problem (given x, is $x \in P$?) in polynomial time, without using linear programming.

Symmetrically, the optimization problem for $\mathcal{V}$-polytopes is easy (just compute the objective value at each extreme point), but optimization for $\mathcal{H}$-polytopes is exactly linear programming.

Some basic questions concerning the two polytope representations are still open. For example, the *polyhedral verification*, or the *convex hull* problem is, given an $\mathcal{H}$-polytope P and a $\mathcal{V}$-polytope Q, to decide whether $P = Q$. Polynomial-time algorithms are known for simple and simplicial polytopes [20], but not for the general case.

2.4 Duality

Consider a linear programming (LP) problem instance whose *feasible region* is the polyhedron $P \subset \mathbf{R}^I$. One can form a matrix $A \in \mathbf{R}^{m \times I}$ and $b \in \mathbf{R}^m$

such that

$$P = \{x \in \mathbf{R}^I \mid A \cdot x \geq b\}.$$

This linear program can then be stated as

$$\begin{array}{ll} \text{minimize} & c \cdot x \\ \text{subject to} & \\ & A \cdot x \geq b. \end{array}$$

In this section we derive a natural notion of a *dual* linear program. Hence, we sometimes refer to the original LP as the *primal*. In the primal LP, the *variables* are indexed by a combinatorially meaningful set I. We may want to similarly index the *constraints*, (currently they are indexed by integers $1, \ldots, m$). We simply write J for the index set of the constraints, and when convenient associate with J a combinatorial interpretation.

One way to motivate the formulation of the dual linear program is by trying to derive valid inequalities for P through linear algebraic manipulations. For each $j \in J$, multiply the constraint indexed by j by some number $y_j \geq 0$, and add all these inequalities to get a new valid inequality. What we are doing here is choosing $y \in \mathbf{R}^J_+$, and deriving

$$(y \cdot A) \cdot x = y \cdot (A \cdot x) \geq y \cdot b. \tag{2.6}$$

If by this procedure we derived the inequality

$$c \cdot x \geq c_0,$$

where $(c, c_0) \in \mathbf{R}^I \times \mathbf{R}$, then clearly c_0 would be a lower bound for an optimal solution to the primal LP. The idea of the dual LP is to obtain the greatest such lower bound. Informally, the dual problem can be stated as:

maximize c_0
subject to
$c \cdot x \geq c_0$
can be derived from $A \cdot x \geq b$
by linear-algebraic manipulations.

More formally, the dual LP is (see (2.6))

$$\begin{array}{lll} \text{maximize} & y \cdot b & \\ \text{subject to} & & \\ & y \cdot A & = \; c \\ & y & \geq \; 0. \end{array}$$

In the next section we derive the *strong duality* theorem, which states that in fact all valid inequalities for P can be derived by these linear-algebraic manipulations.

Suppose the primal LP includes the non-negativity constraints for all the variables — i.e.,

$$\begin{array}{ll} \text{minimize} & c \cdot x \\ \text{subject to} & \\ & A \cdot x \;\geq\; b \\ & x \;\;\;\;\;\; \geq\; 0, \end{array} \tag{2.7}$$

Then, valid inequalities of the form $c \cdot x \geq c_0$ can also be obtained by relaxing the equality constraint $y \cdot A = c$ to $y \cdot A \leq c$ ($\leq$ instead of merely $=$). Hence, the dual LP for this primal is

$$\begin{array}{ll} \text{maximize} & y \cdot b \\ \text{subject to} & \\ & y \cdot A \;\leq\; c \\ & y \;\;\;\;\;\; \geq\; 0. \end{array} \tag{2.8}$$

2.4.1 Properties of Duality

In the remainder of Section 2.4, we focus on the primal-dual pair (2.7) and (2.8). First, observe the following:

THEOREM 2.39 *The dual of the dual is again the primal.*

An easy consequence of the definition of the dual linear problem is the following *weak duality theorem.*

THEOREM 2.40 *If x is feasible for the primal and y is feasible for the dual, then*

$$c \cdot x \geq y \cdot b.$$

In fact, we can state something much stronger than this. A deep result in polyhedral theory says that the linear-algebraic manipulations described in the previous section can produce **all** of the valid inequalities of a polyhedron P. As a consequence, we have the *strong duality theorem.*

THEOREM 2.41 *Suppose the primal and dual polyhedra are non-empty. Then both polyhedra have optimal solutions (by weak duality). Moreover, if x^* is an optimal primal solution and y^* is an optimal dual solution, then*

$$c \cdot x^* = y^* \cdot b.$$

It should be noted that either or both of these polyhedra could be empty, and that if exactly one of these polyhedra is empty, then the objective function of the other is *unbounded.* Also, if either of these polyhedra has an optimal solution, then both of these polyhedra have optimal solutions, for which strong duality holds. This implies a very nice property of linear programming: an optimal solution to the dual provides a *certificate of optimality* for an optimal solution to the primal. That is, merely producing a feasible solution y^* to the dual proves the optimality of a feasible solution x^* to the primal, as long as their objective values are equal (that is, $c \cdot x^* = y^* \cdot b$). Some algorithms for the solution of linear programming problems, such as the simplex method, in fact naturally produce such dual certificates.

To prove the strong duality theorem, according to the previous discussion, it suffices to prove the following:

CLAIM 2.42 (STRONG DUALITY) *If (2.7) has an optimal solution x^*, then there exists a feasible dual solution y^* to (2.8) of equal objective value.*

Proof: Consider an extreme-point solution x^* that is optimal for (2.7). It clearly remains optimal even after all constraints that are not tight, that are either from A or are non-negativity constraints, have been removed from (2.7). Denote the system resulting from the removal of these non-tight constraints by

$$\begin{aligned} A' \cdot x &\geq b' \\ I' \cdot x &\geq 0. \end{aligned}$$

That x^* is optimal can be seen to be equivalent to the fact that there is no improving feasible direction. That is, x^* is optimal for (2.7) if and only if there exists no $r \in \mathbf{R}^I$ such that $x^* + r$ is still feasible and $c \cdot (x^* + r) < c \cdot x^*$. (The vector r is called the *improving feasible direction.*) The problem of finding an improving feasible direction can be cast in terms of linear programming, namely

$$\begin{array}{ll} \text{minimize} & c \cdot r \\ \text{subject to} & \\ & \begin{aligned} A' \cdot r &\geq 0 \\ I' \cdot r &\geq 0. \end{aligned} \end{array} \tag{2.9}$$

Clearly, 0 is feasible for (2.9). Since x^* is optimal for the LP (2.7), 0 is the optimum for the "improving" LP (2.9).

Stack A' and I' to form the matrix $\overline{A}$. Denote the rows of $\overline{A}$ by $\{\overline{a}^1, \overline{a}^2, \ldots, \overline{a}^k\}$. Next we resort to the Farkas Lemma:

LEMMA 2.43 *The minimum of the LP* (2.9) *is* 0 *if and only if c is in the cone generated by $\{\overline{a}^1, \overline{a}^2, \ldots, \overline{a}^k\}$.*

Since x^* is optimal, the minimum for LP (2.9) is 0. Hence, by the Farkas Lemma, c is in the cone generated by $\{\overline{a}^1, \overline{a}^2, \ldots, \overline{a}^k\}$. Thus, there exist multipliers $\overline{y}_1, \overline{y}_2, \ldots, \overline{y}_k \geq 0$ such that

$$c = \sum_{i=1}^{k} \overline{y}_i \overline{a}^i.$$

We now construct our dual certificate y^* from $\overline{y}$. First throw out any $\overline{y}_i$s that arise from the non-negativity constraints of I'. Renumber the rest of the $\overline{y}_i$s so that they correspond with the rows of A. We now define y^* by

$$y_i^* = \begin{cases} 0 & \text{if constraint } i \text{ of } A \text{ is not tight at } x^*, \\ \overline{y}_i & \text{otherwise.} \end{cases} \tag{2.10}$$

One can see that y^* is feasible for (2.8). From weak duality, we then have the chain

$$c \cdot x^* \geq (y^* \cdot A) \cdot x^* = y^* \cdot (A \cdot x^*) \geq y^* \cdot b. \tag{2.11}$$

It is easy to verify that the above chain of inequalities is in fact tight, so we actually have $c \cdot x^* = y^* \cdot b$, completing our proof. □

Proof of the Farkas Lemma: If c is in the cone of the vectors $\overline{a}^i$, then the inequality $c \cdot r \geq 0$ can be easily derived from the inequalities $\overline{a}^i \cdot r \geq 0$ (since c is then a non-negative combination of the vectors $\overline{a}^i$). But 0 is feasible for (2.9), so the minimum of (2.9) is then 0.

Now suppose c is not in the cone of the vectors $\overline{a}^i$. It is then geometrically intuitive that there is a separating hyperplane $r' \cdot a \geq r'_0$ passing through the origin (the $\vec{0}$-vector) such that all the vectors $\overline{a}^i$ (and thus the entire cone) are on one side (halfspace) of the hyperplane, and c is strictly on the other side. In fact, $r'_0 = 0$ since the origin lies on this hyperplane. We place the vector r' so that it is on the side of the hyperplane that contains the cone generated by the vectors $\overline{a}^i$. Then r' is feasible for (2.9), but $r' \cdot c < 0$, demonstrating that the minimum of (2.9) is less than 0. □

Let x^* be feasible for the primal and y^* feasible for the dual. Denote the entry in the j-th row and i-th column of the constraint matrix by $a_{j,i}$. Then, we have the *complementary slackness theorem*:

THEOREM 2.44 *The points x^* and y^* are optimal solutions for the primal and dual if and only if*

$$y_j^*(\sum_{i \in I} a_{j,i} x_i^* - b_j) = 0 \qquad \forall j \in J,$$

and

$$x_i^*(c_i - \sum_{j \in J} a_{j,i} y_j^*) = 0 \qquad \forall i \in I.$$

In other words, for x^*, y^* to be optimal solutions, for each $i \in I$, either the i-th primal variable is 0 or the dual constraint corresponding to this variable is *tight*. Moreover, for each $j \in J$, either the j-th dual variable is 0 or the primal constraint corresponding to this variable is tight.

EXERCISE 2.45 *Consider the linear program*

$$\begin{array}{ll} \textit{minimize} & c \cdot x \\ \textit{subject to} & \\ & A \cdot x \ \geq \ b \\ & I' \cdot x \ \geq \ 0, \end{array}$$

where the matrix $\begin{bmatrix} A \\ I' \end{bmatrix}$ *is square and I' consists of a subset of the set of rows of an identity matrix. Suppose x^* is the unique optimal solution to this linear program that satisfies all constraints at equality. Construct a dual solution y^* that certifies the optimality of x^*.*

EXERCISE 2.46 *(Due to R. Freund [35].) Given the linear program $Ax \geq b$, describe a linear program from whose optimal solution it can immediately be seen which of the inequalities are always satisfied at equality. (Hint: Solving a linear program with $n^2 + n$ variables gives the answer. Is there a better solution?)*

2.4.2 Primal, Dual Programs in the Same Space

We can obtain a nice picture of duality by placing the feasible solutions to the primal and the dual in the same space. In order to do this, let us modify the primal and dual linear programs by adding to them each other's variables and a single innocent constraint to both of them:

$$\begin{array}{ll} \text{minimize} & c \cdot x \\ \text{subject to} & \\ & A \cdot x \ \geq \ b \\ & x \ \geq \ 0 \\ & c \cdot x - y \cdot b \ \leq \ 0, \\ \text{maximize} & y \cdot b \\ \text{subject to} & \\ & y \cdot A \ \leq \ c \\ & y \ \geq \ 0 \\ & c \cdot x - y \cdot b \ \leq \ 0. \end{array}$$

The y variables are in just one primal constraint in the revised primal LP. Hence, this constraint is easy to satisfy and does not affect the x variable values in any optimal primal solution. Likewise for the x variables in the revised dual LP.

Now comes some interesting geometry. Denote the optimal primal and dual objective function values by z_p and z_d, as if you were unaware of strong duality.

THEOREM 2.47 (WEAK DUALITY) *If the primal and dual linear programs are feasible, then the region*

$$\{(x, y) \in \mathbf{R}^{I \cup J} \mid 2z_p \geq c \cdot x + b \cdot y \geq 2z_d\},$$

and in particular the hyperplane

$$\{(x, y) \in \mathbf{R}^{I \cup J} \mid c \cdot x + b \cdot y = z_p + z_d\},$$

separates the revised primal and revised dual polyhedra.

THEOREM 2.48 (STRONG DUALITY) *If the primal and dual linear programs are feasible, then the intersection of the revised primal polyhedron and the revised dual polyhedron is non-empty.*

2.4.3 Duality and the Ellipsoid Method

The ellipsoid method takes as input a polytope P (or more generally a bounded convex body) that is either full-dimensional or empty, and gives as output either an $x \in P$ or an assurance that $P = \emptyset$. In the typical case where P may not be full-dimensional, with extra care, the ellipsoid method will still find an $x \in P$ or an assurance that $P = \emptyset$. With even more care, the case where P may be unbounded can be handled too.

The ellipsoid method, when combined with duality, provides an elegant (though not practical) way of solving linear programs. Finding optimal primal and dual solutions to the linear programs (2.7) and (2.8) is equivalent (see also Chvátal [7], Exercise 16.1) to finding a point in the polyhedron P defined by

$$P := \{(x, y) \in \mathbf{R}^{I \cup J} \mid A \cdot x \geq b,\ y \cdot A \leq c,\ x, y \geq 0,\ c \cdot x - y \cdot b \leq 0\}.$$

Admittedly, P is probably far from full-dimensional, and it may not be bounded either. But, consider the set of ray directions of P, given by

$$\begin{aligned} R \ := \ & \{(s, t) \in \mathbf{R}^{I \cup J} \mid A \cdot s \geq 0,\ t \cdot A \leq 0,\ s, t \geq 0,\ c \cdot s, y \cdot t = 0, \\ & s(I) + t(J) = 1\}. \end{aligned}$$

(For the theory of the derivation of this kind of result, see Nemhauser and Wolsey [30].) Since the polytope R gives the ray directions of P, P is bounded if $R = \emptyset$. Conversely, if $R \neq \emptyset$, then P is unbounded or empty.

Suppose we are in the lucky case where $R = \emptyset$, and thus P is a polytope (the alternative is to artificially bound P). We also choose to ignore the complications arising from P being almost certainly flat. The ellipsoid method first finds an ellipsoid $\mathcal{E}$ that contains P and determines its center x. We run a procedure required by the ellipsoid algorithm (and called the *separation algorithm*) on x. The separation algorithm either determines that $x \in P$ or finds a closed halfspace that contains P but not x. The boundary of this halfspace is called a *separating hyperplane*. From this closed halfspace and $\mathcal{E}$, both of which contain P, the ellipsoid method determines a provably smaller ellipsoid $\mathcal{E}'$ that contains P. It then determines the center x' of $\mathcal{E}'$ and repeats the process. The sequence of ellipsoids shrinks quickly enough so that in a number of iterations polynomial in terms of $|I \cup J|$, we can find an $x \in P$ or determine that $P = \emptyset$, as required.

The alternative to combining duality with the ellipsoid method for solving linear programs is to use the ellipsoid method together with binary search on the objective function value. This is, in fact, a more powerful approach than the one discussed above, because it only requires a linear programming formulation and a separation algorithm, and runs in time polynomial in $|I|$, regardless of $|J|$. Thus the ellipsoid method (if used for the "primal" problem, with binary search) can optimize over polytopes like the matching polytope $PM(n)$ in spite of the fact that $PM(n)$ has exponentially many facets (in terms of n). The size of an explicit description in and of itself simply has no effect on the running time of the ellipsoid algorithm.

2.4.4 Sensitivity Analysis

Sensitivity analysis is the study of how much changing the input data or the variable values affects the optimal LP solutions.

Reduced Cost. Let x^* and y^* be optimal solutions to (2.7)-(2.8). The *reduced cost* c_e^{red} of a variable x_e is defined to be

$$c_e^{red} := c_e - (y^* \cdot A)_e.$$

Note that our dual constraints require that $c_e^{red} \geq 0$. Moreover $c_e^{red} > 0$ only when $x_e^* = 0$ by complementary slackness. The reduced cost is designed to put a lower bound on the increase of the cost of the primal solution when it changes from the solution x^*, where $x_e^* = 0$, to a feasible solution x^+, where $x_e^+ = 1$.

THEOREM 2.49 *If x^+ is a feasible primal solution, and $e \in E$ such that $x_e^* = 0$, but $x_e^+ = 1$, then $c \cdot x^+ \geq c \cdot x^* + c_e^{red}$.*

Proof: Complementary slackness implies $c \cdot x^+ - y^* \cdot b \geq c_e^{red}$. □

The reduced cost is of use when the primal is a relaxation of a 0-1 program or integer program. In particular, if $c \cdot x^* + c_e^{red}$ (can be rounded up for integral c) exceeds the cost of a known feasible integer solution, then the variable indexed by e can never appear in any integer optimal solution. Hence, this variable can be permanently removed from all formulations.

Another use of reduced cost is when the (perhaps "exponentially many") x_e variables are not all explicitly present in the linear program, and one is using *variable generation* (or *column generation*) to add them into the formulation when this results in a lower cost solution. Variable generation is analogous to the ideas of separation, whether using the ellipsoid method or a simplex LP method.

Assume that $|I|$ grows exponentially ($O(2^n)$) with respect to n, but $|J|$ grows polynomially. If a variable x_e is not currently in the n-th formulation, then implicitly, $x_e^* = 0$. If $c_e^{red} < 0$, then there is a chance that adding x_e into the LP formulation will lower the cost of the resulting optimal solution. A valuable *pricing algorithm* (or *dual separation algorithm*) is one that in $O(\log^k |I|)$ steps, either finds an $e \in I$ such that $c_e^{red} < 0$, or an assurance that there is no such $e \in I$. In this latter case, the current solution x^* is in fact optimal for the entire problem! With such a pricing algorithm, the ellipsoid method guarantees that these LP problem instances can be solved in polynomial time with respect to n.

Finally, there is an analogous dual concept of reduced cost, which can be easily derived.

Changing Right Hand Sides. Now we address how the optimal LP solutions are affected by changing the right hand side b or the cost c. We will address the case when b is altered only, since the case when c is altered can be easily derived from this. Suppose we add δ to the right hand side for $j \in J$. The dual solution y^* that was once optimal is still feasible. However, the dual objective function has changed from b to some b' (differing only in row index j). So,

$$b' \cdot y^* = b \cdot y^* + \delta y_j^* = c \cdot x^* + \delta y_j^*.$$

Since y^* is known only to be feasible for the new dual, the optimal solution to this dual can only be better. Since the dual problem is a maximization problem, the value of the new optimal dual solution is thus at least δy_j^* more than the value of the old optimal dual solution y^*. Hence, the new optimal primal solution value is also at least δy_j^* more than that of the old optimal primal solution x^* (now potentially infeasible). In other words, the cost of the best primal solution has gone up by at least δy_j^*.

2.5 Compact Optimization and Separation

2.5.1 Compact Optimization

Consider as an example the following optimization problem: given the integer k and a linear objective function c, maximize $c \cdot x$ under the constraint that the L^1 norm $\|x\|_1$ of $x \in \mathbf{R}^I$ is at most k (by definition, $\|x\|_1 = \sum_{i \in I} |x_i|$). This optimization problem at first leads to a linear programming formulation with $2^{|I|}$ constraints, namely for every vector $a \in \{-1, 1\}^I$, we require that $a \cdot x \leq k$. Introducing $|I|$ new variables makes it possible to reduce the number of constraints to $2|I| + 1$ instead. For each $i \in I$, the variable y_i will represent the absolute value $|x_i|$ of the i-th component of x. This is achieved by imposing the constraints $y_i \geq x_i$ and $y_i \geq -x_i$. Then the $2^{|I|}$ original constraints are replaced by a single one:

$$\sum_{i \in I} y_i \leq k.$$

It turns out that introducing additional variables can make it easier to describe the polyhedron using linear inequalities. In this section, we denote the *additional variables* (those not needed to evaluate the objective function) by y, and the *decision variables* (those needed to evaluate the objective function) by x.

Suppose we have a linear objective representation for each instance of a combinatorial optimization problem. We informally define an approach to describing the convex hull of feasible solutions for such a linear objective representation. A *linear description* is an abstract mathematical description or algorithm that for each instance n, can quickly explicitly produce all facet-defining inequalities and equations valid for the linear objective representation of the n-th instance. Here, "quickly" is in terms of the number of these facet-defining inequalities and equations and the number of variables for an instance. Given a linear description and a particular instance n, we can thus explicitly produce the matrix $A_n \in \mathbf{R}^{J_n \times (I_n \cup I'_n)}$ and right hand side $b_n \in \mathbf{R}^{J_n}$ so that the convex hull P_n of feasible solutions is given by the x variables (Section 2.5.2) of

$$P_n = \{(x, y) \in \mathbf{R}^{I_n \cup I'_n} \mid A_n \begin{bmatrix} x \\ y \end{bmatrix} \geq b_n\}. \tag{2.12}$$

A *compact linear description* is a linear description where the polyhedron lies in a space of polynomial (in terms of n) dimension and is the intersection of polynomially many (in terms of n) closed halfspaces. We call optimization using a compact linear description *compact optimization.* Since linear programming is polynomially solvable, a compact linear description guarantees that this combinatorial optimization problem is in P. Many combinatorial optimization problems in P have compact linear descriptions. A notorious possible exception to this (as we have already indicated in Section 2.3.2) is the perfect matching problem.

2.5.2 Projection and Lifting

Consider the polyhedron P_n defined by (2.12). Recall that the y variable values have no effect on the objective function value for our problem instance. In fact, the only role these y variables have is to make it easier to form a linear description for our problem.

Suppose one wanted to extract the feasible polyhedron $\mathrm{Proj}_x(P_n)$ that results from just the x decision variables when the y variables are removed from the vector representation. The process of doing this is called projection, and the resulting polyhedron $\mathrm{Proj}_x(P_n)$ is called the *projection* of the polyhedron P_n onto the space of the x variables. This new polyhedron is in fact a projection in a geometric sense, satisfying

$$\mathrm{Proj}_x(P_n) = \{x \in \mathbf{R}^{I_n} \mid \exists y \in \mathbf{R}^{I'_n} \text{ s.t. } (x,y) \in P_n\}.$$

A fundamental property of projection is that

$$\begin{array}{ll} \text{minimize} & c \cdot x \\ \text{subject to} & (x,y) \in P_n, \end{array}$$

and

$$\begin{array}{ll} \text{minimize} & c \cdot x \\ \text{subject to} & x \in \mathrm{Proj}_x(P_n) \end{array}$$

have the same minimizers (when the minimizers of P_n are projected onto the space of the x variables).

But what are the facet-defining closed halfspaces and hyperplanes in a description of $\mathrm{Proj}_x(P_n)$? Let us break up A_n by columns into A_x and A_y, the submatrices that act on the x variables and the y variables respectively. By strong duality, one can see that the valid inequalities for $\mathrm{Proj}_x(P_n)$ (involving only the x variables) are precisely

$$\begin{array}{lcl} (\lambda \cdot A_n) \cdot x & \geq & \lambda \cdot b_n, \\ \text{such that} & & \\ \lambda \cdot A_y & = & 0, \\ \lambda & \geq & 0, \\ \sum_{i \in J_n} \lambda_i & = & 1, \end{array} \tag{2.13}$$

where this last sum is an arbitrary scaling of each inequality by a positive multiple.

THEOREM 2.50 *The facet-defining inequalities and equations of the polyhedron* $\mathrm{Proj}_x(P_n)$ *correspond exactly with the extreme points of (2.13).*

The concept opposite to projection is that of *lifting*, where one adds more variables, inter-relating them to the original variables, to form a lifted linear description in a higher dimensional space. As already seen, lifting is useful in producing non-obvious compact linear descriptions for combinatorial optimization problems (when the obvious linear descriptions are exponential in terms of n).

We describe two examples of compact linear programs possible only thanks to the introduction of new variables and constraints.

Example: Parity polytope [43]. Let $PP(n)$ denote the convex hull of zero-one vectors over $\mathbf{R}^n$ with an odd number of ones, that is, $PP(n) = \mathrm{conv}\{x \in \{0,1\}^n \mid \sum_i x_i \text{ is odd}\}$. This polytope has exponentially many facets [18], but by introducing new variables the number of constraints can be reduced significantly. The idea is to write $x \in PP(n)$ as a convex combination $\sum_{k \text{ odd}} \lambda_k y^k$, where y^k is a convex combination of extreme points of $PP(n)$ with k ones. We represent $\lambda_k y^k$ by the variable z^k in the linear program, obtaining

$$
\begin{array}{llll}
\text{minimize} & c \cdot x & & \\
\text{subject to} & & & \\
\sum_{k \text{ odd}} \lambda_k & = & 1 & \\
\sum_{k \text{ odd}} z_i^k & = & x_i & \forall i \\
\sum_i z_i^k & = & k\lambda_k & \text{for odd } k \\
0 & \leq & z_i^k \leq \lambda_k & \forall i, k.
\end{array}
\tag{2.14}
$$

In fact, a similar formulation exists for any *symmetric 0-1 function* $f : \{0,1\}^n \to \{0,1\}$. (A function f is symmetric, if its value doesn't change when its arguments are reordered.) In the 0-1 case, the value of a symmetric function depends only on the number of non-zero coordinates in the argument. Thus in the formulation (2.14), whenever there is a reference to the set of odd integers, it should be replaced with the set I such that $f(x) = 1$ iff the number of nonzero coordinates of x belongs to I.

Example: Arborescences in directed graphs. Given a vertex r of a directed graph $G = (V, E)$, an *arborescence rooted at* r is a subgraph of G that contains a directed path from r to v for each $v \in V$, but no (undirected) cycles. If we represent arborescences in G by their incidence vectors, then the convex hull of all such incidence vectors is the set of solutions to

$$
\begin{array}{lll}
x(\delta^+(S)) & \geq 1 & \forall S \subset V \text{ such that } r \in S \\
x(E) & = |V| - 1 & \\
x & \geq 0, &
\end{array}
\tag{2.15}
$$

where $\delta^+(S)$ denotes the set of directed edges of G leaving S [10]. The constraints can be intuitively be thought of as requiring that every cut separating a vertex of G from the root r is crossed by at least one edge. This idea is the basis of an extended formulation. For each v, we require that the solution support a path from r to v. This path will be represented by a flow of value 1 from r to v.

$$
\begin{array}{llll}
\text{minimize} & c \cdot x & & \\
\text{subject to} & & & \\
x_e & \geq & y_e^t & \forall e \in E,\ \forall t \in V \setminus \{r\} \\
y^t(\delta^-(v)) & = & y^t(\delta^+(v)) & \forall t, v \in V \setminus \{r\},\ v \neq t \\
y^t(\delta^-(t)) & = & 1 & \forall t \in V \setminus \{r\} \\
y^t(\delta^-(r)) & = & 0 & \forall t \in V \setminus \{r\} \\
y^t(\delta^+(r)) & = & 1 & \forall t \in V \setminus \{r\} \\
x(E) & = & |V| - 1 & \\
y & \geq & 0 &
\end{array}
\tag{2.16}
$$

Now by the max-flow-min-cut theorem [11], each of the cuts in the constraints of LP (2.15) has value at least 1 iff for every $v \neq r$, there is an r-v flow of value at least 1. In other words, the projection of LP (2.16) to the space spanned by the x-variables is exactly the r-arborescence polytope defined by LP (2.15). The number of constraints in the flow-based formulation is polynomial in the size of G. See Pulleyblank [33] for a discussion and references.

2.5.3 Compact Separation

Consider an optimization problem that can be phrased as

$$\begin{array}{ll} \text{minimize} & c \cdot x \\ \text{subject to} & \\ & x \in P_n^{\mathcal{A}} (\subset \mathbf{R}^{I_n}), \end{array} \tag{2.17}$$

such as we saw earlier (Definition 2.34) in Section 2.3.2. Here, n is an integer parameter indicating the size of the instance, and $P_n^{\mathcal{A}}$ is a polyhedron whose variables are indexed by I_n and constraints come from a class $\mathcal{A}$ of valid inequalities. Unfortunately, the number of inequalities describing $P_n^{\mathcal{A}}$ will usually grow exponentially with n. In this case, directly solving this problem through linear programming takes exponential time. But, as we saw in the last two sections, a compact linear description may still be possible. That is, if we introduce additional variables y, we are sometimes able to express the polyhedron $P_n^{\mathcal{A}}$ as:

$$P_n^{\mathcal{A}} = \text{Proj}_x\{(x, y) \in \mathbf{R}^{I_n \cup I'_n} \mid A_n \cdot \begin{bmatrix} x \\ y \end{bmatrix} \geq b_n\}, \tag{2.18}$$

where the total number of variables and constraints grows polynomially with respect to n, as was done in (2.12). This gives a polynomial-time solution to our problem using linear programming, showing that this problem is in $\mathcal{P}$.

Recall the separation problem defined in Section 2.3.2 for our problem (2.17). In this section, we learned that (2.17) is in $\mathcal{P}$ if and only if the separation problem for the class $\mathcal{A}$ of inequalities is polynomially solvable. This raises the following natural question:

QUESTION 2.51 *Can we use the polynomial solvability of the separation problem to construct a compact linear description for our problem?*

A partial answer to this question was given by Martin [27] and by Carr and Lancia [5].

ANSWER 2.52 *If (and only if) the separation problem can be solved by compact optimization then one can construct a compact linear description of our problem from a compact linear description of the separation problem.*

In other words, if we can phrase the separation algorithm as a compact linear program, then we can use this to optimize compactly.

When one can solve the separation problem by compact optimization, this is called *compact separation*. We first argue that compact separation is a plausible case. Consider (2.17) again. Denote by Q_n the set of points $(a, a_0) \in \mathbf{R}^{I_n} \times \mathbf{R}$ corresponding to valid inequalities $a \cdot x \geq a_0$ of $P_n^{\mathcal{A}}$ (with (a, a_0) scaled so that $-1 \leq (a, a_0) \leq 1$). It follows from the theory behind projection (described in Section 2.5.2) and duality that Q_n is a (bounded) polyhedron!

Hence, the separation problem for the class $\mathcal{A}$ of inequalities can be phrased as a linear program of the form

$$\begin{array}{ll} \text{minimize} & x^* \cdot a - a_0 \\ \text{subject to} & \\ & (a, a_0) \in Q_n, \end{array} \tag{2.19}$$

where, x^* is the fractional point we wish to separate from $P_n^{\mathcal{A}}$. There is a violated inequality if and only if the minimum objective function value is less than 0. In this case, if the minimizer is (a^*, a_0^*), then

$$a^* \cdot x \geq a_0^*$$

is a valid inequality for $P_n^{\mathcal{A}}$ that is violated by x^*.

Unfortunately, the polyhedron Q_n often has exponentially many constraints with respect to n. But it may be possible to obtain compact separation by examining either Q_n or the workings of a polynomial separation algorithm. We thus strive to obtain

$$Q_n = \text{Proj}_{(a,a_0)}\{(a, a_0, w) \in \mathbf{R}^{I_n} \times \mathbf{R} \times \mathbf{R}^{J_n} \mid B_n\,[a\,a_0\,w]^\tau \geq d_n\}, \tag{2.20}$$

where there is a polynomial number of variables and constraints with respect to n.

We now state the main theorem of this section.

THEOREM 2.53 *[27, 5] Compact optimization is possible for $P_n^{\mathcal{A}}$ if (and only if) compact separation is possible for the class $\mathcal{A}$ of valid inequalities of $P_n^{\mathcal{A}}$.*

Proof: We only give the "if" direction here. Let Q_n be the polyhedron of valid inequalities $a \cdot x \geq a_0$ of $P_n^{\mathcal{A}}$. Assuming compact separation is possible, let (2.21) be the LP that achieves this, with a polynomial number of variables and constraints with respect to n.

$$\begin{array}{ll} \text{minimize} & x^* \cdot a - a_0 \\ \text{subject to} & \\ & B_n \begin{bmatrix} a \\ a_0 \\ w \end{bmatrix} \geq d_n \end{array} \tag{2.21}$$

Here, x^* violates an inequality in $\mathcal{A}$ if and only if the minimum of (2.21) is less than 0. Hence, x^* satisfies all such inequalities iff the minimum of (2.21) is at least 0. Partition B_n by columns into $[B_n^a B_n^{a_0} B_n^w]$. Consider the dual linear program:

$$
\begin{array}{ll}
\text{maximize} & y \cdot d_n \\
\text{subject to} & \\
& y \cdot B_n^a = x^* \\
& y \cdot B_n^{a_0} = -1 \\
& y \cdot B_n^w = 0 \\
& y \geq 0.
\end{array}
\tag{2.22}
$$

By strong duality, we now have that x^* satisfies all such inequalities iff the maximum of (2.22) is at least 0. When we remove the star (!) from x^*, we see that x satisfies all the inequalities of $\mathcal{A}$ iff the linear system

$$
\begin{array}{l}
y \cdot d_n \geq 0 \\
y \cdot B_n^a = x \\
y \cdot B_n^{a_0} = -1 \\
y \cdot B_n^w = 0 \\
y \geq 0
\end{array}
\tag{2.23}
$$

has a feasible solution. Thus, $P_n^{\mathcal{A}}$ is given by

$$P_n^{\mathcal{A}} = \text{Proj}_x\{(x,y) \in \mathbf{R}^{I_n \cup J_n} \mid (x,y) \text{ satisfies (2.23)}\},$$

which is of the form (2.20). Hence, compact optimization of (2.17) is given by

$$
\begin{array}{ll}
\text{minimize} & c \cdot x \\
\text{subject to} & \\
& (x,y) \text{ satisfies (2.23)},
\end{array}
\tag{2.24}
$$

which concludes our proof. □

2.6 Integer Polyhedra, TDI Systems and Dynamic Programming

In this section we are concerned with problems formulated in terms of integer programs. In order to solve such a problem in polynomial time we may apply a linear programming algorithm to the linear relaxation of the integer program. This approach sometimes works because polyhedra P associated with several combinatorial optimization problems are integer polyhedra (have only integer extreme points). In such a case, any polynomial-time algorithm for linear programming that outputs an extreme point solution implies a polynomial-time algorithm for the problem. How can we tell whether a given polyhedron is integer? It is easy to see that the problem of determining whether a given polyhedron is integer is in Co-NP, but it is not known if it is in NP. However, in order to optimize a linear objective function over the polyhedron, we do not need to know that it is integer. Since the known polynomial-time algorithms for linear programming can be used to find an optimal extreme point, it follows that an integer programming problem can be solved in polynomial time when the polyhedron of its linear relaxation is integer.

There do exist several typical classes of integer polyhedra and they come with characterizations that allow one to prove their integrality, or at least, that an integer optimum can be found efficiently.

2.6.1 Total Unimodularity

An important special case of integer polyhedra is the class defined by *totally unimodular* matrices. In this case the matrix of the polyhedron nearly guarantees all by itself that the polyhedron is integer.

DEFINITION 2.54 *A matrix A is totally unimodular, if for each integer vector b, the polyhedron $\{x \mid x \geq 0, Ax \leq b\}$ has only integer vertices.*

It can easily be seen that all the entries of a totally unimodular matrix belong to $\{0, -1, 1\}$.

There exist several other conditions that can be used as definitions of total unimodularity. We only list a few.

THEOREM 2.55 *For a matrix A with entries from $\{0, -1, 1\}$, the following conditions are equivalent:*

1 *For each integral vector b, the polyhedron $\{x \mid x \geq 0, Ax \leq b\}$ has only integral vertices.*

2 *The determinant of each square submatrix of A belongs to $\{0, -1, 1\}$.*

3 *Each subset of columns (rows) of A can be split into two parts A_1 and A_2 so that the difference of the sum of columns (rows) in A_1 and the sum of columns (rows) in A_2 has all entries in $\{0, -1, 1\}$.*

Total unimodularity is preserved under typical matrix operations such as transposition, multiplication of a row or column by -1, interchange or duplication of rows or columns, or appending a basis vector.

The typical example of a totally unimodular matrix is the vertex-edge incidence matrix of a bipartite graph. To prove the total unimodularity, split the rows (corresponding to vertices) into two sets according to a bipartition of the graph. Then for each column (edge), the difference between the sums of the rows in the two subsets is 0 (one endpoint in each set), and the third condition of Theorem 2.55 is satisfied. It is not difficult to prove the converse, either, namely that the vertex-edge incidence matrix of a nonbipartite graph is not totally unimodular.

2.6.2 Total Dual Integrality

As long as the right-hand side vector b is integral, a totally unimodular matrix A will define an integral polyhedron $Ax \leq b$. In order to generalize the class of integral polyhedra we consider, we may include the right-hand side in the specification of an instance. The system $Ax \leq b$ (with rational A and b) is *totally dual integral*, if it is feasible, and for every integral objective function c, the dual problem

$$\min\{y \cdot b \mid y \geq 0,\ yA = c\} \tag{2.25}$$

has an integer optimal solution y^*, when the dual is feasible.

THEOREM 2.56 *If $Ax \leq b$ is totally dual integral and b is integer, then $Ax \leq b$ defines an integer polyhedron.*

From Theorem 2.56 it follows that one can optimize efficiently over polyhedra defined by totally dual integral systems. However, a stronger result holds and this is possible even without requiring the integrality of b [35].

THEOREM 2.57 *If $Ax \leq b$ is a totally dual integral system and c an integral objective function, then it is possible to find an integral solution to the dual problem* (2.25) *in polynomial time.*

Totally dual integral systems do not behave as nicely as totally unimodular matrices when subjected to transformations. In fact, *any* rational system $Ax \leq b$ can be made totally dual integral by dividing both sides of the equation with an appropriately chosen (potentially large) integer k [35]. (Of course, if b was integral before, it is unlikely to remain integral upon dividing by such a k.) Another way of stating this property of rational systems is that regardless of the objective function (as long as it is integral), there is an integer k such that in the optimum solution to the dual problem (2.25) each entry is an integral multiple of $1/k$.

However, there is a simple and useful operation that does preserve total dual integrality.

THEOREM 2.58 *If $Ax \leq b$ is a totally dual integral system, then so is the system obtained from it by making any inequality an equality.*

2.6.3 Dynamic Programming and Integrality

In this section we describe a very general source of totally dual integral systems, integer polytopes and compact linear formulations of combinatorial optimization problems. The main source for this material is the paper of Martin et al. [26].

Discrete Dynamic Programming. A combinatorial minimization problem (maximization can be handled analogously) can be solved using dynamic programming, if for each instance of the problem there exists a family of instances S (called the family of *subproblems*, or *states*) and a decomposition relation $D \subseteq 2^S \times S$ such that the following two conditions hold:

1 (*Optimality principle.*) For each $(A, I) \in D$, an optimal solution to I can be constructed from any collection that includes an optimal solution to every instance in A.

2 (*Acyclicity.*) There is an ordering $\sigma : S \longrightarrow [m]$ (where $m = |S|$) compatible with the decomposition relation: given $I' \in S$, $A \subseteq S$ such that $(A, I') \in D$, the condition $\sigma(I) < \sigma(I')$ holds for all $I \in A$.

The dynamic programming algorithm solves the given problem I_m by solving recursively each of the subproblems I such that $I \in A$ for some $A \subseteq S$ that satisfies $(A, I_m) \in D$. Then the algorithm builds an optimal solution to I_m from optimal solutions to the collection of subproblems A' that minimizes (over the selection of A) the cost $\sum_{I' \in A} c(I') + c(A, I_m)$. Here, $c(I)$ is the cost of an optimal solution to I and $c(A, I_m)$ denotes the cost of building a solution to I_m from the solutions of the subproblems in A.

A simple example of this paradigm in action is provided by the *shortest path problem*: given a graph $G = (V, E)$ with a nonnegative cost (or length) function $c : E \longrightarrow R_+$ defined on the set of edges, and given two vertices s and t of G, find an s-t path of minimum cost. The collection of subproblems consists of pairs (v, i), where $v \in V$ and $i \in [n]$ $(n = |V|)$. Solutions to the subproblem (v, i) are all s-v paths no more than i edges long. Then for every pair of vertices u and v such that $uv \in E$ and for every $i \leq n$, there will be an element $(\{(u, i)\}, (v, i + 1))$, in the relation D such that the optimal cost for $(v, i + 1)$ is equal to the optimal cost for the pair (u, i) plus c_{uv}. Consequently, the associated cost $c(\{(u, i)\}, (v, i + 1))$ equals the cost c_{uv} of the edge uv.

The shortest path problem possesses a special property, namely the relation D again defines a graph on the set of subproblems, and in fact, solving the dynamic programming formulation is equivalent to solving a shortest path problem on this *decision graph.* This has been used by several authors [21, 17, 8]. We follow the paper of Martin et al. [26] in describing a more general *hypergraph model.* In this context, the elements of the relation D will be called *decisions* or *hyperarcs*.

Another example is provided by the parity polytope. We gave a linear description of the parity polytope in Section 2.5.1, following Yannakakis [43]. Here is an alternative description motivated by dynamic programming.

Let $P_n = \{x \in \{0, 1\}^n \mid \sum_i x_i \text{ is odd}\}$, as in Section 2.5.1. We introduce additional variables p_i, $i = 1, \ldots, n$. The idea is to have $p_{i'}$ express the parity of the partial sum $\sum_{i \le i'} x_i$. The first constraint is

$$p_1 = x_1. \tag{2.26}$$

For each $i \in \{2, \ldots n-1\}$, there will be four constraints:

$$\begin{array}{rcl} p_{i+1} & \le & p_i + x_{i+1} \\ p_{i+1} & \le & 2 - (p_i + x_{i+1}) \\ p_{i+1} & \ge & p_i - x_{i+1} \\ p_{i+1} & \ge & x_{i+1} - p_i. \end{array} \tag{2.27}$$

Finally, to ensure the sum of all components of x is odd, the constraint

$$p_n = 1 \tag{2.28}$$

is needed. Upper and lower bounds complete the formulation:

$$0 \le x, p \le 1. \tag{2.29}$$

It is easy to interpret this formulation in terms of the dynamic programming paradigm and infer the integrality of the polyhedron described by the constraints. Note the small numbers of $2n$ variables and $6n-6$ constraints used, as opposed to roughly $n^2/2$ variables and constraints each in the formulation (2.14). However, this alternative formulation is not symmetric with respect to the variables. In fact, the asymmetry seems to have helped make this a smaller formulation. Yannakakis proves [43] that no symmetric linear programming formulation of polynomial size exists for the matching or traveling salesman polytopes and conjectures that asymmetry will not help there. Is he right?

Polyhedral formulations. To construct a polyhedral description of the general discrete dynamic programming problem, let S denote the set of states, ordered by $\sigma : S \to [m]$, and let D be the decomposition relation.

For each $(A, I) \in D$, let there be a binary variable $z_{A,I}$ whose value indicates whether the hyperarc (A, I) is a part of the (hyper)path to the optimum. Let $I_m \in S$ be the subproblem corresponding to the original instance. It must be solved by the dynamic program, hence the constraint

$$\sum_{\{A \mid (A, I_m) \in D\}} z_{A,I_m} = 1. \tag{2.30}$$

If some subproblem is used in the solution of another, then it must be solved too. This leads us to *flow conservation* constraints

$$\sum_{\{A \mid (A,I) \in D\}} z_{A,I} = \sum_{\{(A,J) \in D \mid I \in A\}} z_{A,J}, \quad \text{for all } I \neq I_m. \tag{2.31}$$

The polyhedral description of the dynamic programming problem is completed by the nonnegativity constraints

$$z_{A,I} \geq 0, \quad \text{for all } (A, I) \in D. \tag{2.32}$$

Now the complete formulation is

$$\begin{array}{ll} \text{minimize} & \sum_{(A,I)\in D} c(A, I) z_{A,I} \\ \text{subject to} & \\ & (2.30)\text{–}(2.32). \end{array} \tag{2.33}$$

To relate this to the dynamic programming algorithm, consider the dual formulation

$$\begin{array}{lll} \text{maximize} & v & \\ \text{subject to} & & \\ v - \sum_{J\in A} u_J & \leq c(A, I) & \text{for all } (A, I) \in D \\ & & \quad \text{such that } I = I_m \\ u_I - \sum_{J\in A} u_J & \leq c(A, I) & \text{for all } (A, I) \in D, \\ & & \quad \text{such that } I \neq I_m. \end{array} \tag{2.34}$$

Dual construction.
The dynamic programming procedure directly finds an optimal solution to the dual LP (2.34) by setting the dual variables u_I corresponding to the constraints (2.30)–(2.31) in the order defined by σ: once the value u_I^* has been defined for each I such that $\sigma(I) < i$, the procedure sets

$$u_{I_i}^* = \min_{A \mid (A,I_i)\in D} \{c(A, I_i) + \sum_{J\in A} u_J^*\},$$

and continues until reaching I_m. □

Given u^*, the computation of the dynamic program can be reversed to recursively construct a primal solution z^* as follows.

Primal Construction:

(1) Initialize $z^* = 0, t = 0, A_0 = \{I_m\}$ (A_t is the set of subproblems whose solution is required at depth t).

(2) While $A_t \neq \emptyset$, do steps **(3)–(5)**.

(3) Pick a subproblem $I \in A_t$, remove it from A_t.

(4) Select a set A_{t+1} such that $u_I^* = c(A_{t+1}, I) + \sum_{J\in A_{t+1}} u_J^*$.

(5) If $t > 0$, set $t = t - 1$ and return to step **(2)** (backtrack).

THEOREM 2.59 *Consider u^* and z^* produced by the two constructions above. If the acyclic condition defined in Section 2.6.3.0 holds, then z^* and u^* are optimal solutions to* (2.33) *and* (2.34)*, respectively, and z^* is an integer solution.*

Note that z^* is integer by the primal construction. Hence it is enough to prove that z^* is feasible for the primal (2.33), that u^* is feasible for the dual (2.34), and that the complementary slackness conditions (Theorem 2.44) are satisfied. We leave the details to the reader, but note that in the critical step **(4)** in the primal construction, the set A_{t+1} can always be found because of the way u_I^* has been defined during the dual construction. This same observation helps in establishing complementary slackness: a primal variable is only increased from its original zero value if the corresponding dual constraint is tight.

Theorem 2.59 implies also total dual integrality of the system defining the LP (2.33). However, it is possible that z^* contains some entries of value greater than 1, and so does not correspond to a dynamic programming solution. If constraints are added to LP (2.33), requiring that each entry of z is at most 1, then the LP may not have an integer optimum solution. Thus we must impose additional structure on the dynamic programming problem to ensure that zero-one solutions are extreme points. This structure can be expressed by associating with each state $I \in S$ a nonempty subset R_I of a finite *reference set* R. The partial order defined on the family of reference subsets induced by set inclusion must be compatible with the relation D:

$$R_I \subseteq R_J,$$

for all I, J such that there exists an A such that $I \in A$ and $(A, J) \in D$. The reference subsets must also satisfy the following condition:

$$R_I \cap R_{I'} = \emptyset \quad \text{for all } (A, J) \in D \text{ such that } I, I' \in A.$$

THEOREM 2.60 *For a discrete dynamic programming problem formulated over a set of states S with the decision variables $z_{A,I}$ for $(A, I) \in D$, where there exists a reference set R satisfying the conditions described above, the LP* (2.33) *describes the convex hull of zero-one solutions.*

A few final observations:

1. The resulting formulations are compact.
2. The resulting formulations are typically not symmetric.
3. Usually related to the second statement is the fact that the formulations are typically not over the most natural variable space. A good example is the k-median problem on a tree. There exist dynamic programming algorithms for this problem [39], but a polyhedral formulation in terms

of natural vertex decision variables is not known [9]. Theoretically, it is possible to express the formulation in terms of natural variables by using a linear transformation and then a projection. However, this is often not practical.

2.7 Approximation Algorithms, Integrality Gaps and Polyhedral Embeddings

Many combinatorial optimization problems are NP-hard. In such a case, we cannot hope to have a complete polyhedral description that would be solvable in polynomial time and we must resort to relaxations (Section 2.3.1). In many cases, an optimal solution to the relaxation of a polytope can be used as a starting point in the derivation of a point belonging to the polytope. In a combinatorial optimization setting, the motivation is to try and use the optimal LP solution to find a good—though not necessarily optimal—solution to the problem in question. Such a procedure is usually called *LP rounding* because from an optimal (fractional) solution to the LP relaxation it produces an integral solution to the problem.

Consider a combinatorial minimization problem. Denote by $z^*(I)$ the cost of an optimal solution to the instance I. An algorithm A that produces a feasible solution of cost $z_A(I)$ such that

$$\frac{z_A(I)}{z^*(I)} \leq \alpha$$

for all instances I is called an *α-approximation algorithm* for the given problem. In other words, an approximation algorithm provides a guarantee on the quality of solutions it produces.

An approximation algorithm based on LP rounding usually (but not always [23]) implies a bound on the maximum ratio between the cost z^* of an optimal solution to the problem and the cost z^*_{LP} of an optimal solution to the relaxation. Given an LP relaxation of a combinatorial optimization problem, the value

$$\sup_I \frac{z^*(I)}{z^*_{LP}(I)}$$

is called the *integrality gap* of the relaxation. The integrality gap will be a constant in most of our examples, but in many cases it is a function of the instance size.

2.7.1 Integrality Gaps and Convex Combinations

A standard example of the integrality gap is provided by the *vertex cover* problem. Given a graph $G = (V, E)$ with a cost function $c : V \to \mathbf{R}_+$, a set of $U \subset V$ is a vertex cover of G, if for every edge $e \in E$, at least one endpoint of

e belongs to U. The cost of the vertex cover U is defined as $c(U) = \sum_{v \in U} c_v$. Any minimum-cost vertex cover of G is an optimal solution to the vertex cover problem on G. The vertex cover problem is NP-hard and so we do not know a polyhedral description, but it turns out that the relaxation (2.35) is useful in designing an approximation algorithm.

$$\begin{array}{ll} \text{minimize} & \sum_{v \in V} c_v x_v \\ \text{subject to} & \\ & \begin{array}{rcll} x_u + x_v & \geq & 1 & \forall uv \in E \\ x & \geq & 0 & \end{array} \end{array} \tag{2.35}$$

Let x^* be an optimal solution to the relaxation (2.35) and consider $2x^*$. Since $x_u + x_v \geq 1$ for every $uv \in E$, at least one of x_u and x_v is at least $1/2$. For each edge $e \in E$, choose an endpoint v_e of e such that $v_e \geq 1/2$. Let $U = \{v_e \mid e \in E\}$. Then define a 0-1 vector x^U by

$$x_v^U = \begin{cases} 1, & v \in U \\ 0, & v \notin U. \end{cases}$$

Now x^U is the incidence vector of a vertex cover, and since

$$x^U \leq \lfloor 2x^* \rfloor \leq 2x^*,$$

it follows that $2x^*$ dominates a convex combination (in this case a trivial combination, consisting only of x^U) of integral solutions. The cost of the solution x^U then satisfies

$$c \cdot x^U \leq 2c \cdot x^*.$$

One consequence of this argument is that the integrality gap of the LP (2.35) is at most 2. Another is that the algorithm that computes and outputs x^U is a 2-approximation algorithm.

How does one find a rounding procedure? The vertex cover case is simple enough that the rounding can be guessed directly and the guarantee proven easily, but in general the question of whether there is a rounding procedure may be very difficult. The procedure described for vertex cover can be generalized to all relaxations with a nonnegative cost function. In situations when there is more than one integral solution with a nonzero coefficient in the convex combination, to guarantee a bound on the integrality gap we use the additional observation that the cheapest among these solutions costs no more than the convex combination itself. In fact, the following can be shown [3, 6]:

THEOREM 2.61 *For a minimization problem with a nonnegative cost function, the integrality gap is bounded above by α if (and only if) the corresponding multiple αx^* of any extreme-point solution to the relaxation dominates (component-wise) a convex combination of integral solutions.*

Thus there is no loss of generality in considering only the rounding procedures of this form. In Sections 2.7.2 and 2.7.5 we show two more examples of this principle: those of metric STSP and edge-dominating set problems.

Integrality gaps of many combinatorial optimization problems have been studied in the literature. A notable example is the traveling salesman problem where the cost satisfies the triangle inequality (that is, defines a metric on the underlying graph), in both undirected and directed versions. For the undirected case (that is, metric STSP), the integrality gap is between $4/3$ and $3/2$ [13, 19]. In the directed (asymmetric) case the best currently known bounds are $4/3$ and $\log n$ [6]. For the metric STSP, it is conjectured and widely believed that the gap is $4/3$. Many of the integrality gap results have been motivated by the search for approximation algorithms [41].

2.7.2 An STSP Integrality Gap Bound

Closely related to Hamilton cycles in combinatorial optimization are *spanning Eulerian multigraphs*. An *Eulerian (multi)graph* is a graph $T = (V, E(T))$ where every vertex in V has even degree (and parallel edges are allowed in $E(T)$). Such a graph is *spanning* if every vertex in V is an endpoint of some edge of $E(T)$ (and thus at least 2 such edges). If the cost function satisfies the triangle inequality, then any such multigraph can be converted to a Hamilton cycle of no greater cost by taking a so-called *Euler tour* (visiting each edge in $E(T)$ exactly once), and *shortcutting*, creating a new edge to bypass each vertex $v \in V$ that has already been visited, thus replacing the two edges entering and leaving v on this subsequent occasion. This relationship between Eulerian graphs and Hamilton cycles yields the following result:

THEOREM 2.62 *The integrality gap for the metric STSP is at most α if (and only if) for each extreme point x^* of the subtour relaxation, αx^* can be expressed as a convex combination of spanning Eulerian multigraphs.*

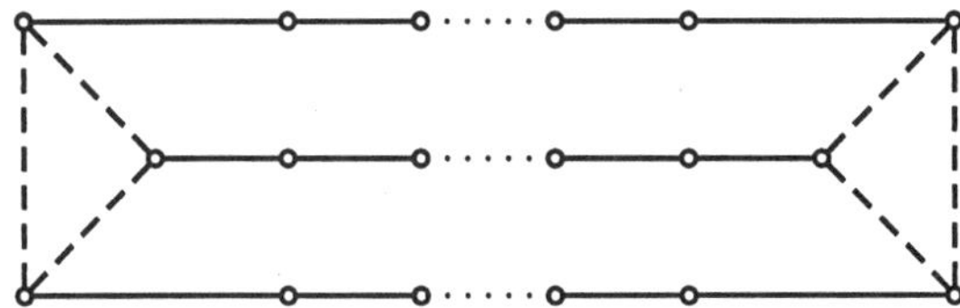

Figure 2.1. Diagram of an STSP extreme point. Each of the three horizontal paths has k edges. The solid lines represent edges with value 1, dashed with value $\frac{1}{2}$.

Consider the extreme point in Figure 2.7.2. By applying the only if part of Theorem 2.62, one can show that the integrality gap of metric STSP is at least

4/3 (an appropriate choice of a cost function will obtain this result as well). Wolsey [42], and later Shmoys and Williamson [38], show this gap to be at most 3/2. We now outline an elegant argument for this in the spirit of Theorem 2.62. Let $n = |V|$. Recall that SEP denotes the subtour polytope (Section 2.3.1).

LEMMA 2.63 *Given any extreme point x^* of SEP, $\frac{n-1}{n}x^*$ can be expressed as a convex combination of (incidence vectors of)* spanning trees *(i.e., acyclic, connected spanning graphs).*

Proof: Consider the spanning tree polytope [25]. It can be shown that $\frac{n-1}{n}x^*$ satisfies all the constraints. □

LEMMA 2.64 *Given any extreme point x^* of SEP and any $T \subset V$ with $|T|$ even, $\frac{1}{2}x^*$ can be expressed as a convex combination of* T-joins *(multigraphs whose odd degree vertices are precisely those in T, see Exercise 2.36).*

Proof: Consider the T-join polyhedron [12]. It can be shown that $\frac{1}{2}x^*$ satisfies all the constraints. □

THEOREM 2.65 *Given any extreme point x^* of SEP, $(\frac{3}{2} - \frac{1}{n})x^*$ can be expressed as a convex combination of spanning Eulerian multigraphs.*

Proof: Express $\frac{n-1}{n}x^*$ as a convex combination

$$\sum_i \lambda_i \chi^{E(T_i)} \tag{2.36}$$

of spanning trees, using Lemma 2.63. Consider one such tree T_i and define T_i^{odd} to be the odd degree nodes of this tree. If an incidence vector of a T_i^{odd}-join is added to $\chi^{E(T_i)}$, we obtain an incidence vector of an Eulerian graph. Hence, using Lemma 2.64, we can express $\chi^{E(T_i)} + \frac{1}{2}x^*$ as a convex combination

$$\sum_j \mu_{i,j} \chi^{E(K_{i,j})} \tag{2.37}$$

of spanning Eulerian multigraphs. But, $(\frac{n-1}{n} + \frac{1}{2})x^*$ can be rewritten using (2.36) as a convex combination

$$\sum_i \lambda_i (\chi^{E(T_i)} + \frac{1}{2}x^*).$$

Therefore, $(\frac{n-1}{n} + \frac{1}{2})x^*$ can be expressed as the convex combination

$$\sum_{i,j} \lambda_i \mu_{i,j} \chi^{E(K_{i,j})}$$

of spanning Eulerian multigraphs by using (2.37), as required. □

COROLLARY 2.66 *The integrality gap of metric STSP is at most* 3/2.

2.7.3 Half-integrality

The linear relaxation (2.35) of vertex cover in fact satisfies a strong, very interesting property.

THEOREM 2.67 *[29] Every extreme point of the polyhedron defined by the LP* (2.35) *has coordinates that all belong to* $\{0, 1/2, 1\}$*, that is, the polyhedron is half-integral.*

Proof: Let x^* be an optimal solution to the LP (2.35) and define

$$\begin{aligned} U_{-1} &= \{v \in V \mid 0 < x_v^* < 1/2\}, \\ U_1 &= \{v \in V \mid 1/2 < x_v^* < 1\}. \end{aligned}$$

Then modify the values of x_v^* for those v contained in either U_{-1} or U_1: define $y_v^1 = x_v^* + k\epsilon$ and $y_v^2 = x_v^* - k\epsilon$ if $v \in U_k$, for $k \in \{-1, 1\}$. Define $y_v^1 = y_v^2 = x_v$ for $v \notin U_{-1} \cup U_1$. Then $x^* = (y^1 + y^2)/2$. If either U_{-1} or U_1 is nonempty, ϵ may be chosen small enough that both y^1 and y^2 are feasible, and thus $U_{-1} = U_1 = \emptyset$ for an extreme point x^*. □

Half-integrality is common in combinatorial optimization. We give another example in Section 2.7.5 and refer the reader to Hochbaum's survey [16] for additional results.

2.7.4 Rounding Reductions

Occasionally, approximation algorithms may be composed by reductions that preserve approximation guarantees. For example, suppose we can transform an instance I_A of problem A to an instance I_B of problem B so that there is a one-to-one correspondence between solutions of I_A and those of I_B such that if s_A is a solution to I_A and s_B the corresponding solution to I_B, their costs satisfy $\alpha c(s_A) \geq c(s_B)$. Then a β-approximation algorithm for B implies an $\alpha\beta$ approximation algorithm for A.

In Section 2.7.5 we present such a reduction that uses linear relaxations and rounding. The problems involved are the *weighted edge-dominating set* (EDS) and *edge-cover* (EC). The algorithm starts with an optimal solution of a relaxation of EDS and multiplies it by 2. The resulting point can be interpreted as a solution to a relaxation of EC, and after multiplication by $4/3$ expressed (after reducing the values of some of the variables) as a convex combination of integral solutions.

Other similar results exist [2, 22]. The integrality gap or the approximation guarantee is sometimes the product of those achieved by the intermediate steps (Section 2.7.5) and sometimes the sum (Section 2.7.2). Even better results are sometimes possible with careful analysis [4].

2.7.5 Weighted edge-dominating sets

Given a graph $G = (V, E)$ with a cost function $c : E \to \mathbf{R}_+$ defined on the edge set, an *edge-dominating set* of G is any subset $F \subseteq E$ of edges such that for every $e \in E$, at least one edge adjacent to e is contained in F. The cost of an edge-dominating set is defined as the sum of the costs of its edges. An optimal solution to the edge-dominating set (EDS) instance defined by G is then a minimum-cost edge-dominating set.

For edges $e, f \in E$, write $e \sim f$ if e and f are adjacent. A simple linear relaxation for EDS is as follows.

$$\begin{array}{lrcll} \text{minimize} & \sum_{e\in E} c_e x_e & & & \\ \text{subject to} & & & & \\ & \sum_{f\sim e} x_f & \geq & 1 & \forall e \in E \\ & 0 \leq x & \leq & 1 & \end{array} \tag{2.38}$$

Let x^* be an optimal solution to (2.38). Then for each edge $uv \in E$,

$$\max\{2x^*(\delta(u)), 2x^*(\delta(v))\} \geq 1 + x^*_{uv}.$$

For each edge $e \in E$, let v_e be the endpoint that achieves the maximum above. Then let $U = \{v_e \mid e \in E\}$, and $y_e = \min\{2x^*_e, 1\}$ for all $e \in E$. The point y is now an element of the polyhedron

$$\begin{array}{rcll} 0 \leq y_e & \leq & 1 & \forall e \in E \\ y(\delta(v)) & \geq & 1 & \forall v \in U. \end{array} \tag{2.39}$$

An important observation is that any integral solution to (2.39) is also an edge-dominating set for G. We shall prove that an extreme point optimal solution to (2.39) can be efficiently rounded to an edge-dominating set while increasing the cost by a factor at most $4/3$, implying an $8/3$-approximation algorithm (and integrality gap) for EDS.

THEOREM 2.68 *Let y^* be an extreme point of the polyhedron* (2.39). *Then y^* is half-integral and the edges of value $1/2$ induce disjoint odd cycles in G.*

The proof of Theorem 2.68 is longer, but based on the same ideas as that of Theorem 2.67 (half-integrality of vertex cover).

A final fact needed to justify the performance guarantee:

LEMMA 2.69 *Let C be an odd cycle of edges with value $1/2$. Let y_C denote the vector of fractional edge-values of C. Then $(4/3)y_C$ dominates a convex combination of edge-covers of C.*

Proof: Let $C = (V_C, E_C)$, where $V_C = \{v_1, \ldots, v_k\}$ and $E_C = \{v_1v_2, v_2v_3, \ldots, v_kv_1\}$. Define integral edge-covers $D_1, \ldots, D_k$ as follows. For each i, D_i

will have $(k+1)/2$ edges. For $i < k$, let D_i contain e_i and e_{i+1}. Then add $(k-3)/2$ edges to D_i so that e_i and e_{i+1} are the only adjacent pair of edges in D_i. In D_k, the adjacent edges are e_k and e_1.

Let y^i denote the incidence vector of D_i. Then

$$y^0 = \frac{1}{k}(y^1 + y^2 + \cdots + y^k)$$

is a convex combination of edge-covers, and $y_i^0 = (k+1)/(2k)$ for all i. Since $4/3 \geq (k+1)/k$ for all $k \geq 3$ (and the shortest odd cycle is a triangle), it follows that $(4/3)y_C \geq y^0$. □

Note that, because the sum of coefficients in any convex combination is 1, it follows that at least one of the edge-covers $D_1, \ldots, D_k$ has cost no more than $4/3$ times the cost of y_C.

Algorithm:

(1) Solve the LP-relaxation (2.38). Let x^{LP} be an optimal solution.

(2) Let $U = \emptyset$. For each edge $e = uv \in E$, if $x^{LP}(\delta(u)) \geq 1/2$, add u to U, otherwise add v to U.

(3) Solve the relaxed edge-cover linear program, where the constraints $y(\delta(u)) \geq 1$ are included only for $u \in U$, and the objective function is $\min \sum_{e \in E} c_e y_e$. Find an optimal solution y^* which is an extreme point.

(4) Let $D = \{e \in E \mid y_e^* = 1\}$. Find the least cost edge-cover of each odd cycle in y^*, and add its edges to D.

(5) Output D.

Note that D is an edge-dominating set because the edge-covering constraints for U require at least one endpoint of each edge in E to be covered (dominated) by an edge in D.

THEOREM 2.70 *The edge-dominating set D that the above algorithm outputs has cost no more than* $8/3$ *times the cost of an optimal edge-dominating set.*

THEOREM 2.71 *For any extreme point* x^{LP} *of* (2.38), $(8/3)x^{LP}$ *dominates a convex combination of edge-dominating sets. Thus the integrality gap of the LP* (2.38) *is at most* $8/3$.

A more careful analysis shows that the integrality gap of the relaxation (2.38) is in fact at most 2.1. In fact, the integrality gap is *exactly* 2.1 [**?**]: consider the complete graph K_{5n} on $5n$ vertices and partition it into n disjoint copies of K_5, named $G_1, \ldots, G_n$. Let $c_e = 1$ for each edge e induced by some G_i, and $c_e = 11n$ for every other edge e. Now any integral edge-dominating set has cost at least $3n - 1$, but on the other hand assigning $x_e = 1/7$ to all the edges of G_i for each i results in a feasible solution of cost $10n/7$.

It is possible to strengthen the linear relaxation (2.38) by adding inequalities for *hypomatchable sets* [?]. The integrality gap of the resulting relaxation is 2, and an algorithm with approximation guarantee 2 follows as well. Further improvements would imply improvements also for vertex cover, which is an outstanding open problem.

As a final remark, we note that the algorithm described above, as well as many others based on the same paradigm, can actually be implemented without explicitly solving the initial LP relaxation, and instead based on (more efficient) algorithms for approximating the "intermediate problem" (vertex cover).

Acknowledgment

This chapter arose partially from notes for a course taught at the University of New Mexico. Sandia is a multiprogram laboratory operated by Sandia Corporation, a Lockheed Martin Company, for the U.S. Department of Energy under Contract DE-AC04-94AL85000.

References

[1] E. Balas and M. Fischetti. Polyhedral theory for the asymmetric traveling salesman problem. In *The traveling salesman problem and its variations*, pages 117–168. Kluwer, 2002.

[2] D. Bienstock, M. Goemans, D. Simchi-Levi, and David Williamson. A note on the prize-collecting traveling salesman problem. *Mathematical Programming, Series A*, 59(3):413–420, 1993.

[3] S. Boyd and R. D. Carr. A new bound for the radio between the 2-matching problem and its linear relaxation. *Mathematical Programming, Series A*, 86(3):499–514, 1999.

[4] R. D. Carr, T. Fujito, G. Konjevod, and O. Parekh. A $2\frac{1}{10}$ approximation algorithm for a generalization of the weighted edge-dominating set problem. *Journal of Combinatorial Optimization*, 5:317–326, 2001.

[5] R. D. Carr and G. Lancia. Compact vs. exponential-size LP relaxations. *Operations Research Letters*, 30(1):57–65, 2002.

[6] R. D. Carr and S. Vempala. Towards a 4/3-approximation for the asymmetric traveling salesman problem. In *Proceedings of the 11th Annual ACM-SIAM Symposium on Discrete Algorithms*, pages 116–125, 2000.

[7] V. Chvátal. *Linear programming*. Freeman, 1983.

[8] E. V. Denardo. *Dynamic programming: theory and practice*. Prentice-Hall, 1978.

[9] S. de Vries, M. E. Posner, and R. Vohra. Polyhedral properties of the K-median problem on a tree. No 1367 in Discussion Papers, Northwestern University, Center for Mathematical Studies in Economics and Management Science, 42 pages, 2003.

[10] J. Edmonds. Optimum branchings. *Journal of Research of the National Bureau of Standards*, 71B:233–240, 1967.

[11] L. R. Ford and D. R. Fulkerson. *Flows in Networks*. Princeton University Press, Princeton, NJ, 1962.

[12] A. M. H. Gerards. Matching. In *Network models*, volume 7 of *Handbooks in Operations Research*, pages 135–224. Elsevier, 1995.

[13] M. Goemans. Worst-case comparison of valid inequalities for the TSP. *Mathematical Programming*, 69:335–349, 1995.

[14] M. Grötschel, L. Lovász, and A. Schrijver. *Geometric Algorithms and Combinatorial Optimization*. Springer Verlag, 1988.

[15] M. Grötschel and L. Lovász. Combinatorial optimization. In *Handbook of combinatorics*, volume 2, pages 1541–1597. Elsevier, 1995.

[16] D. S. Hochbaum. Approximating covering and packing problems: set cover, vertex cover, independent set, and related problems. In *Approximation algorithms for NP-hard problems*, pages 94–143. PWS, 1997.

[17] T. Ibaraki. Solvable classes of discrete dynamic programming. *Journal of Mathematical Analysis and Applications*, 43:642–693, 1973.

[18] R. G. Jeroslow. On defining sets of vertices of the hypercube by linear inequalities. *Discrete Mathematics*, 11:119–124, 1975.

[19] M. Jünger, G. Reinelt, and G. Rinaldi. The traveling salesman problem. In *Network models*, volume 7 of *Handbooks in Operations Research*, pages 225–329. Elsevier, 1995.

[20] V. Kaibel and M. Pfetsch. Some algorithmic problems in polytope theory. In *Algebra, Geometry and Software Systems*, pages 23–47. Springer Verlag, 2003. arXiv:math.CO/0202204v1.

[21] R. M. Karp and M. Held. Finite state processes and dynamic programming. *SIAM Journal of Applied Mathematics*, 15:693–718, 1967.

[22] J. Könemann, G. Konjevod, O. Parekh, and A. Sinha. Improved approximation algorithms for tour and tree covers. In *Proceedings of APPROX 2000*, volume 1913 of *Lecture Notes in Computer Science*, pages 184–193, 2000.

[23] G. Konjevod, R. Ravi, and A. Srinivasan. Approximation algorithms for the covering Steiner problem. *Random Structures & Algorithms*, 20:465–482, 2002.

[24] B. Korte and J. Vygen. *Combinatorial optimization. Theory and algorithms.* Springer Verlag, 2000.

[25] T. L. Magnanti and L. A. Wolsey. Optimal trees. In *Network models*, volume 7 of *Handbooks in Operations Research*, pages 503–615. Elsevier, 1995.

[26] R. K. Martin, R. Rardin, and B. A. Campbell. Polyhedral characterizations of discrete dynamic programming. *Operations Research*, 38(1):127–138, 1990.

[27] R. K. Martin. Using separations algorithms to generate mixed integer model reformulations. *Operations Research Letters*, 10(3):119–128, 1991.

[28] D. Naddef. Polyhedral theory and branch-and-cut algorithms for the symmetric TSP. In *The traveling salesman problem and its variations*, pages 29–116. Kluwer, 2002.

[29] G. L. Nemhauser and L. E. Trotter. Properties of vertex packing and independence system polyhedra. *Mathematical Programming*, 6:48–61, 1974.

[30] G. Nemhauser and L. Wolsey. *Integer and Combinatorial Optimization*. Wiley, 1988.

[31] C. Papadimitriou. *Computational Complexity*. Addison-Wesley, 1994.

[32] O. Parekh. Edge-dominating and hypomatchable sets. In *Proceedings of the 13th Annual ACM-SIAM Symposium on Discrete Algorithms*, pages 287–291, 2002.

[33] W. R. Pulleyblank. Polyhedral combinatorics. In *Optimization*, volume 1 of *Handbooks in Operations Research*, pages 371–446. Elsevier, 1989.

[34] W. R. Pulleyblank. Matchings and extensions. In *Handbook of combinatorics*, volume 1, pages 179–232. Elsevier, 1995.

[35] A. Schrijver. *Theory of Linear and Integer Programming*. Wiley, 1986.

[36] A. Schrijver. Polyhedral combinatorics. In *Handbook of combinatorics*, volume 2, pages 1649–1704. Elsevier, 1995.

[37] A. Schrijver. *Combinatorial Optimization: Polyhedra and Efficiency*. Springer Verlag, 2003.

[38] D. B. Shmoys and D. P. Williamson. Analyzing the Held-Karp TSP bound: a monotonicity property with application. *Information Processing Letters*, 35:281–285, 1990.

[39] A. Tamir. An $O(pn^2)$ algorithm for the p-median and related problems on tree graphs. *Operations Research Letters*, 19(2):59–64, 1996.

[40] T. Terlaky and S. Zhang. Pivot rules for linear programming: a survey on recent theoretical developments. *Annals of Operations Research*, 46:203–233, 1993.

[41] V. V. Vazirani. *Approximation algorithms*. Springer Verlag, 2001.

[42] L. A. Wolsey. Heuristic analysis, linear programming and branch and bound. *Mathematical Programming Study*, 13:121–134, 1980.

[43] M. Yannakakis. Expressing combinatorial optimization problems by linear programs. *Journal of Computer and System Sciences*, 43(3):441–466, 1991.

[44] G. Ziegler. *Lectures on Polytopes*. Springer, 1995.

Chapter 3

CONSTRAINT LANGUAGES FOR COMBINATORIAL OPTIMIZATION

Pascal Van Hentenryck
Computer Science Department, Brown University
pvh@cs.brown.edu

Laurent Michel
Department of Computer Science & Engineering, University of Connecticut
ldm@engr.uconn.edu

Abstract This tutorial reviews recent developments in the design and implementation of constraint languages for combinatorial optimization. In particular, it argues that constraint-based languages are a unifying technology to express combinatorial optimization problems, supporting a variety of solvers including constraint programming, local search, and mathematical programming.

Introduction

Combinatorial optimization problems are ubiquitous in many parts of our societies. From the airline industry to courier services, from supply-chain management to manufacturing, and from facility location to resource allocation, many important decisions are taken by optimization software every day. In general, optimization problems are extremely challenging both from a computational and software engineering standpoint. They cannot be solved exactly in polynomial time and they require expertise in algorithms, applied mathematics, and the application domain. Moreover, the resulting software is often large, complex, and intricate, which makes it complicated to design, implement, and maintain. This is very ironic, since many optimization problems can be specified concisely. The distance between the specification and the final program is thus considerable, which indicates that software tools are seriously lacking in expressiveness and abstractions in this application area.

Because of the nature of optimization problems, no single approach is likely to be effective on all problems, or even on all instances of a single problem. Solving optimization problems has remained a very experimental endeavor: what will, or will not, work in practice is hard to predict. As a consequence, it is of primary importance that all major approaches to optimization be supported by high-level tools automating many of the tedious and complex aspects of these applications. The design of such languages is the primary topic of this chapter.

Historically, languages for expressing combinatorial optmization problems have focused on mathematical programming. Starting from matrix generators, this line of research led to the development of elegant modeling languages (e.g., [5, 14]), where collections of linear and nonlinear constraints are expressed using traditional algebraic and set notations. These constraints, and the associated objective functions are then sent to LP or MIP solvers.

The last two decades have witnessed the emergence of *constraint programming* as a fundamental methodology to solve a variety of combinatorial applications, and of *constraint programming languages* to express them. The focus of constraint programming is on reducing the search space by pruning values which cannot appear in any solution, or in any optimal solution. Because of the distinct nature of this paradigm, it is not surprising that constraint programming languages have contributed many innovations in the modeling of combinatorial optimization problems. Typically, constraint programming languages feature rich languages for expressing and combining constraints, the concept of combinatorial constraints, as well as the ability to specify search procedures at a high level of abstraction. Moreover, there have been several attempts to unify constraint programming and mathematical modeling languages (e.g., [15, 45, 46]).

Surprisingly, neighborhood search, one of the oldest optimization techniques, has been largely ignored until recently as far as modeling support is considered (e.g., [10, 43, 27, 54]). This is a serious gap in the repertoire of tools for optimization, since neighborhood search is the technique of choice for a variety of fundamental applications. In the last three years, the COMET project has attempted to address this need. COMET is an object-oriented language supporting a constraint-based architecture for neighborhood search and featuring novel declarative and control abstractions. Interestingly, COMET programs often feature declarative components which closely resemble those of constraint programs, although the actual interfaces, functionalities, and search procedures differ considerably. Moreover, COMET decreases the size of neighborhood search programs significantly and enhances compositionality, modularity, and reuse for this class of applications.

The purpose of this chapter is to review these recent developments. Its main goal is to show that constraint languages constitute a natural vehicle to express mathematical programming, neighborhood search, and constraint programming applications. These languages may dramatically reduce the distance between the specification and the computer solution of an application, enhance compositionality, modularity, and reuse of combinatorial algorithms, and provide a platform for building hybrid solutions integrating these complementary paradigms. The main focus of the chapter is on constraint programming and neighborhood search, since most readers are likely familiar with mathematical modeling languages. However, the chapter also indicates, whenever appropriate, the potential impact of recent developments for mathematical programming.

The rest of this chapter is organized as follows. It starts with a brief overview of constraint languages for combinatorial optimization. It then studies constraint programming languages and constraint-based languages for neighborhood search. For each language class, the chapter describes the architecture of the languages, the solution of several applications, and a presentation of scheduling abstractions. The chapter concludes with a discussion of the similarities and differences of these language classes. The chapter uses OPL [45] and COMET [28, 50, 51] for illustration purposes, but it is important to note that most of the concepts are, or could be, supported in a variety of other languages and libraries. The references in the chapter are not meant to be comprehensive; they simply provide links where additional information is available.

3.1 Constraint Languages for Combinatorial Optimization

The fundamental design rationale behind constraint languages is the recognition that constraints are a natural vehicle for expressing combinatorial optimization problems. Indeed, constraints are concise and declarative abstractions to specify the properties to be satisfied by solutions.

Constraints. One of the main innovations in constraint languages is what "constraint" means in this context. In constraint languages, constraints should not be understood as a set of (linear) arithmetic constraints as is typically the case in modeling languages. Rather, constraints capture complex combinatorial substructures and can be combined through a variety of operators such as disjunctions and cardinality operators. The idea that constraints may capture complex relationships between variables was integral part of constraint programming languages early on [12, 47, 11, 44]. Indeed, the primary role of constraints in these languages was to reduce the search space by removing, from the domains of variables, values that could not appear in any solution. Complex relationships allowed for more compact and more efficient models.

Combinatorial constraints became an important research topic in the 1990s [3, 37]. The prototypical example is the *alldifferent*$(x_1, \ldots, x_n)$ constraint, which holds if the variables $x_1, \ldots, x_n$ are given all different values. With this novel modeling tool at hand, applications are typically modeled by combining a number of constraints, each of which capturing a combinatorial structure found. Combinatorial constraints filter the domains of the variables by using dedicated algorithms which exploit combinatorial properties (e.g., the *alldifferent* constraint uses matching and connected component algorithms).

Search. A second innovation of constraint languages is their ability to specify search procedures. This feature, which also departs radically from traditional modeling languages, was natural in constraint languages given their roots of logic programming, a class of nondeterministic programming languages where search can be expressed concisely. Once again, the ability to express search procedures makes it possible to exploit the problem structure to guide the search and remove symmetries, two critical components in many applications. As a consequence, an application in a constraint language was typically modeled by a declarative component specifying the problem constraints and a search component specifying how to search for solutions. Search components in early constraint languages only specified the search tree, not how to explore it. Typically, the search tree was explored through depth-first search, once again a consequence of their roots from logic programming. The late 1990s saw the emergence of control abstractions to specify how to search the trees. These meta-search specifications, often called search strategies in constraint programming, made it possible to express a variety of advanced search techniques such as limited discrepancy search [18].

Domain-Specific Abstractions. A third innovation are domain-specific extensions for some important applications areas. These extensions built on top of the constraint and search components just described and were motivated by the fact that some application areas, such as scheduling and routing, are typically expressed in terms of high-level concepts such as activities, resources, visits, and customers. Many constraint languages now feature these high-level modeling concepts, which encapsulate variables, combinatorial constraints, and support for search procedures.

Advantages of Constraint Languages. Constraint programming languages offer a variety of benefits. They make it easy to express many complex applications by combining combinatorial constraints to model the applications and specifying high-level search procedures. Moreover, the languages promote a programming style which is extremely modular, separating the declarative and search components and favoring reusability. In particular, it is possible in constraint languages to modify one of the components without affecting the other and to write code which can be reused across applications.

Constraint and Mathematical Programming. Constraint programming (CP) models are, in general, very different from integer programming (IP) statements, yet they are not bound to a particular solver technology: In fact, these models are best viewed as high-level specifications making combinatorial substructures of an application explicit. These high-level models can be "compiled down" into a variety of solvers (e.g., CP and IP solvers) and/or they can be the basis for the integration of the technologies. In this last case, combinatorial constraints not only encapsulate filtering algorithms; they also generate linear relaxations and tighten them using cuts (e.g., [19, 36]). Pure IP solvers may then exploit these combinatorial substructures to obtain tighter relaxations and/or to design more effective search procedures as suggested in [9]. This will not be discussed in much detail in this chapter due to space reasons, but this research direction is certainly worth exploring.

Neighborhood Search. Perhaps one of the most surprising developments of recent years is the recognition that constraint languages also apply, and bring substantial benefits, to neighborhood search. Typically, scientific papers on neighborhood search rarely describe algorithms in terms of constraints. They talk about neighborhoods, evaluation functions, and heuristics (and meta-heuristics). It is also not clear what is reusable across applications and how these algorithms can be modularized. As a consequence, the gap between an application and a neighborhood search algorithm is generally quite substantial, since they operate at different conceptual levels. What made the link between constraint languages and neighborhood search possible is the concept of *differentiable constraints*. The key idea is that neighborhood search applications can also be modeled by expliciting combinatorial structures through constraints, which now maintain incremental data structures to explore the neighborhood efficiently. More precisely, each combinatorial structure can be queried to evaluate the effect of local moves and to update their data structures incrementally when a move is executed. As a consequence, it becomes possible to write neighborhood search algorithms whose declarative components closely resemble those of typical constraint programs. Moreover, the resulting algorithms clearly separate constraints, heuristics, and meta-heuristics, and offer the same modularity and reuse as constraint programming languages.

3.2 Constraint Programming

This section is a brief overview of constraint programming. It describes its architecture and illustrates its modelling style and computational approach on several applications. It uses the modeling language OPL [45].

3.2.1 The Architecture

Constraint programming features a two-level architecture integrating a *constraint* and a *search* component as shown in Figure 3.1. The constraint component, often called the constraint store, collects the constraints at every computation step. Operating around the constraint store is a search component which assigns variables, or adds constraints, in nondeterministic ways. At this conceptual level, the architecture is similar in essence to integer programming systems. The main originality in constraint programming lies in the constraint-store organization.

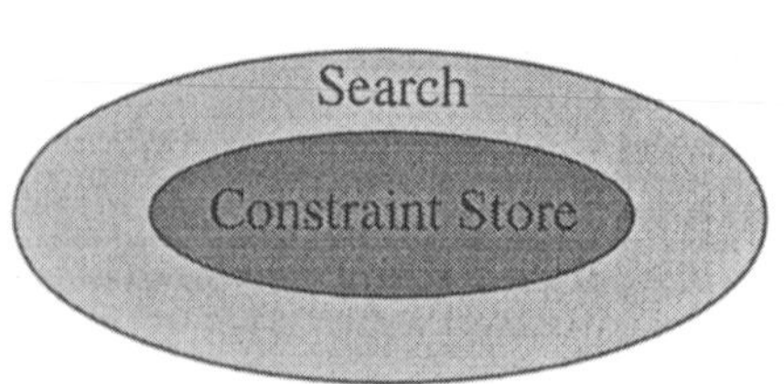

Figure 3.1. The CP Architecture

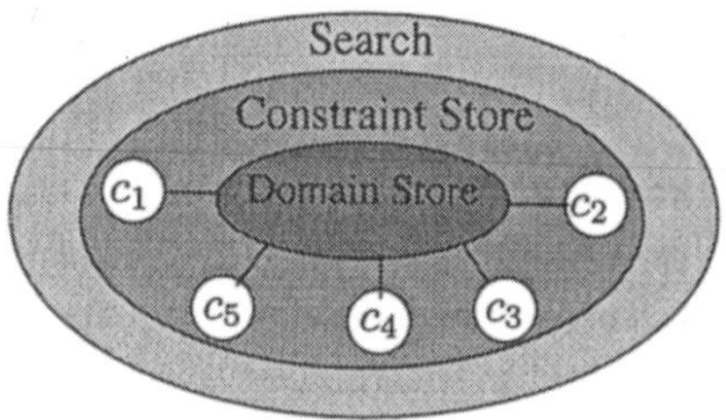

Figure 3.2. The CP Constraint Store

Constraints. Figure 3.2 depicts the architecture of the constraint store. The core of the architecture is the *domain store* that associates a domain with each variable. The domain of a variable represents its set of possible values at a given computation state (e.g., at a node in the tree search). Gravitating around these domains are the various constraints, which have no knowledge of each other. Associated with each constraint is a constraint-satisfaction algorithm whose primary role is to perform two main tasks:

1 to determine if the constraint is consistent with the domain store, i.e., if there exist values for the variables satisfying the constraint;

2 to apply a filtering algorithm to remove, from the domains of the constraint variables, values that do not appear in any of its solutions.

In addition, the constraint-satisfaction algorithm may add new constraints to the constraint store.

The constraint solver can then be viewed as a simple iterative algorithm whose basic step consists of selecting a constraint and applying its constraint-satisfaction algorithm. The algorithm terminates when a constraint is inconsistent with respect to the domain store or when no more domain reductions are possible. Note that, on termination, there is no guarantee that there exists a solution to the constraint store. The constraints may all have a local solution with respect to the domain store, but these solutions may be incompatible globally. This architecture, which is now the cornerstone of most modern constraint-programming systems, was pioneered by the CHIP system [13, 44, 49]) which included a solver for discrete finite domains based on constraint-satisfaction techniques (e.g., [24, 31, 56]).

Unlike linear and integer programming, constraint-programming systems support *arbitrarily complex constraints*. These constraints are not restricted to linear constraints, or even nonlinear constraints, and they may impose complex relations between their variables. For instance, a disjunctive scheduling constraint imposes that its activities do not overlap in time. As a consequence, it is useful, and correct, to think of a constraint as representing *an interesting subproblem* (combinatorial substructure) of an application. It is one of the fundamental issues in constraint programming to isolate classes of constraints that are widely applicable and amenable to efficient implementation.

Ideally, constraint-satisfaction algorithms should be complete (i.e., it should remove all inconsistent values) and run in polynomial time. Such complete algorithms enforce arc consistency [24]. However, enforcing arc consistency may sometimes prove too hard (i.e., it may be an NP-hard problem), in which case simpler consistency notions are defined and implemented. This is the case, for instance, in scheduling algorithms.

Search. The search component of a constraint program specifies the search tree to explore in order to find solutions. Typically, the search tree, often called an *and/or tree* in artificial intelligence terminology, describes how to assign values to decision variables. Constraint programming offers a variety of control abstractions to simplify that process. Their main benefit is to abstract the tree exploration (e.g., through backtracking) by nondeterministic constructs, automating one of the most tedious aspects of constraint programs. For instance, the OPL snippet

```
forall(v in V)
   tryall(d in D)
      a[v] = d;
```

nondeterministically assigns values in `D` to the variables `a[v]` (`v` $\in$ `D`). More precisely, the `forall` instruction corresponds to an *and-node* and executes its body for each value `i` in `Size`. The `tryall` instruction corresponds to an *or-node* and is nondeterministic, i.e., it specifies a choice point with a number of alternatives and one of them must be selected.

In practical applications, great care must be taken to select the next variable to assign and the order in which these values are tried, since these choices may have a significant impact on the size of the search tree explored by the algorithm. The `forall` and `tryall` instructions support dynamic orderings of the variables and the values, using information from the constraints store.

The search snippet presented earlier describes the search space to explore implicitly in order to find solutions. However, it does not prescribe how to explore the tree. By default, constraint programming languages use depth-first search. However, on a variety of applications, other search strategies prove more effective. Limited Discrepancy Search (LDS) [18] is a search strategy relying on a good heuristic for the problem at hand. Its basic idea is to explore the search tree in waves and each successive wave allows the heuristic to make more mistakes. Wave 0 simply follows the heuristic. Wave 1 explores the solutions which can be reached by assuming that the heuristic made one mistake. More generally, wave i explores the solutions which can be reached by assuming that the heuristic makes i mistakes.

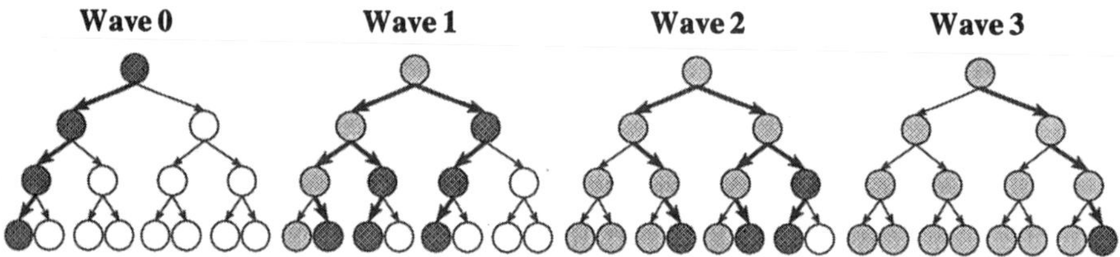

Figure 3.3. The Successive Waves of LDS.

Figure 3.3 illustrates these waves graphically on a binary search tree where the heuristic always selects the left branch first. By exploring the search tree according to the heuristic, LDS may reach good solutions (and thus an optimal solution) much faster than depth-first and best-first search for some applications. Its strength is its ability to explore diverse parts of the search tree containing good solutions which are only reached much later by depth-first search. Modern constraint languages now support high-level abstractions to specify search strategies and a variety of other search constructs. See, for instance, [22, 35, 40, 53] for more information.

3.2.2 The Queens Problems

We now illustrate constraint programming on a simple example. The n-queens problem consists of placing n queens on a chessboard of size $n \times n$ so that no two queens lie on the same row, column, or diagonal. Since no two queens can be placed on the same column, a simple program of this problem consists of associating a queen with each column and searching for an assignment of rows to the queens so that no two queens are placed on the same row or on the same diagonals. Figure 3.4 depicts a program for the n-queens problem in the modeling language OPL.

The OPL program illustrates the structure typically found in constraint programs: the declaration of the data, the declaration of decision variables, the

```
int n = 512;
range Size = 1..n;
var Size queen[Size];
int neg[i in Size] = -i;
int pos[i in Size] = i;

solve {
   forall(i in Domain) {
      alldifferent(queen);
      alldifferent(queen,neg);
      alldifferent(queen,pos);
   }
};
search {
   forall(i in Size ordered by increasing dsize(queen[i]))
      tryall(v in Size)
         queen[i] = v;
};
```

Figure 3.4. A Simple N-Queens Constraint Program

statement of constraints, and the search procedure. The program first declares an integer `n`, and a range `Size`. It then declares an array of n variables, all of which take their values in the range `1..n`. In particular, variable `queen[i]` represents the row assigned to the queen placed in column `i`. The next two instructions declare arrays of constants which are used to state the constraints.

The `solve` instruction defines the problem constraints, i.e., that no two queens should attack each other. It indicates that the purpose of this program is to find an assignment of values to the decision variables that satisfies all constraints. The basic idea in this program is to generate, for all $1 \leq i < j \leq n$, the constraints

```
queen[i] <> queen[j]
queen[i] - i <> queen[j] - j
queen[i] + i <> queen[j] + j
```

where the symbol <> means "not-equal." This is achieved by the "global" `alldifferent` constraint [37] with proper offsets. As mentioned earlier, combinatorial constraints of this type are critical in large and complex applications, since they encapsulate efficient pruning algorithms for substructures arising in many applications.

The rest of the program specifies the search procedure. Its basic idea is to consider each decision variable in sequence and to generate, in a nondeterministic way, a value for each of them. If, at some computation stage, a failure occurs (i.e., the domain of a variable becomes empty), the implementation backtracks and tries another value for the last queen assigned. Note that the `forall` instruction features a dynamic ordering of the variables: at each iteration, it selects the variable with the smallest domain, implementing the so-called *first-fail principle*.

Figure 3.5. The 5-Queens Problem After One Choice

Figure 3.6. The 5-Queens Problem After Two Choices

To illustrate the computational model, consider the five-queens problem. Constraint propagation does not reduce the domains of the variables initially. OPL thus generates a value, say 1, for one of its variables, say `queen[1]`. After this nondeterministic assignment, constraint propagation removes inconsistent values from the domains of the other variables, as depicted in Figure 3.5. The next step of the generation process tries the value 3 for `queen[2]`. OPL then removes inconsistent values from the domains of the remaining queens (see Figure 3.6). Since only one value remains for `queen[3]` and `queen[4]`, these values are immediately assigned to these variables and, after more constraint propagation, OPL assigns the value 4 to `queen[5]`. A solution to the five-queens problem is thus found with two choices and without backtracking, i.e., without reconsidering any of the choices.

Observe the clean separation between the constraints and the search procedure description in this program. Although these two components are *physically separated* in the program, they *cooperate* in solving the problem. Whenever a variable is assigned a value, a constraint propagation step is initiated, which prunes the search space. This pruning in turn affects the control behavior through backtracking or through the heuristics.

3.2.3 Car Sequencing

Car sequencing is a challenging combinatorial optimization application [12, 34] which illustrates several interesting features of constraint programming. It also shows the use of redundant (surrogate) constraints to prune the search space more effectively. The problem can be described as follows. Cars in production are placed on an assembly line moving through various units that install options such as air-conditioning and moonroofs. The assembly line is thus best viewed as a sequence of slots and each car must be allocated to a single slot. The car assignment must satisfy the constraints on the production units, which must place the options on the cars as the assembly line is moving. These capacity constraints are formalized using constraints of the form *r outof s*, indicating that, out of each sequence of *s* cars, at most *r* cars can be fitted with the option. The car-sequencing problem amounts to finding an assignment of cars to the slots that satisfies the capacity constraints. The search space in this problem is made up of the possible values for the assembly-line slots.

```
1.  int nbCars = ...;
2.  int nbOptions = ...;
3.  int nbSlots = ...;
4.  range Cars = 1..nbCars;
5.  range Options = 1..nbOptions;
6.  range Slots = 1..nbSlots;
7.  struct Cap { int l;int u;};
8.  int demand[Cars] = ...;
9.  int option[Options,Cars] = ...;
10. Cap capacity[Options] = ...;
11. int optionDemand[i in Options]=sum(j in Cars) demand[j]*option[i,j];
12. var Cars slot[Slots];
13. var int setup[Options,Slots] in 0..1;
14. solve {
15.    cardinality(demand,Cars,slot);
16.    forall(o in Options,s in [1..nbSlots - capacity[o].u + 1])
17.       sum(j in [s..s+capacity[o].u-1]) setup[o,j] <= capacity[o].l;
18.    forall(o in Options,s in Slots)
19.       setup[o,s] = option[o,slot[s]];
20.
21.    forall(o in Options,i in [1..optionDemand[o]])
22.       sum(s in [1..nbSlots - i * capacity[o].u]) setup[o,s] >=
23.        optionDemand[o] - i * capacity[o].l;
24. }
```

Figure 3.7. The Car-Sequencing Model in OPL

Figure 3.7 describes a model for the car sequencing problem and Figure 3.8 depicts a very simple data set. The model first declares the number of car types, options, and slots, as well as their associated ranges. A structure type is then declared to specify the capacity constraints of the form *l outof u*. The next declarations specify the demand for each car type, the options required by each car type, and the capacity of the production units for the options. The declaration on line 11 defines an array of integers representing the total demand for each option, i.e., it computes the total number of cars requiring each of the

```
nbCars = 6;
nbOptions = 5;
nbSlots = 10;
demand = [1,1,2,2,2,2];
options = [
    [1,0,0,0,1,1],
    [0,0,1,1,0,1],
    [1,0,0,0,1,0],
    [1,1,0,1,0,0],
    [0,0,1,0,0,0]];
capacity = [<1,2>,<2,3>,<1,3>,<2,5>,<1,5>];
```

Figure 3.8. Instance Data for the Car-Sequencing Problem

options. This information is useful in defining redundant constraints that speed up the computation.

The model uses two sets of variables: the `slots` variables, which specify the car types assigned to the slots, and the `setup` variables which, given an option and a slot, specify whether the car type in that slot requires the option. The main output of the model is the `slot` variables: The `setup` variables are used to express the model easily. There are various types of constraints in this model. Line 15 specifies a cardinality which ensures that the requested cars are produced. More precisely, constraint `cardinality(demand,Cars,slot)` holds if

$$\#\{\texttt{s} \in \texttt{Slots} \mid \texttt{slot}[\texttt{s}] = \texttt{c}\} = \texttt{demand}[\texttt{c}]$$

forall `c` in `Cars`. The second set of constraints on lines 16-17 specify the capacity constraints on the production units in terms of the `setup` variables. A constraint is created for each option and each window of width u on the assembly line; it states that the number of cars requiring the option in the window cannot exceed the bound l. The third set of constraints in lines 18-19 are particularly interesting. Their purpose is to link the `slot` and `setup` variables using the ubiquitous *element* constraint which makes it possible to index an array with decision variables. A constraint

```
setup[o,s] = option[o,slot[s]];
```

specifies that `setup[o,s]` is 1 if and only if the car assigned to slot `s`, i.e., `slot[s]`, requires option `o`. This constraint is multidirectional and reduces the domains of one of its variables, once information on the other becomes available. The fourth and last group of constraints in lines 21-23 are redundant constraints stated for each option. Intuitively, the constraints exploit the *l outof u* constraint by observing that, on the last $k * u$ slots of the assembly line, at most $k * l$ cars can be fitted with the option for any value of k. Consequently, with a total demand *d*, the complementary prefix of the assembly line must have at least $d - k * l$ cars with the option.

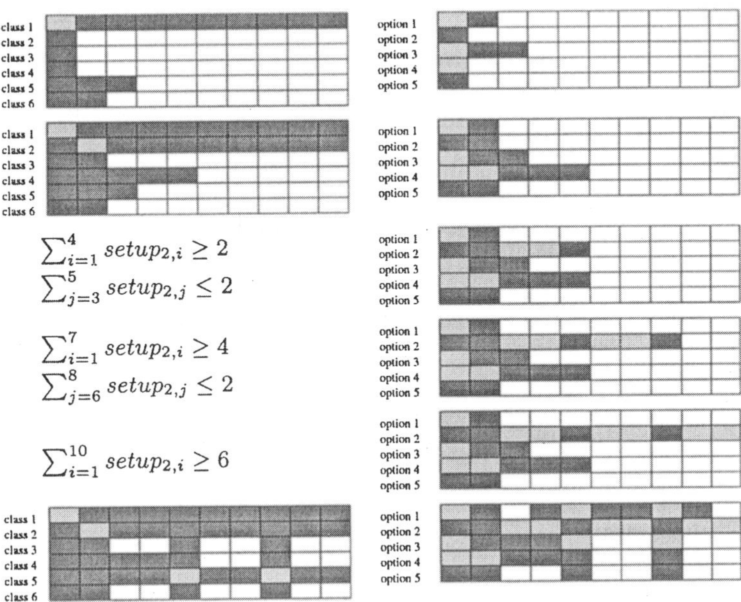

Figure 3.9. Illustrating the First two Choices and their Propagation.

Figure 3.9 illustrates the first two choices during the search and the effect of propagation on the very simple data set. The left column depicts the 10 `slot` variables with one row per value in its domain. A light gray entry shows that this is the chosen value, while a dark gray entry indicates that the value has been removed from the domain. The right column depicts the `setup` variables. A white cell indicates that the domain still contains both value $\{true, false\}$. A light gray corresponds to $true$ while dark gray is $false$. The labeling begins with assigning value 1 to `slot[1]`. No other slot variable can be assigned class 1, since only 1 car of class 1 must be produced. Hence class 1 is removed from the domains of the other slot variables. Since class 1 requires options $\{1, 3, 4\}$, the corresponding entries in the first column of the setup matrix become true (light gray) (and $\{2, 5\}$ are false). Finally, since option 1 obeys the *1 outof 2* rule, the entry `setup[1,2]` must be false. The same reasoning applies to option 3, forcing `option[3,2]` and `option[3,3]` to become false as well. Moreover, since classes 5 and 6 require option 1, the variable `slot[2]` cannot be assigned one of these classes. Similarly, since class 5 requires option 3, `slot[3]` cannot be assigned a car of that class. These values are removed from the domains of these variables as indicated in the top left picture.

It is interesting to observe the consequences of the second choice which assigns a car of class 2 to `slot[2]`. The third, fourth and fifth blocks in the right column of Figure 3.9 depict the pruning resulting from the capacity and redundant constraints, which are shown on the left. Intuitively, this propagation fixes all the `setup` variables for option 2, since 6 cars requiring option 2 must be produced and the capacity constraint is of type `2 outof 3`. The sixth block describes the last round of constraint propagation, which is initiated by the demand constraints on class 5 and the redundant constraints on option 1. This propagation leads directly to a solution without backtracking and with only two choices.

Constraint programming is an effective tool for solving car-sequencing applications [12, 38]. It often finds solutions to large instances and it can prove infeasible when no solution exists.

3.2.4 Scheduling

An interesting feature of constraint programming is its ability to support high-level abstractions for specific application domains such as scheduling and vehicle routing. These abstractions make it possible to design models that are closer to problem specifications. In constraint programming, these abstractions encapsulate decision variables and combinatorial constraints, and provide support for search procedures tailored to the application domain. This section illustrates such vertical extensions for scheduling and present a simple model for job-shop scheduling. See, for instance, [52] for more details on scheduling applications in OPL.

Activities. The central concept for scheduling applications is the *activity*. An activity can be thought of as an object containing three data items: a starting date, a duration, and an ending date, together with the duration constraint stating that the ending date is the starting date plus the duration. In many applications, the duration of an activity is known and the activity is declared together with its duration as in the OPL snippet

```
Activity a(4);
Activity b(5);
```

which declares two activities `a` and `b` with duration 4 and 5 respectively.

Precedence Constraints. Activities can be linked by various precedence constraints. For instance, a precedence constraint between activities `a` and `b` can be stated as

```
a.precedes(b);
```

Disjunctive Resources. Typically, activities require some resources for their execution and they can be of different kinds. A disjunctive resource is a resource which can only be used by one activity at the time. In other words, all activities requiring the resource cannot overlap in time. Typically, disjunctive resources encapsulate the edge-finder algorithm which shrinks the starting and ending dates of activities. The OPL snippet

```
UnaryResource machines[1..10];
```

declares an array of 10 disjunctive resources. The instruction

```
a.requires(machines[1])
```

specifies that activity a requires `machine[1]` during its execution.

Cumulative Resources. Cumulative resources generalize disjunctive resources and model situations where a resource is available in multiple units. The schedule must ensure that the demands of activities requiring a cumulative resource at time t do not exceed the resource capacity at that time. The OPL snippet

```
DiscreteResource worker(10);
Activity framing(5);
```

shows how to declare a cumulative resource `worker` of capacity 10 in OPL. The instruction `framing.requires(worker,4)` can be used to specify that activity `framing` requires 4 units of `worker` during its execution.

Search. The scheduling abstractions also features some support for search procedures exploiting the nature of these applications. For instance, in jobshop scheduling, the search procedure consists of ordering the activities on each disjunctive resource. The nondeterministic OPL instruction `tryRankFirst(d,a)` can be used for that purpose: It tries to schedule a first on resource d and, on backtracking, it imposes that a be not scheduled first.

3.2.5 Jobshop Scheduling

Jobshop scheduling consists of scheduling a number of jobs to minimize makespan, i.e., the completion time of the project. Each job is a sequence of tasks, linked by precedence constraints. Moreover, each task requires a machine and no two tasks executing on the same machine can overlap in time. Figure 3.10 depicts an OPL model which explicits both the constraint and the search components. The search component could in fact be omitted in OPL.

The Constraint Component. The model assumes the existence of ranges `JobRange`, `TaskRange` and `MachineRange` and use them to declare the activities and disjunctive resources in lines 1-2. Line 3 declares a dummy activity to

```
1.  Activity act[j in JobRange,t TaskRange](duration[j,t]);
2.  UnaryResouce res[MachineRange];
3.  Activity makespan(0);
4.  minimize makespan.end
5.  subject to {
6.     forall(j in Jobs)
7.        act[j,nbTasks].precedes(makespan);
8.     forall(j in JobRange) {
9.        forall(t in TaskRange :  t <> 1)
10.          act[j,t].precedes(act[j,t-1]);
11.    forall(j in JobRange,t in TaskRange)
12.       act[j,t].requires(res[machine[j,t]));
13. }
14. search {
15.    forall(r in MachineRange ordered by increasing localSlack(res[r]))
16.       while not res[r].isRanked() do
17.          select(t in TaskRange:  not a[t].isRanked(res[r])
18.                   ordered by increasing dmin(a[t].start))
19.          tryRankFirst(res[r],a[t]);
20. }
```

Figure 3.10. A Constraint Programming Model for Jobshop Scheduling

represent the makespan. Line 4 states the objective function while lines 6-12 state the problem constraints. In particular, lines 6-7 specify that the makespan follows the last tasks of each job, lines 9-10 state the precedence constraints, and lines 11-12 specify the disjunctive constraints.

The Search Component. The search procedure (lines 15-19) orders the activities on the same resource. Its key idea is to select each resource successively and to rank its activities. To rank a resource, the search procedure selects an unranked task whose earliest possible starting date is minimal and tries to schedule it first using the instruction `tryRankFirst` presented earlier. This process is repeated until the resource is ranked. Note also that the search procedure ranks first the resources that are most constrained, i.e., those left with the smallest slack.

Once again, it is useful to emphasize that constraint programming is an effective approach to solve a variety of scheduling applications. See [1] for a review of scheduling applications through constraint programming and [45] for a variety for some scheduling models in OPL.

3.3 Neighborhood Search

We now turn to neighborhood search and we show that much of the style, and many of the benefits, of constraint programming, carry over to this funda-

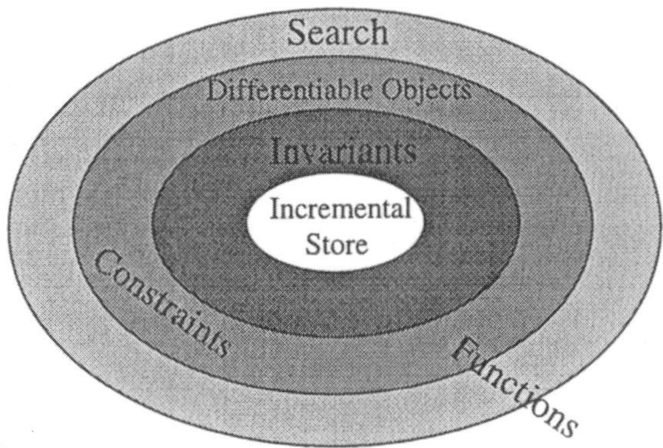

Figure 3.11. The Comet Architecture.

mentally different search paradigm. In particular, we show that neighborhood search can also be supported by a layered architecture, which cleanly separates modeling and search, and supports a rich language for constraints and search. The technical details and the computational model are significantly different however, due to the nature of the underlying approach to combinatorial optimization. The section uses the novel programming language COMET for illustration purposes.

3.3.1 The Architecture

The architecture is shown in Figure 3.11. It consists of a declarative and a search component organized in three layers, which we now review.

Invariants. The kernel of the architecture is the concept of *invariants* (or one-way constraints) over algebraic and set expressions [26]. Invariants are expressed in terms of incremental variables and specify a relation which must be maintained under assignments of new values to its variables. For instance, the code fragment

```
inc{int} s(m) <- sum(i in 1..10) a[i];
```

declares an incremental variable `s` of type int (in a solver `m`) and an invariant specifying that `s` is always the summation of `a[1],...,a[10]`. Each time, a new value is assigned to an element `a[i]`, the value of `s` is updated accordingly (in constant time). Note that the invariant specifies the relation to be maintained incrementally, not how to update it. Incremental variables are always associated with a local solver (`m` in this case). This makes it possible to use a very efficient implementation by dynamically determining a topological order in which to update the invariants. As we will see later, COMET supports a wide variety of algebraic, graph, and set invariants.

Differentiable Objects. Once invariants are available, it becomes natural to support the concept of *differentiable objects*, a fundamental abstraction for local search programming. *Differentiable objects maintain a number of properties (using invariants or graph algorithms) and can be queried to evaluate the effect of local moves on these properties.* They are fundamental because many neighborhood search algorithms evaluate the effect of various moves before selecting the neighbor to visit. Two important classes of differentiable objects are constraints and functions. A differentiable constraint maintains properties such as its satisfiability, its violation degree, and how much each of its underlying variables contribute to the violations. It can be queried to evaluate the effect of local moves (e.g., assignments and swaps) on these properties. A differentiable function maintains the value of a (possibly) complex function and can also be queried to evaluate the variation in the function value under local changes.

Differentiable objects capture combinatorial substructures arising in many applications and are appealing for two main reasons. On the one hand, they are high-level modeling tools which can be composed naturally to build complex neighborhood search algorithms. As such, they bring into neighborhood search some of the nice properties of modern constraint satisfaction systems. On the other hand, they are amenable to efficient incremental algorithms that exploit their combinatorial properties. The use of combinatorial constraints is implicitly emerging from a number of research projects: It was mentioned in [55] as future research and used, for instance, in [7, 16, 32] as building blocks for satisfiability problems. Combinatorial functions play a similar role for optimization problems, where it is important to evaluate the variation of complex objective functions efficiently.

The `AllDifferent` constraint, which we encountered earlier, is an example of differential object. In its simplest form, `AllDifferent(a)` receives an array `a` of incremental variables and holds if all its variables are given different values. The `AllDifferent` constraint maintains a variety of properties incrementally. They include its violation degree, i.e., how much it is violated, as well as the set of variables which occur in such violations. Observe that the same combinatorial constraint, which is natural in many applications, is being used in two very different ways: to prune the search space in constraint programming and to maintain incremental properties in COMET.

In COMET, differentiable constraints are objects implementing the interface `Constraint`, an excerpt of which is shown in Figure 3.12. The first three methods give access to incremental variables that maintain important properties of the constraint: its satisfiability, its violation degree, and the violations of its variables. The other two methods shown in Figure 3.12 are particularly interesting as they evaluate the effect of assignments and swaps on the violation degree of the constraint. Method `getAssignDelta(var,val)` returns the variation of the violation degree when variable `var` is assigned value `val`.

```
interface Constraint {
   inc{int} true();
   inc{int} violationDegree();
   inc{int} violations(inc{int} var);
   int getAssignDelta(inc{int} var,int val);
   int getSwapDelta(inc{int} v,inc{int} w);
   ...
}
```

Figure 3.12. The Interface `Constraint`

Similarly, method `getSwapDelta(v,w)` returns the variation of the violation degree when the values of variables `v` and `w` are swapped. Although this paper presents several differentiable constraints available in COMET, observe that COMET is an open language where users can add their own constraints by implementing the interface `Constraint`. (Similar considerations apply to differentiable functions.)

Constraints can be composed naturally using *constraints systems*. A constraint system groups a collection of constraints together and maintains the satisfiability and violation degree of the set incrementally. The violation degree of a set of constraints is the summation of the violation degrees of its constraints. Being constraints themselves, constraint systems are differentiable and can also be queried to evaluate the effect of local moves. Constraint systems make neighborhood search algorithms more *compositional* and easier to extend. In particular, they allow new constraints to be added in the declarative component of an algorithm without changing the search component.

These two layers, invariants and differentiable objects, constitute the declarative component of the architecture.

Search. The third layer of the architecture is the search component which aims at simplifying the neighborhood exploration and the implementation of heuristics and meta-heuristics, two other critical aspects of neighborhood search algorithms. The search component features high-level constructs and abstractions to foster and increase separation of concerns. They include multidimensional selectors, events, the `neighbor` construct, and checkpoints. Each of them aims at separating various concerns arising in neighborhood search algorithms. For instance, events makes it possible to separate heuristics and meta-heuristics, the `neighbor` construct disconnect the neighborhood exploration from its use, and checkpoints factorize large neighborhood search.

One of the main issues addressed by these abstractions is the temporal disconnection between the definition of a behavior and its use which typically plagues the implementation of neighborhood search algorithms. This issue arises in meta-heuristics and in applications whose neighborhoods are heterogeneous and consists of various types of moves (e.g., [4, 8, 20, 23]). COMET abstractions heavily rely on first-class closures to address this temporal disconnection and to implement events, neighbors, and checkpoints. We now review closures and the `neighbor` construct. More details on the search abstractions of COMET can be found in [50].

Closures. A closure is a piece of code together with its environment. Closures are ubiquitous in functional programming languages, where they are first-class citizens. They are rarely supported in object-oriented languages however. To illustrate the use of closures in COMET, consider the class `DemoClosure` (on the left) and the code snippet (on the right)

```
class DemoClosure {
  DemoClosure() {}
  Closure print(int i) {
    return new closure
      {cout << i << endl;}
  }
}
```

```
DemoClosure demo();
Closure c1 = demo.print(9);
Closure c2 = demo.print(5);
call(c2);
call(c1);
call(c2);
```

Method `print` receives an integer `i` and returns a closure which, when executed, prints `i` on the standard output. The snippet on the right shows how to use closures in COMET: It displays 5, 9, and 5 on the standard output. Observe that closures are first-class citizens: They can be stored in data structures, passed as parameters, and returned as results. The two closures created in the example above share the same code (i.e., `cout << i << endl`), but their environments differ. Both contain only one entry (variable `i`), but they associate the value 9 (closure `c1`) and the value 5 (closure `c2`) to this entry. When a closure is created, its environment is saved and, when a closure is executed, the environment is restored before, and popped after, execution of its code. Closures can be rather complex and have environments containing many parameters and local variables, as will become clear later on.

Neighbors. The `neighbor` construct was motivated by complex applications in scheduling and routing which employ neighborhoods composed of several classes of heterogeneous moves. The difficulty in expressing their search components come from the temporal disconnection between move selection and execution. In general, a tabu-search or a greedy local search algorithm first scans the neighborhood to determine the best move, before executing the selected move. However, in these complex applications, the exploration cannot be expressed using a (multidimensional) selector, since the moves are heterogeneous and obtained by iterating over different sets. As a consequence, an implementation would typically create classes to store the information necessary to characterize the different types of moves. Each of these classes would inherit from a common abstract class (or would implement the same interface). During the scanning phase, the algorithm would create instances of these classes to represent selected moves and store them in a selector whenever appropriate. During the execution phase, the algorithm would extract the selected move and apply its `execute` operation. The drawbacks of this approach are twofold. On the one hand, it requires the definition of a several classes to represent the moves. On the other hand, it fragments the code, separating the *evaluation* of a move from its *execution* in the program source. As a result, the program is less readable and more verbose.

COMET supports a `neighbor` construct, which relies heavily on closures and eliminates these drawbacks. It makes it possible to specify the move evaluation and execution in one place and avoids unnecessary class definitions. More important, it significantly enhances compositionality and reuse, since the various subneighborhoods do not have to agree on a common interface or abstract class and are independent of the move selection heuristics.

The `neighbor` construct is of the form

```
neighbor(δ,N) M
```

where `M` is a move, δ is its evaluation, and `N` is a neighbor selector, i.e., a container object to store one or several moves and their evaluations. COMET supports a variety of such selectors and users can define their own, since they all implement a common interface. For instance, a typical neighbor selector for tabu-search maintains the best move and its evaluation. The `neighbor` instruction queries selector `N` to find out whether it accepts a move of quality δ, in which case the closure of `M` is submitted to `N`. The construct will be illustrated subsequently on a scheduling application.

3.3.2 The Queens Problem

We now reconsider the queens problem for neighborhood search. Figure 3.13 depicts a COMET program which shares many features with the constraint program in Figure 3.4, although the two programs implement fundamentally different algorithms. The algorithm in Figure 3.13 implements the min-conflict heuristic [30]. It starts with an initial random configuration. Then, at each iteration, it chooses the queen violating the largest number of constraints and moves this queen to a position minimizing its violations. This step is iterated until a solution is found. Once again, the COMET program associates a queen with every column and it uses `queen[i]` to denote the row of the queen on column `i`. As before, the program starts by declaring the size of the board and a range. It then declares a local solver, which will hold the incremental variables, the invariants, and the constraints. It also declares a uniform distribution. The next instruction

```
inc{int} queen[i in Size](m,Size) := distr.get();
```

```
int n = 512;
range Size = 1..n;
LocalSolver m();
UniformDistribution distr(Size);

inc{int} queen[i in Size](m,Size) := distr.get();
int neg[i in Size] = -i;
int pos[i in Size] = i;

ConstraintSystem S(m);
S.post(new AllDifferent(queen));
S.post(new AllDifferent(queen,neg));
S.post(new AllDifferent(queen,pos));
inc{set{int}} conflicts(m) <- argMax(q in Size) S.violations(queen[q]);
m.close();

while (S.violationDegree())
  select(q in conflicts)
    selectMin(v in Size)(S.getAssignDelta(queen[q],v))
      queen[q] := v;
```

Figure 3.13. The Queens Problem in Comet.

declares the incremental variables, which are initialized randomly and receive the local solver in their constructors. Incremental variables are central in COMET: They are used in invariants and constraints, and changes to their values trigger events and update all affected invariants and constraints. Note the use of the assignment operator `:=`, which assigns a value of type T to an incremental variable of type `inc{T}`. By contrast, the operator = assigns references.

The next block of instructions describes the declarative component of the application. Instruction

```
ConstraintSystem S(m);
```

declares a constraint system S, while the instructions

```
S.post(new AllDifferent(queen));
S.post(new AllDifferent(queen,neg));
S.post(new AllDifferent(queen,pos));
```

add the problem constraints to S. Observe that these are the very same constraints as in the constraint programming solution. The final instruction of the declarative component

```
inc{set{int}} conflicts(m) <- argMax(q in Size) S.violations(queen[q]);
```

declares an incremental variable `conflicts` whose values are of type "set of integers" and imposes an invariant ensuring that `conflicts` always represents the set of queens that violate the most constraints. Observe that

```
S.violations(queen[q])
```

returns an incremental expression representing the number of violations in S. Operator `argMax(i in S) E` simply returns the set of values v in S which maximizes E. The last instruction `m.close()` closes the model, which enables COMET to construct a dependency graph in order to update the constraints and invariants under changes to the incremental variables.

Observe that the declarative component only specifies the properties of the solutions, as well as the data structures to maintain. It does not specify how to update the constraints, e.g., the violations of the variables, or how to update the conflict set. These are performed by the COMET runtime system, which uses optimal incremental algorithms in this case. The final part of the program

```
while (S.violationDegree())
   select(q in conflicts)
      selectMin(v in Size)(S.getAssignDelta(queen[q],v))
         queen[q] := v;
```

states the search strategy. It performs local moves until the violation degree of the system is zero, in which case all constraints are satisfied. Each iteration randomly selects a queen q in the conflict set, selects a value v for queen q that minimizes the number of violations, and assigns v to `queen[q]`. The selection of a new row for `queen[q]`

```
selectMin(v in Size)(S.getAssignDelta(queen[q],v))
```

uses the ability to query the constraint system to determine the impact of assigning v to queen q. This query can be performed in constant time, thanks to the invariants maintained in each of the constraints.

There are a couple of observations to make at this point. First, the search and declarative components are clearly separated in the program, as was the case with the constraint programming solution. It is thus easy to modify one of them without affecting the other. For instance, it is possible to add new constraints to the constraint system without any change to the search component. Similarly, it is easy to replace the search component without affecting the declarative component. Although the two components are physically separated in the program code, they closely collaborate during execution. The search component uses the declarative component to guide the search, while the assignment `queen[q] := v` starts a propagation phase which updates all invariants and constraints. Once again, these are the same appealing features as constraint programming. Finally, the resulting COMET program operates at a high level of abstraction.

The declarative component is close to a formal specification of the problem, while the search component is a very abstract description of the search strategy. In fact, this search component is not problem-specific and could almost be reused as such in other contexts.

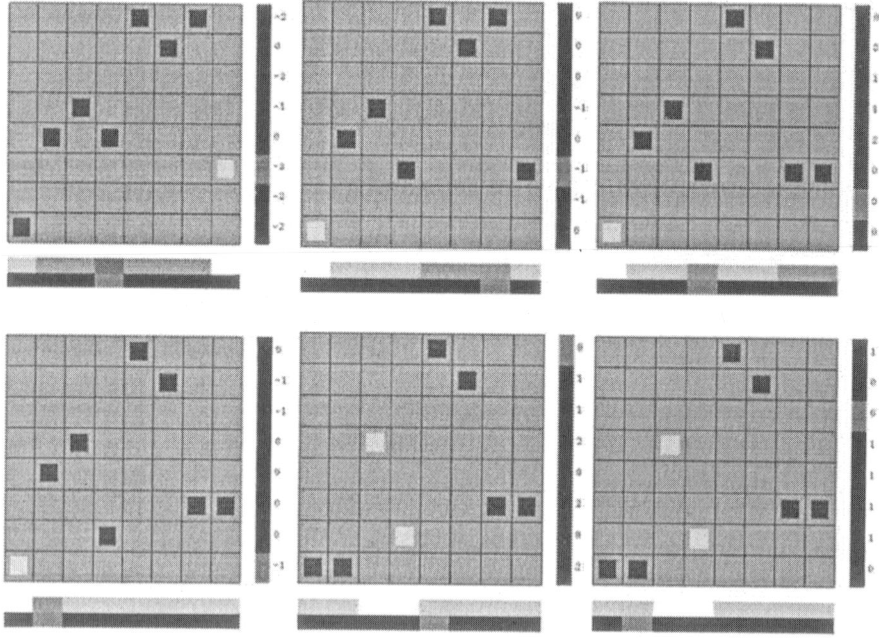

Figure 3.14. The First Six Steps of the Comet Program

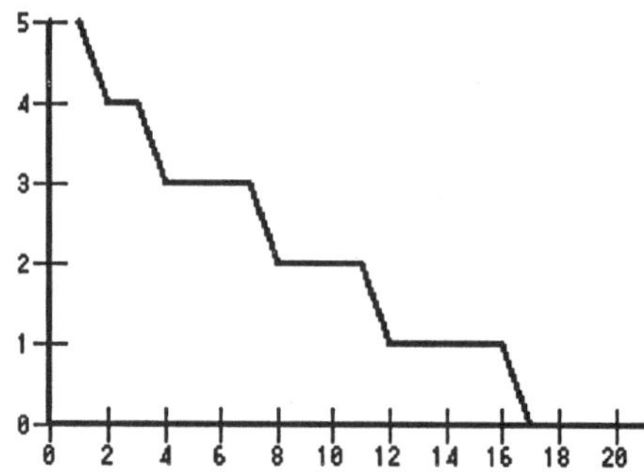

Figure 3.15. The Number of Violations over Time

Figure 3.14 illustrates the computation model and depicts the first six iterations of the algorithm. Figure 3.15 also shows a plot of the number of violations over time. Each box represents the board and is adorned at the bottom with a color-coded bar showing which queen are currently violating constraints. The color intensity indicates the violation level. Darker tabs correspond to more violations. Below the color-coded bar, a tab shows which queen is picked for the next transition and the tab to the right of the board shows the row the selected queen should be assigned to. Additionally, the column of numbers on the right of each board displays the gain that will result from the transition. In essence, the figure displays many of the properties maintained by the COMET program incrementally. In the first step, queen 4 is selected and randomly moved to position 6 reducing the number of violations by 2. (It could have been moved to positions $\{1,3,7,8\}$). In the second step, queen 7 is moved to position 6, reducing the number of violations by one. Note that, in the fifth iteration, the algorithm decides not to move the selected queen as this is, locally, one of the most profitable moves. As a side note, observe that the result of forbidding such moves is not beneficial in general. Indeed, the algorithm is then forced to consider moves which degrade the value of the objective, it oscillates around 1 and 2 violations, and it takes much longer to terminate.

Table 3.1. Experimental Results for N-Queens.

Size	State(s)	Solve(s)	Total(s)	Iter
8	0.01	0.02	0.03	249
16	0.01	0.00	0.01	21
32	0.02	0.01	0.03	56
64	0.04	0.02	0.06	80
128	0.10	0.08	0.18	238
256	0.17	0.12	0.29	206
512	0.35	0.37	0.72	304
1024	0.71	1.64	2.35	628
2048	1.43	6.07	7.50	1100
4096	2.95	22.56	25.51	2092
8192	6.16	85.91	92.07	4040
16384	14.69	333.89	348.58	7968
32768	42.10	1320.43	1362.53	15899

Experimental Results. Table 3.1 reports some experimental results for the program. The program runs in $O(n)$ space and each iteration takes time $O(n)$, since it is necessary to find the value minimizing the number of conflicts. In general, the propagation step takes constant time because the conflict set is small and does not change very much. The table reports the total CPU time in seconds on a 1.1Ghz PC running Linux, as well as the time for stating the constraints and for the search. It also reports the number of iterations. The results clearly indicate that the implementation is extremely competitive with low-level encodings of the algorithm [7] and is quadratic in time experimentally. In summary, COMET makes it possible to design a very efficient algorithm at the same level of abstraction as typical constraint programs for the task. Moreover, although the underlying algorithms are quite different, the two solutions share the same structure, compositionality, and modularity. Note that constraint programming solutions take quadratic space, which reduces their applicability to large number of queens.

3.3.3 Car Sequencing

Our second application is the car-sequencing problem introduced in Section 3.2.3. The main objective is to present the core of a neighborhood search solution and to show combinatorial constraints and the separation between the declarative and search components on a more realistic application. The complete COMET program closed an open problem and is extremely efficient compared to other approaches. See [28] for more details on this application.

The Comet Program. Figure 3.16 depicts the core of the Comet program. It assumes that the data has been read and preprocessed, which takes about 20 lines of code and is not shown here. The fragment simply assumes that ranges `Configs`, `Slots`, and `Options` specify the types of cars to sequence, the slots of the assembly line, and their possible options. There are as many slots on the assembly line as there are cars to produce. For every car `c`, the array element `car[c]` specifies the configuration of car `c`. For every option `o`, array element `option[o]` denotes the set of cars requiring `o`, while `cap[o].l` and `cap[o].u` specify its capacity constraint, i.e., they specify that no more than `cap[o].l` cars in each sequence of size `cap[o].u` may take option `o`. The neighborhood search algorithm is a tabu-search procedure that minimizes the number of violations of the capacity constraints until a valid sequencing is found. Its basic idea is to select the most violated slot and to swap it with another slot in order to minimize the number of conflicts.

The Declarative Component. The declarative component of the algorithm states the capacity constraints. More specifically, it declares an array `slot` of incremental variables: `slot[i]` denotes the car type assigned to slot `i` in the assembly line. It then declares the constraint system `S` and posts all capacity constraints in `S`. The capacity constraints are of the form `SequenceAtmost(slot,O,n,m)`, where `O` is the set of car types requiring an option `o` and *n outof m* is the capacity constraint of `o`. These constraints hold whenever every subsequence `slot[i],...,slot[i+m]` has atmost `n` cars whose types are in `O`. These constraints are differentiable and they capture

```
1.    int tLen = 10;
2.    RandomPermutation perm(Cars);

3.    LocalSolver m();
4.    inc{int} slot[Slots](m,Configs) := cars[perm.get()];
5.    ConstraintSystem S(m);
6.    forall(o in Options)
7.      S.post(new SequenceAtmost(slot,options[o],cap[o].l,cap[o].u));
8.    inc{int} violations = S.violationDegree();
9.    m.close();

10.   int it = 0;
11.   int best = violations;
12.   int tabu[Slots,Slots] = -1;
13.   while (violations) {
14.     int old = violations;
15.     selectMax(s1 in Slots)(S.violations(slot[s1]))
16.       selectMin(s2 in Slots:  slot[s1]!=slot[s2] && tabu[s1,s2]<=it)
17.                (S.getSwapDelta(slot[s1],slot[s2])) {
18.         slot[s1] :=:  slot[s2];
19.         if (violations >= old) {
20.           tabu[s1,s2] = it + tLen;
21.           tabu[s1,s2] = it + tLen;
22.         }
23.       }
24.     }
25.     it = it + 1;
26.   }
```

Figure 3.16. Car Sequencing in Comet

a combinatorial subproblem that arise in many sequencing and scheduling applications, including sport-scheduling problems. The violation degree of such a constraint is the number of subsequences

$$\texttt{slot[i],...,slot[i+m]}$$

that have more than n cars whose types are in O. The declarative part also declares an incremental variable `violations` which maintains the violation degree of S. Observe the simplicity and high-level nature of the declarative component.

The Search Component. The search component is a tabu-search procedure with diversification and intensification, which are not shown here for space reasons. Its key idea is to select a slot appearing in the most conflicts and to swap it with the slot which leads to the least violations. Lines 18-21 are the core of the search component. Observe how the `selectMin` instruction queries the constraint system with calls `S.getSwapDelta(slot[s1],slot[s2])` to evaluate the effect of swapping `slot[s1]` and `slot[s2]`. Once the two slots are selected, the instruction `slot[s1] :=: slot[s2]` performs the swap and updates the invariants and differentiable objects.

Experimental Results. Table 3.2 gives the experimental results. We took the satisfiable instances from the CSP lib library [17]. These instances were generated by B. Smith and are very difficult in general (Problem 26 was open). We also took a satisfiable version of Problem 21 which is also open. The algorithm solved all satisfiable instances and closes Problem 26. We report average, minimum, and maximum CPU Time in seconds over 10 runs on a 800mhz PC running Linux. The COMET program is extremely effective on this problem and is probably one of the most efficient algorithms available at this point. It is important to stress that this performance was obtained by a concise and high-level program.

Table 3.2. Experimental Results for Car-Sequencing.

Problem	4	8	26	41	21a
Time(Min)	0.41	2.3	1.02	1.31	0.91
Time(Max)	0.64	56.05	21.21	20.85	3.27
Time(Av.)	0.497	25.765	5.057	7.126	1.935

3.3.4 The Progressive Party Problem

We now describe a COMET solution to the Progressive Party (PP) problem, which is often used as a benchmark for comparing approaches and algorithms in combinatorial optimization. The PP problem features a variety of heterogeneous constraints, which makes it an interesting exercise in modeling.

The Problem. The progressive party problem can be described as follows (as is traditional, we assume that host boats have been selected). We are given a set of boats (the hosts), a set of parties (the guests), and a number of time periods. Each host has a limited capacity and each party has a given size. The problem consists of assigning a host to every guest for each time period in order to satisfy three sets of constraints. First, the capacity of the hosts cannot be exceeded by the parties. Second, every party must visit a different host at each time period. Finally, no two parties can meet more than once over the course of the party.

Figure 3.17 depicts the core of the COMET program. It receives as inputs two ranges `Hosts` and `Parties` that denote the set of hosts and the set of parties respectively. In addition, it receives two arrays of integers `cap` and `party` where `cap[h]` is the capacity of host `h` and `party[g]` is the size of guest g.

```
LocalSolver m();
UniformDistribution distr(Hosts);
inc{int} boat[Guests,Periods](m,Hosts) := distr.get();
ConstraintSystem S(m);
forall(g in Guests) {
  inc{int} r[p in Periods] = boat[g,p];
  S.post(new AllDifferent(r),2);
}
forall(p in Periods) {
  inc{int} c[g in Guests] = boat[g,p];
  S.post(new WeightedAtmost(c,party,cap),2);
}
forall(i in Guests, j in Guests :  j > i) {
  inc{int} ri[p in Periods] = boat[i,p];
  inc{int} rj[p in Periods] = boat[j,p];
  S.post(new MeetAtmost(ri,rj,1));
}
inc{int} vd = S.violationDegree();
m.close();
int tabu[Guests,Periods,Hosts] = -1;
Solution sol(m); int best = vd;int tLen = 2;
int it = 0; int stable = 0;int stableLimit = 2000;
while (vd) {
  int old = vd;
  selectMax(g in Guests,p in Periods)(S.violations(boat[g,p]))
    selectMin(h in Hosts:tabu[g,p,h] <= it ||
                         S.getAssignDelta(boat[g,p],h) + vd < best)
              (S.getAssignDelta(boat[g,p],h)) {
      tabu[g,p,boat[g,p]] = it + tLen;
      boat[g,p] := h;
      if (vd < old && tLen > 2) tLen = tLen - 1;
      if (vd >= old && tLen < tbl) tLen = tLen + 1;
    }
  if (vd < best) {
    best = vd; stable = 0;
    sol = new Solution(m);
  } else {
    if (stable == stableLimit) {
      sol.restore();
      stable = 0; it = it + tLen;
    }
    it = it + 1; stable = stable + 1;
  }
}
```

Figure 3.17. The Progressive Party Problem in COMET

Once again, the neighborhood search algorithm for the PP problem features a declarative and a search component.

The Declarative Component. The declarative component primarily declares the incremental variables and the problem constraints. It features a variety of combinatorial constraints and illustrates how to associate weights with constraints, a common technique in neighborhood search (e.g., [42]). The instructions

```
inc{int} r[p in Periods] = boat[g,p];
S.post(new AllDifferent(r),2);
```

specify that a party g never visits a host more than once over the course of the party. Observe that the `post` method which specifies a weight of 2 for the constraint. The instructions

```
inc{int} c[g in Parties] = boat[g,p];
S.post(new WeightedAtmost(c,party,cap),2);
```

specify the capacity constraints for period p. Constraint

```
WeightedAtmost(c,party,cap)
```

is a generalized cardinality constraint [48, 2] which holds if

$$\sum_{\texttt{i}\in\texttt{Parties}} \texttt{party[i]} * (\texttt{c[i]} = \texttt{j}) \leq \texttt{cap[j]}$$

for all $\texttt{j} \in$ `Hosts`. Once again, this constraint is differentiable and captures a subproblem arising in many applications. Finally, the instructions

```
inc{int} ri[p in Periods] = boat[i,p];
inc{int} rj[p in Periods] = boat[j,p];
S.post(new MeetAtmost(ri,rj,1));
```

specify that no two guest parties meet more than once. `MeetAtmost(a,b,l)` is also a cardinality constraint which holds if

$$\#\{i \in \texttt{R} \mid \texttt{a}[i] = \texttt{b}[i]\} \leq l$$

where R is the range of arrays a and b.

The Search Component. The search component is a tabu-search procedure minimizing the violation degree of the constraint system. It first selects a party g and a time period t such that variable `boat[g,t]` appears in the most conflicts. It then selects a non-tabu host for the pair `(g,t)` that minimizes the violation degree of the system. The host selection features an aspiration criteria

```
S.getAssignDelta(boat[g,p],h) + vd < best
```

which overrides the tabu status when the assignment improves upon the best "solution" found so far. The search also includes an *intensification* process. The key idea behind the intensification is to restart from the best found solution when the number of violations has not decreased for a long time. The search uses the solution concept of COMET for implementing the intensification in a simple manner. The instruction

```
Solution sol(m);
```

declares a solution which stores the current value of all decision variables, i.e., incremental variables whose values are not specified by an invariant. Each time a better solution is found, the instruction

```
sol = new Solution(m);
```

makes sure that `sol` now maintains the best solution found so far. After a number of iterations without improvements, the instruction `sol.restore()` restores the best available solution and resumes the search from this state. Once again, observe the simplicity of the search component which can be described very concisely.

Table 3.3. Experimental Results for the PP Problem.

Hosts/Periods	6	7	8	9	10
1-12,16	0.98	1.64	5.13	90.19	
1-13	0.61	0.90	1.17	4.41	21.00
1,3-13,19	0.90	1.53	5.28	253.92	
3-13,25,26	1.21	1.81	7.02	82.66	
1-11,19,21	4.50	24.35			
1-9,16-19	6.20	161.16			

Experimental Results. Before presenting the results, it is useful to emphasize that the program provides a very high-level and natural modeling of the application. Yet, as the results show, the algorithm compares extremely well with low-level codes developed for this application. The experiments use the same host configurations as in [55] to evaluate the algorithm. For each of these configurations, we consider problems with 6 periods (as in [55]) but also with 7, 8, 9, and 10 time periods. Table 3.3 reports median CPU Time in seconds for 10 runs of the algorithm. As can be seen, the algorithms easily outperforms the approach in [55].

3.3.5 Scheduling

This section reviews some of the scheduling abstractions of COMET. These abstractions make it possible to specify declarative components for scheduling applications that closely resemble those of constraint programming. Moreover, they enable search procedures to be written at a high conceptual level using scheduling concepts. Because of the nature of constraint programming and neighborhood search, the computational model underlying the scheduling abstractions, and the search procedures, are fundamentally different. But the benefits of the abstractions are similar. They reduce the distance between the model and the implementation, they give rise to concise and elegant models, and they foster compositionality and reuse.

The Computational Model. *The integration of scheduling abstractions within* COMET *raised interesting challenges due to the fundamentally different nature of neighborhood search algorithms for scheduling.* Indeed, in constraint programming, the high-level modeling abstractions encapsulate global constraints such as the edge finder and provide support for search procedures dedicated to scheduling. In contrast, neighborhood search algorithms move from (possibly infeasible) schedules to their neighbors in order to reduce infeasibilities or to improve the objective function. Moreover, neighborhood search algorithms for scheduling typically do not perform moves which assign the value of some decision variables, as is the case in many other applications. Rather, they walk from schedules to schedules by adding and/or removing sets of precedence constraints.[1] This is the case in algorithms for job-shop scheduling where makespan (e.g., [23, 33]) or total weighted tardiness (e.g., [21]) is minimized, flexible job-shop scheduling where activities have alternative machines on which they can be processed (e.g., [25]), and cumulative scheduling where resources are available in multiple units (e.g., [6]) to name only a few.

COMET's main contribution in this area is a novel computational model for the abstractions that captures the specificities of scheduling by neighborhood search. The core of the computational model is *an incremental precedence graph*, which specifies a candidate schedule at every computation step and can be viewed as a complex incremental variable. Once the concept of precedence graph is isolated, scheduling abstractions, such as resources and tardiness functions, become *differentiable objects* which maintain various properties and how they evolve under various local moves.

This model has a number of benefits. From a programming standpoint, neighborhood search algorithms are short and concise, and they are expressed in terms of high-level concepts which have been shown robust in the past.

[1]We use *precedence constraints* in a broad sense to include distance constraints.

In fact, their declarative components closely resemble those of constraint programs, although their search components radically differ. From a computational standpoint, the computational model naturally integrates with the architecture of COMET, allows for efficient incremental algorithms, and induces a reasonable overhead. From a language standpoint, the computational model suggests novel modeling abstractions which explicit the structure of scheduling applications even more. These novel abstractions make scheduling applications more compositional and modular, fostering the main modeling benefits of COMET and complementing its control abstractions.

Scheduling Abstractions. This brief overview is not meant to be comprehensive, but to convey the spirit of the abstractions. More details can be found in [51]. Scheduling applications in COMET are organized around the traditional concepts of schedules, activities, and resources. The fragment

```
Schedule sched(mgr);
Activity a(sched,4);Activity b(sched,5);
a.precedes(b);
sched.close();
```

introduces the most basic concepts. It declares a schedule `sched`, two activities `a` and `b` of duration 4 and 5 respectively, and a precedence constraint between `a` and `b`. This excerpt highlights the high-level similarity between the declarative components of COMET and constraint-based schedulers. In constraint-based scheduling, these instructions create domain-variables for the starting dates of the activities and the precedence constraints reduce their domains. In COMET, these instructions specify a *candidate schedule* satisfying the precedence constraints. For instance, the above snippet assigns starting dates 0 and 4 to activities `a` and `b`. The expression `a.getESD()` can be used to retrieve the starting date of activity `a` which typically vary over time as the neighborhood search moves from candidate schedules to their neighbors.

Schedules in COMET always contain two basic activities of zero duration: the source and the sink. The source precedes all other activities, while the sink follows every other activity. The availability of the source and the sink often simplifies the design and implementation of neighborhood search algorithms.

Many neighborhood search algorithms rely on the job structure to specify their neighborhood, which makes it natural to include jobs as a modeling object for scheduling. A job is simply a sequence of activities linked by precedence constraints. The structure of jobs is specified in COMET through precedence constraints. For instance, the following

```
Schedule sched(mgr);
Job j(sched);
Activity a(sched,4); Activity b(sched,5);
```

```
4.   a.precedes(b,j);
5.   sched.close();
```

specifies a job `j` with two activities `a` and `b`, where `a` precedes `b`. This code also highlights an important feature of COMET: Precedence constraints can be associated with modeling objects such as jobs and resources (see line 4). This functionality simplifies the description of neighborhood search algorithms which may retrieve subsets of precedence constraints easily. Since each activity belongs to at most one job, COMET provides methods to access the job predecessors and successors. For instance, the expression `b.getJobPred()` returns the job predecessor of `b`, while `j.getFirst()` returns the first activity in job `j`.

Cumulative Resources. Cumulative Resources are traditionally used to model the processing requirements of activities. For instance, the instruction

```
CumulativeResource cranes(sched,5);
```

specifies a pool of 5 cranes, while the instruction `a.requires(cranes,2)` specifies that activity `a` requires 2 cranes during its execution. Once again, COMET reuses traditional modeling concepts from constraint programming, and the novelty is in their functionalities. Resources in COMET are not instrumental in pruning the search space: They are differentiable objects which maintain invariants and data structures to define the neighborhood. In particular, a cumulative resource maintains violations induced by the candidate schedule. A violation appears at time t where the demands of the activities executing on r at t in the candidate schedule exceeds the capacity of the resource. Cumulative resources can also be queried to return sets of tasks responsible for a given violation. As mentioned, precedence constraints can be associated with resources, e.g., `a.precedes(b,crane)`, a functionality illustrated later in the chapter.

Disjunctive Resources. Disjunctive resources are special cases of cumulative resources with unit capacity. Activities requiring disjunctive resources cannot overlap in time and are strictly ordered in feasible schedules. Neighborhood search algorithms for applications involving only disjunctive resources (e.g., various types of jobshop and flexible scheduling problems) typically move in the space of feasible schedules by reordering activities on a disjunctive resource. As a consequence, COMET provides a number of methods to access the (current) disjunctive sequence. For instance, method `d.getFirst()` returns the first activity in the sequence of disjunctive resource `d`, while method `a.getSucc(d)` returns the successor of activity `a` on `d`. COMET also provides a number of local moves for disjunctive resources which can all be viewed as the addition and removal of precedence constraints. For instance, the move `d.moveBackward(a)` swaps activity a with its predecessor on disjunctive resource d. This move removes three precedence constraints and adds three new ones. Note that such a move does not always result in a feasible schedule: activity a must be chosen carefully to avoid introducing cycles in the precedence graph.

Objective Functions. One of the most innovative scheduling abstractions featured in COMET is the concept of objective functions. At the modeling level, the key idea is to specify the "global" structure of the objective function explicitly. At the computational level, objective functions are differentiable objects which incrementally maintain invariants and data structures to evaluate the impact of local moves. The ubiquitous objective function in scheduling is of course the makespan which can be specified as follows:

```
Makespan makespan(sched);
```

Once declared, an objective function can be evaluated (i.e., `makespan.eval()`) and queried to determine the impact of various moves. For instance, the expression `makespan.evalAddPrecedenceDelta(a,b)` evaluates the makespan variation obtained by adding precedence $\mathtt{a} \rightarrow \mathtt{b}$. Similarly, the effect on the makespan of swapping activity a with its predecessor on disjunctive resource d can be queried using `makespan.evalMoveBackwardDelta(a,d)`.

The makespan maintains a variety of interesting information besides the total duration of the schedule. In particular, it maintains the latest starting date of each activity, as well as the *critical* activities, which appear on a longest path from the source to the sink in the precedence graph. These information are generally fundamental in defining neighborhood search and heuristic algorithms for scheduling. They can also be used to estimate quickly the impact of a local move. For instance, the expression

```
makespan.estimateMoveBackwardDelta(a,d)
```

returns an approximation to the makespan variation when swapping activity a with its predecessor on disjunctive resource d.

Although the makespan is probably the most studied objective function in scheduling, there are many other criteria to evaluate the quality of a schedule. One such objective is the concept of tardiness which has attracted increasing attention in recent years. The instruction

```
Tardiness tardiness(sched,a,dd);
```

declares an objective function which maintains the tardiness of activity a with respect to its due date dd, i.e., $\max(0, \mathtt{e} - \mathtt{dd})$ where e is the finishing date of activity a in the candidate schedule. Once again, a tardiness object is differentiable and can be queried to evaluate the effect of local moves on its value. For instance, the instruction `tardiness.evalMoveBackwardDelta(a,d)` determines the tardiness variation which would result from swapping activity a with its predecessor on disjunctive resource d.

The objective functions share the same differentiable interface, thus enhancing their compositionality and reusability. In particular, they combine naturally to build more complex optimization criteria. For instance, the snippet

```
1.  Tardiness tardiness[j in Job](sched,job[j].getLast(),dd[j]);
2.  ScheduleObjectiveSum totalTardiness(sched);
3.  forall(j in Job)
4.    totalTardiness.add(tardiness[j]);
```

defines an objective function `totalTardiness`, another differentiable function, which specifies the total tardiness of the candidate schedule. Line 1 defines the tardiness of every job `j`, i.e., the tardiness of the last activity of `j`. Line 2 defines the differentiable object `totalTardiness` as a sum of objective functions. Lines 3 and 4 adds the job tardiness functions to `totalTradiness` to specify the total tardiness of the schedule. Queries on the aggregate objective `totalTardiness`, e.g., `totalTardiness.evalMoveBackwardDelta(a,d)`, are computed by querying the individual tardiness functions and aggregating the results. It is easy to see how similar code could define, for instance, maximum tardiness as an objective function.

3.3.6 Jobshop Scheduling

This section describes the core of a jobshop scheduling algorithm in COMET. More precisely, it outlines the implementation of the effective tabu-search algorithm of Dell'Amico and Trubian (DT) [23] which features a complex and heterogeneous neighborhood. COMET has been applied to many other scheduling applications, including the minimization of total weighted tardiness in a jobshop [51] and cumulative scheduling [29].

The Declarative Component. Figure 3.18 depicts the declarative component of the jobshop scheduling algorithm. Line 1 declares a disjunctive schedule and lines 3-5 create the activities, disjunctive resources, and jobs respectively. Line 6 specifies the objective function. The remaining part of the declarative component states the constraints in a way which is essentially similar to the constraint program in Figure 3.10. Once again, observe the independence of the declarative and search components.

The Search Component. The main complexity in implementing the DT algorithm is its search procedure. Algorithm DT uses the neighborhood

```
void state() {
   sched = new DisjunctiveSchedule(mgr);
   act = new Activity[i in ActRange](sched,duration[i]);
   res = new DisjunctiveResource[MachineRange](sched);
   job = new Job[JobRange](sched);
   obj = new Makespan(sched);

   forall(a in ActRange)
      act[a].requires(res[machine[a]]);
   forall(j in JobRange)
      forall(t in TaskRange:  t != TaskRange.low())
         jobAct[j,t-1].precedes(jobAct[j,t]);
   sched.close();
   mgr.close();
}
```

Figure 3.18. The Declarative Component of Jobshop Scheduling

$NC = RNA \cup NB$, where *RNA* is a neighborhood swapping vertices on a critical path (critical vertices) and *NB* is a neighborhood where a critical vertex is moved toward the beginning or the end of its critical block. More precisely, *RNA* considers sequences of the form $\langle p, v, s\rangle$, where v is a critical vertex and p, v, s are successive tasks on the same machine, and explores all permutations of these three vertices. Neighborhood *NB* considers a maximal sequence $\langle v_1, \ldots, v_i, \ldots, v_n\rangle$ of critical vertices on the same machine. For each such subsequence and each vertex v_i, it explores the schedule obtained by placing v_i at the beginning or at the end of the block, i.e.,

$$\langle v_i, v_1, \ldots, v_{i-1}, v_{i+1}, \ldots, v_n\rangle \vee \langle v_1, \ldots, v_{i-1}, v_{i+1}, \ldots, v_n, v_i\rangle$$

Since these schedules are not necessarily feasible, *NB* actually considers the leftmost and rightmost feasible positions for v_i (instead of the first and last position). *NB* is connected, which is an important theoretical property of neighborhoods.

```
void executeMove() {
   MinNeighborSelector N();
   exploreNeighborhood(N);
   if (N.hasMove())
      call(N.getMove());
}
void exploreNeighborhood(NeighborSelector N) {
   set{int} ca = setof(v in Activities) obj.isStronglyCritical(act[v]);
   exploreRNA(N,ca);
   exploreNB(N,ca);
}
```

Figure 3.19. Neighborhood *NB* in COMET.

The main complexity in implementing algorithm DT lies in the temporal disconnection between the move declarations, selections, and executions. The `neighbor` construct of COMET precisely addresses this issue. The top-level method of the search procedure is depicted in Figure 3.19. Method `executeMove` creates a selector, explores the neighborhood, and executes the best move (if any). The selector `N` is of type `MinNeighborSelector` which means that it selects the move with the smallest evaluation. This code nicely illustrates the compositionality of the approach. In particular, it demonstrates how new neighborhoods can be added without modifying existing code, since the subneighborhoods do not have to agree on a common interface or abstract class. The implementation of `exploreRNA` and `exploreNB` is of course where the `neighbor` construct is used.

```
1.  void exploreNB(NeighborSelector N,set{int} Criticals) {
2.   forall(v in Criticals) {
3.      Activity a = act[v];
4.      DisjunctiveResource res = a.getResource();
5.      int lm = obj.leftMostCriticalShift(a,res);
6.      int rm = obj.rightMostCriticalShift(a,res);
7.      Activity p = a.getPred(res);
8.      Activity s = a.getSucc(res);
9.      if (lm > 0) {
10.        int eval = obj.evalMoveBackwardDelta(a,res,lm);
11.        if (acceptNBLeft(eval,v))
12.           neighbor(eval,N)
13.              res.moveBackward(a,lm);
14.     }
15.     if (rm > 0) {
16.        int eval = obj.evalMoveForwardDelta(a,res,rm);
17.        if (acceptNBRight(eval,v))
18.           neighbor(eval,N)
19.              res.moveForward(a,rm);
20.     }
21.  }
22.}
```

Figure 3.20. Neighborhood *NB* in COMET.

Figure 3.20 gives the implementation of `exploreNB`: method `exploreRNA` is similar in spirit, but somewhat more complex, since it involves 5 different moves and additional conditions to ensure feasibility (lines 5-6). The code for method `exploreNB` is particularly interesting. It iterates over all critical activities (lines 1-2) and computes how much each such activity can be shifted left (resp. right) while staying in a critical block and maintaining feasibility. Lines 10-14 considers the left shift if one exists (line 9). Line 10 evaluates the effect on the makespan of shifting activity `a` by `lm` positions on its disjunctive resource. If this move is accepted (e.g., if it is not tabu), then the `neighbor` construct is executed (lines 12-13). Its effect is to submit the move `res.moveBackward(a,lm)` with its evaluation `eval` to the selector. The move is not executed; it is simply stored as a closure and will only be executed if it is the best move after complete exploration of the neighborhood (line 5 in Figure 3.19). The rest of the method describes the symmetric treatment for the right shift.

The neighborhood exploration is particularly elegant (in our opinion). Although a move evaluation and its execution take place at different execution times, the `neighbor` construct makes it possible to specify them together, significantly enhancing clarity and programming ease. The move evaluation and execution are textually adjacent and the logic underlying the neighborhood is not made obscure by introducing intermediary classes and methods. Compositionality is another fundamental advantage of the code organization. As mentioned earlier, new moves can be added easily, without affecting existing code. Equally or more important perhaps, the approach separates the neighborhood definition (method `exploreNeighborhood`) from its use (method `executeMove` in the DT algorithm). This makes it possible to use the neighborhood exploration in many different ways without any modification to its code. For instance, a semi-greedy strategy, which selects one of the k-best moves, only requires to use a semi-greedy selector. Similarly, the method can be used to collect all neighbors which is useful in intensification strategies based on elite solutions [33].

Table 3.4. Computational Results on the Tabu-Search Algorithm (DT)

	abz5	abz6	abz7	abz8	abz9	mt10	mt20
DT	139.5	86.8	320.1	336.1	320.8	155.8	160.1
DT*	6.2	3.8	14.2	15.1	14.2	6.9	7.1
KS	7.8	8.2	20.7	23.1	20.3	8.7	16.4
KS*	4.6	4.8	12.2	13.6	11.9	5.1	9.6
CO	5.9	5.7	11.7	9.9	9.0	6.7	9.8
	orb1	orb2	orb3	orb4	orb5		
DT	157.6	136.4	157.3	156.8	140.1		
DT*	7.0	6.0	7.0	6.9	6.2		
KS	9.2	7.8	9.3	8.5	8.1		
KS*	5.4	4.6	5.5	5.0	4.8		
CO	5.6	4.8	5.6	6.3	6.5		

Experimental Results. Table 3.4 presents some preliminary experimental results on jobshop scheduling. It compares various implementations of the tabu-search algorithm DT. In particular, it compares the original results [23], a C++ implementation [39], and the COMET implementation. Table 3.4 presents the results corresponding to Table 3 in [23]. Since DT is actually faster on the LA benchmarks (Table 4 in [23]), these results are representative. In the table, DT is the original implementation on a 33mhz PC, DT* is the scaled times on a 745mhz PC, KS is the C++ implementation on a 440 MHz Sun Ultra, KS* are the scaled times on a 745mhz PC, and CO is the COMET implementation on a 745mhz PC. Scaling was based on the clock frequency, which is favorable to slower machines (especially for the Sun). The times corresponds to the average over multiple runs (5 for DT, 20 for KS, and 50 for CO). Results for COMET are for the JIT compiler but include garbage collection. The results clearly indicate that COMET can be implemented to be competitive with specialized programs. Note also that the C++ implementation is more than 4,000 lines long, while the COMET program has about 400 lines.

3.4 Comet in Context

It is worth summarising the results presented in this paper. *The main message is that, although they support fundamentally different types of algorithms, constraint programming and* COMET *share a common architecture which promotes modularity, compositionality, reuse, and separation of concerns.* The architecture combines declarative and search components which express the problem constraints at a high level of abstraction and allow for concise descriptions of search algorithms. As a result, programs in constraint programming languages and COMET often exhibit a similar organization and modeling components which are relatively close. The search components are quite distinct in general, because of the nature of the underlying algorithms.

Table 3.5. Contrasting CP and NS

Issue	CP	NS
Variables	logical/domain	incremental
Constraints/objective	numeric	numeric
	combinatorial	combinatorial
Search	tree search	graph exploration
	nondeterministic	randomized
	strategies	meta-heuristics
Architecture	layered	layered
Constraints/objective	pruning/relaxation	differentiability
Search	choice points	closures
	backtracking	inverse functions

There are of course fundamental technical difference between constraint programming and COMET, some of which are captured in Table 3.5. The top part of the table discusses conceptual differences, while the bottom part addresses operational concerns.

At the conceptual level, *the key distinction lies in the nature of the variables.* Constraint programming languages for combinatorial optimization are based on logical variables. The values of these variables are unknown initially and it is the purpose of the computation to find values that satisfy all constraints. COMET, in contrast, relies on incremental variables, i.e., variables which maintain both an old and a new state to facilitate incremental computation. Once these respective variables are in place, the constraints are essentially similar. They capture properties of the solutions, they may be numeric or combinatorial, and they may be composed naturally. As mentioned earlier, search is rather different in both frameworks. Constraint programming heavily relies on a tree-search exploration model, and uses nondeterministic constructs to specify the search tree and search strategies to describe how to explore it. COMET supports graph exploration procedures which rely on randomization, checkpointing, and meta-heuristics. Interestingly, COMET supports the concept of invariants. Invariants provide an intermediate layer between variables and constraints, simplifying the implementation of constraints and other differentiable objects. Such intermediate layers will become increasingly important as constraint programming matures. In fact, invariants could already be useful for a variety of purposes in constraint programming, although this has not been explored to date.

At the operational level, the main commonality is the overall architecture which relies on data-driven computations. But the differences are more striking. Constraints in constraint programming embody pruning algorithms, while they encapsulate incremental algorithms in COMET. Search in constraint programming relies on choice points, backtracking, and trailing,[2] while COMET uses closures and inverse functions derived from incremental algorithms.

In summary, recent research on COMET seems to indicate that constraint programming and neighborhood search can be supported by high-level languages and libraries with similar abstraction levels, compositionality, and programming style. Obviously, the actual abstractions and their operational semantics are quite distinct, which is to be expected given the nature of the underlying paradigms. But, over the course of the research, novel concepts, as well as novel uses of old ones, have emerged to show remarkable similarities in the way these two paradigms can be supported.

It is also interesting to observe that the CP column would directly apply to MIP languages as well. Indeed, constraint languages for mixed integer programming may follow a similar architecture and OPL was a first step in that direction. What differs is how pruning, relaxation, and search take place, since they exploit linear programming algorithms. However, despite this difference

[2] Some implementations also use copying [41].

in the underlying technology, the compositional architecture, the idea of combinatorial constraints, and the novel search constructs may bring significant benefits to this area as well. Constraint and modeling languages have evolved significantly over the last two decades and it seems likely that this trend will continue to accommodate the significant progress in the underlying paradigms and technologies.

Acknowledgment

This work was partially supported by NSF ITR Awards DMI-0121495 and ACI-0121497.

References

[1] P. Baptiste, C. Le Pape, and W. Nuijten. *Constraint-Based Scheduling*. Kluwer Academic Publishers, 2001.

[2] N. Beldiceanu and M. Carlsson. Revisiting the cardinality Operator. In *ICLP-01*, Cyprus, 2001.

[3] N. Beldiceanu and E. Contejean. Introducing Global Constraints in CHIP. *Journal of Mathematical and Computer Modelling*, 20(12):97–123, 1994.

[4] R. Bent and P. Van Hentenryck. A Two-Stage Hybrid Local Search for the Vehicle Routing Problem with Time Windows. *Transportation Science*, 2001. (To Appear).

[5] J. Bisschop and A. Meeraus. On the Development of a General Algebraic Modeling System in a Strategic Planning Environment. *Mathematical Programming Study*, 20:1–29, 1982.

[6] Amedeo Cesta, Angelo Oddi, and Stephen F. Smith. Iterative flattening: A scalable method for solving multi-capacity scheduling problems. In *AAAI/IAAI*, pages 742–747, 2000.

[7] C. Codognet and D. Diaz. Yet Another Local Search Method for Constraint Solving. In *AAAI Fall Symposium on Using Uncertainty within Computation*, Cape Cod, MA., 2001.

[8] B. De Backer, V. Furnon, P. Shaw, P. Kilby, and P. Prosser. Solving Vehicle Routing Problems Using Constraint Programming and Metaheuristics. *Journal of Heuristics*, 6:501–523, 2000.

[9] I.R. de Farias, E.L. Johnson, and G.L. Nemhauser. Branch-and-Cut for Combinatorial Optimization Problems without Auxiliary Binary Variables. *Knowledge Engineering Review*, 16:25–39, 2001.

[10] L. Di Gaspero and A. Schaerf. *Optimization Software Class Libraries*, chapter Writing Local Search Algorithms Using EasyLocal++. Kluwer, 2002.

[11] M. Dincbas, H. Simonis, and P. Van Hentenryck. Solving a Cutting-Stock Problem in Constraint Logic Programming. In *Fifth International Conference on Logic Programming*, Seattle, WA, August 1988.

[12] M. Dincbas, H. Simonis, and P. Van Hentenryck. Solving the Car Sequencing Problem in Constraint Logic Programming. In *ECAI-88*, August 1988.

[13] M. Dincbas, P. Van Hentenryck, H. Simonis, A. Aggoun, T. Graf, and F. Berthier. The Constraint Logic Programming Language CHIP. In *Proceedings of the International Conference on Fifth Generation Computer Systems*, Tokyo, Japan, December 1988.

[14] R. Fourer, D. Gay, and B.W. Kernighan. *AMPL: A Modeling Language for Mathematical Programming*. The Scientific Press, San Francisco, CA, 1993.

[15] R. Fourer and D.M. Gay. Extending an Algebraic Modeling Language to Support Constraint Programming. *Informs Journal on Computing*, 14(4), 2002.

[16] P. Galinier and J.-K. Hao. A General Approach for Constraint Solving by Local Search. In *CP-AI-OR'00*, Paderborn, Germany, March 2000.

[17] I.P. Gent and T. Walsh. CSPLib: a benchmark library for constraints. In *CP'99 (Short Paper)*, Alexandra, VA, October 1999.

[18] W.D. Harvey and M.L. Ginsberg. Limited Discrepancy Search. In *Proceedings of the 14th International Joint Conference on Artificial Intelligence*, Montreal, Canada, August 1995.

[19] J.N. Hooker. *Logic-Based Methods for Optimization: Combining Optimization and Constraint Satisfaction*. John Wiley and Sons, 2000.

[20] Kindervater, G. and Savelsbergh, M.W. Vehicle routing: Handling edge exchanges. In E. Aarts and J.K. Lenstra, editors, *Local Search in Combinatorial Optimization*, pages 482–520. Wiley-Interscience Series in Discrete Mathematics and Optimization, John Wiley & Sons Ltd, England, 1997.

[21] W. Kreipl. A large step random wlak for minimizing total weighted tardiness in a job shop. *Journal of Scheduling*, 3:125–138, 2000.

[22] F. Laburthe and Y. Caseau. SALSA: A Language for Search Algorithms. In *Fourth International Conference on the Principles and Practice of Constraint Programming (CP'98)*, Pisa, Italy, October 1998.

[23] Dell'Amico M. and Trubian M. Applying Tabu Search to the Job-Shop Scheduling Problem. *Annals of Operations Research*, 41:231–252, 1993.

[24] A.K. Mackworth. Consistency in Networks of Relations. *Artificial Intelligence*, 8(1):99–118, 1977.

[25] M. Mastrolilli and L.M. Gambardella. Effective neighborhood functions for the flexible job shop problem. *Journal of Scheduling*, 3(1):3–20, 2000.

[26] L. Michel and P. Van Hentenryck. Localizer: A Modeling Language for Local Search. *Informs Journal on Computing*, 11(1):1–14, 1999.

[27] L. Michel and P. Van Hentenryck. Localizer. *Constraints*, 5:41–82, 2000.

[28] L. Michel and P. Van Hentenryck. A constraint-based architecture for local search. In *Conference on Object-Oriented Programming Systems, Languages, and Applications.*, pages 101–110, Seattle, WA, USA, November 4-8 2002. ACM.

[29] L. Michel and P. Van Hentenryck. Iterative Relaxations for Iterative Flattening in Cumulative Scheduling. In *14th International Conference on Automated Planning & Scheduling (ICAPS'04)*, Whistler, BC, Canada, 2004.

[30] S. Minton, M.D. Johnston, and A.B. Philips. Solving Large-Scale Constraint Satisfaction and Scheduling Problems using a Heuristic Repair Method. In *AAAI-90*, August 1990.

[31] U. Montanari. Networks of Constraints : Fundamental Properties and Applications to Picture Processing. *Information Science*, 7(2):95–132, 1974.

[32] A. Nareyek. *Constraint-Based Agents*. Springer Verlag, 1998.

[33] E. Nowicki and C. Smutnicki. A Fast Taboo Search Algorithm for the Job Shop Problem. *Management Science*, 42(6):797–813, 1996.

[34] B.D. Parrello, W.C. Kabat, and L. Wos. Job-Shop Scheduling Using Automated Reasoning: A Case Study of the Car-Sequencing Problem. *Journal of Automated Reasoning*, 2(1):1–42, 1986.

[35] L. Perron. Search Procedures and Parallelism in Constraint Programming. In *Fifth International Conference on the Principles and Practice of Constraint Programming (CP'99)*, Alexandra, VA, October 1999.

[36] P. Refalo. Tight Cooperation and its Application in Piecewise Linear Optimization. In *Fifth International Conference on the Principles and Practice of Constraint Programming (CP'99)*, Alexandra, VA, October 1999.

[37] J-C. Régin. A filtering algorithm for constraints of difference in CSPs. In *AAAI-94, proceedings of the Twelth National Conference on Artificial Intelligence*, pages 362–367, Seattle, Washington, 1994.

[38] J.-C. Regin and J.-F. Puget. A Filtering Algorithm for Global Sequencing Constraints. In *Third International Conference on the Principles and Practice of Constraint Programming (CP'97)*, Lintz, Austria, October 1997.

[39] K. Schmidt. Using Tabu-search to Solve the Job-Shop Scheduling Problem with Sequence Dependent Setup Times. ScM Thesis, Brown University, 2001, 2001.

[40] C. Schulte. Programming Constraint Inference Engines. In *Proceedings of the Third International Conference on Principles and Practice of Constraint Programming*, volume 1330, pages 519–533, Schloss Hagenberg, Linz, Austria, October 1997. Springer-Verlag.

[41] Christian Schulte. Comparing trailing and copying for constraint programming. In Danny De Schreye, editor, *Proceedings of the Sixteenth International Conference on Logic Programming*, pages 275–289, Las Cruces, NM, USA, November 1999. The MIT Press.

[42] B. Selman, H. Kautz, and B. Cohen. Noise Strategies for Improving Local Search. In *AAAI-94*, pages 337–343, 1994.

[43] P. Shaw, B. De Backer, and V. Furnon. Improved local search for CP toolkits. *Annals of Operations Research*, 115:31–50, 2002.

[44] P. Van Hentenryck. *Constraint Satisfaction in Logic Programming*. Logic Programming Series, The MIT Press, Cambridge, MA, 1989.

[45] P. Van Hentenryck. *The OPL Optimization Programming Language*. The MIT Press, Cambridge, Mass., 1999.

[46] P. Van Hentenryck. Constraint and Integer Programming in OPL. *Informs Journal on Computing*, 14(4):345–372, 2002.

[47] P. Van Hentenryck and J-P. Carillon. Generality Versus Specificity: An Experience with AI and OR Techniques. In *Proceedings of the American Association for Artificial Intelligence (AAAI-88)*, (St. Paul, MN), August 1988. AAAI, Menlo Park, Calif.

[48] P. Van Hentenryck and Y. Deville. The Cardinality Operator: A New Logical Connective and its Application to Constraint Logic Programming. In *ICLP-91*, pages 745–759, June 1991.

[49] P. Van Hentenryck and M. Dincbas. Domains in Logic Programming. In *AAAI-86*, Philadelphia, PA, August 1986.

[50] P. Van Hentenryck and L. Michel. Control Abstractions for Local Search. In *Ninth International Conference on Principles and Practice of Constraint Programming*, Cork, Ireland, 2003. (Best Paper Award).

[51] P. Van Hentenryck and L. Michel. Scheduling Abstractions for Local Search. In *Proceedings of the First International Conference on the Integration of AI and OR Techniques in Constraint Programming for Combinatorial Optimisation Problems (CP-AI-OR'04)*, Nice, France, 2004.

[52] P. Van Hentenryck, L. Michel, P. Laborie, W. Nuijten, and J. Rogerie. Combinatorial Optimization in OPL Studio. In *Proceedings of the 9th Portuguese Conference on Artificial Intelligence International Conference (EPIA'99)*, pages 1–15, Evora, Portugal, September 1999. (Invited Paper).

[53] P. Van Hentenryck, L. Perron, and J-F. Puget. Search and Strategies in OPL. *ACM Transactions on Computational Logic*, 1(2):1–36, October 2000.

[54] S. Voss and D. Woodruff. *Optimization Software Class Libraries.* Kluwer Academic Publishers, 2002.

[55] J. Walser. *Integer Optimization by Local Search.* Springer Verlag, 1998.

[56] D. Waltz. Generating Semantic Descriptions from Drawings of Scenes with Shadows. Technical Report AI271, MIT, MA, November 1972.

Chapter 4

A TUTORIAL ON RADIATION ONCOLOGY AND OPTIMIZATION

Allen Holder
Department of Mathematics, Trinity University
aholder@trinity.edu

Bill Salter
Radiation Oncology Department and Cancer Therapy and Research Center
University of Texas Health Science Center, San Antonio
bsalter@ctrc.net

Abstract Designing radiotherapy treatments is a complicated and important task that affects patient care, and modern delivery systems enable a physician more flexibility than can be considered. Consequently, treatment design is increasingly automated by techniques of optimization, and many of the advances in the design process are accomplished by a collaboration among medical physicists, radiation oncologists, and experts in optimization. This tutorial is meant to aid those with a background in optimization in learning about treatment design. Besides discussing several optimization models, we include a clinical perspective so that readers understand the clinical issues that are often ignored in the optimization literature. Moreover, we discuss many new challenges so that new researchers can quickly begin to work on meaningful problems.

Keywords: Optimization, Radiation Oncology, Medical Physics, Operations Research

Introduction

The interaction between medical physics and operations research (OR) is an important and burgeoning area of interdisciplinary work. The first optimization model used to aid the design of radiotherapy treatments was a linear model in 1968 [1], and since this time medical physicists have recognized that optimization techniques can support their goal of improving patient care. However, OR experts were not widely aware of these problems until the middle 1990s, and the last decade has witnessed a substantial amount of work focused on medical physics. In fact, three of the four papers receiving the Pierskalla prize from 2000 to 2003 address OR applications in medical physics [14, 25, 54].

The field of medical physics encompasses the areas of Imaging, Health Physics, and Radiation Oncology. These overlapping specialties typically combine when a patient is treated. For example, images of cancer patients are used to design radiotherapy treatments, and these treatments are monitored to guarantee safety protocols. While optimization techniques are useful in all of these areas, the bulk of the research is in the area of Radiation Oncology, and this is our focus as well.

Specifically, we study the design and delivery of radiotherapy treatments. Radiotherapy is the treatment of cancerous tissues with external beams of radiation, and the goal of the design process is to find a treatment that destroys the cancer but at the same time spares surrounding organs. Radiotherapy is based on the fact that unlike healthy tissue, cancerous cells are incapable of repairing themselves if they are damaged by radiation. So, the idea of treatment is to deliver enough radiation to kill cancerous tissues but not enough to hinder the survival of healthy cells.

Treatment design was, and to a large degree still is, accomplished through a trial-and-error process that is guided by a physician. However, the current technological capabilities of a clinic make it possible to deliver complicated treatments, and to take advantage of modern capabilities, it is necessary to automate the design process. From a clinical perspective, the hope is to improve treatments through OR techniques. The difficulty is that there are numerous ways to improve a treatment, such as delivering more radiation to the tumor, delivering less radiation to sensitive organs, or shortening treatment time. Each of these improvements leads to a different optimization problem, and current models typically address one of these aspects. However, each decision in the design process affects the others, and the ultimate goal is to optimize the entire process. This is a monumental task, one that is beyond the scope of current optimization models and numerical techniques. Part of the problem is that different treatment goals require different areas of expertise. To approach the problem in its entirety requires a knowledge of modeling, solving, and analyzing both deterministic and stochastic linear, nonlinear, integer, and global optimization problems. The good news for OR experts is that no matter what niche one studies, there are related, important problems. Indeed, the field of radiation oncology is a rich source of new OR problems that can parlay new academic insights into improved patient care.

Our goals for this tutorial are threefold. First, we discuss the clinical aspects of treatment design, as it is paramount to understand how clinics assess treatments. It is easy for OR experts to build and solve models that are perceived to be clinically relevant, but as every OR expert knows, there are typically many attempts before a useful model is built. The clinical discussions in this tutorial will help new researchers avoid traditional academic pitfalls. Second, we discuss the array of optimization models and relate them to clinical techniques. This will help OR experts identify where their strengths are of greatest value. Third, the bibliography at the end of this tutorial highlights some of the latest work in the optimization and medical literature. These citations will quickly allow new researchers to become acquainted with the area.

4.1 Clinical Practice

As with most OR applications, knowledge about the restrictions of the other discipline are paramount to success. This means that OR experts need to become familiar with clinical practice, and while treatment facilities share many characteristics, they vary widely in their treatment capabilities. This is because there are differences in available technology, with treatment machines, software, and imaging capabilities varying from clinic to clinic. A clinic's staff is trained on the clinic's equipment and rarely has the chance to experiment with alternate technology. There are many reasons for this: treatment machines and software are extremely expensive (a typical linear accelerator costs more than $1,000,000), time restrictions hinder exploration, etc.... A dialog with a clinic is invaluable, and we urge interested readers to contact a local clinic.

We begin by presenting a brief overview of radiation therapy (RT) concepts, with the hope of familiarizing the reader with some of the terminology used in the field, and then describe a "typical" treatment scenario, beginning with patient imaging and culminating with delivery of treatment.

4.1.1 Radiation Therapy Concepts and Terminology

Radiation therapy (RT) is the treatment of cancer and other diseases with ionizing radiation; ionizing radiation that is sufficiently energetic to dislodge electrons from their orbits and send them penetrating through tissue depositing their energy. The energy deposited per unit mass of tissue is referred to as *Absorbed Dose* and is the source of the biological response exhibited by irradiated tissues, be that lethal damage to a cancerous tumor or unwanted side effects of a healthy tissue or organ. Units of absorbed dose are typically expressed as Gy (pronounced Gray) or centiGray (cGy). One Gy is equal to one Joule (J) of energy deposited in one kilogram (kg) of matter.

Cancer is, in simple terms, the conversion of a healthy functioning cell into one that constantly divides, thus reproducing itself far beyond the normal needs of the body. Whereas most healthy cells divide and grow until they encounter another tissue or organ, thus respecting the boundaries of other tissues, cancerous cells continue to grow into and over other tissue boundaries. The use of radiation to "treat" cancer can adopt one of two general approaches.

One delivery approach is used when healthy and cancerous cells are believed to co-mingle, making it impossible to target the cancerous cells without also treating the healthy cells. The approach adopted in such situations is called *fractionation*, which means to deliver a large total dose to a region containing the cancerous cells in smaller, daily fractions. A total dose of 60 Gy, for example, might be delivered in 2 Gy daily fractions over 30 treatment days. Two Gy represents a daily dose of radiation that is typically tolerated by healthy cells but not by tumor cells. The difference between the tolerable dose of tumor and healthy cells is often referred to as a *therapeutic advantage*, and radiotherapy exploits the fact that tumor cells are so focused on reproducing that they lack a well-functioning repair mechanism possessed by healthy cells. By breaking the total dose into smaller pieces, damage is done to tumor cells each day (which they do not repair) and the damage that is done to the healthy cells is tolerated, and in fact, repaired over the 24 hours before the next daily dose. The approach can be thought of as bathing the region in a dose that tumor cells will not likely survive but that healthy cells can tolerate.

The second philosophy that might be adopted for radiation treatment dosage is that of Radio*Surgery*. Radiosurgical approaches are used when it is believed that the cancer is in the form of a solid tumor which can be treated as a distinct target, without the presence of healthy, co-mingling cells. In such approaches it is believed that by destroying all cells within a physician-defined target area, the tumor can be eliminated and the patient will benefit. The treatment approach utilized is that of delivering one fraction of dose (i.e., a single treatment) which is extremely large compared to fractionated approaches. Typical radiosurgical treatment doses might be 15 to 20 Gy in a single fraction. Such doses are so large that all cells which might be present within the region treated to this dose will be destroyed. The treatment approach derives its name from the fact that such methods are considered to be the radiation equivalent to surgery, in that the targeted region is completely destroyed, or ablated, as if the region had been surgically removed.

The physical delivery of RT treatment can be broadly sub-categorized into two general approaches: *brachytherapy* and *external beam* radiation therapy (EBRT), each of which can be effectively used in the treatment of cancer. Brachytherapy, which could be referred to as *internal* radiation therapy, involves a minimally invasive surgical procedure wherein tiny radioactive "seeds" are deposited, or implanted, in the tumor. The optimal arrangement of such seeds, and the small, roughly spherical distribution of dose which surrounds them, has been the topic of much optimization related research. *External* beam radiation therapy involves the delivery of radiation to the tumor, or target, from a source of radiation located outside of the patient; thus the external component of the name. The radiation is typically delivered by a device known as a *linear accelerator*, or linac. Such a device is shown in Figures 4.1 and 4.2. The device is capable

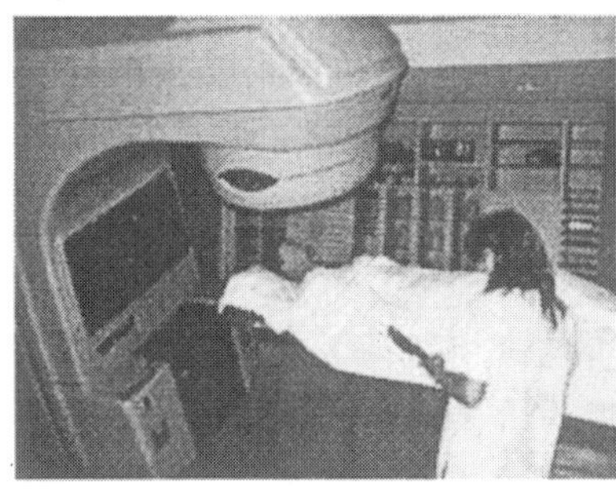

Figure 4.1. A Linear Accelerator

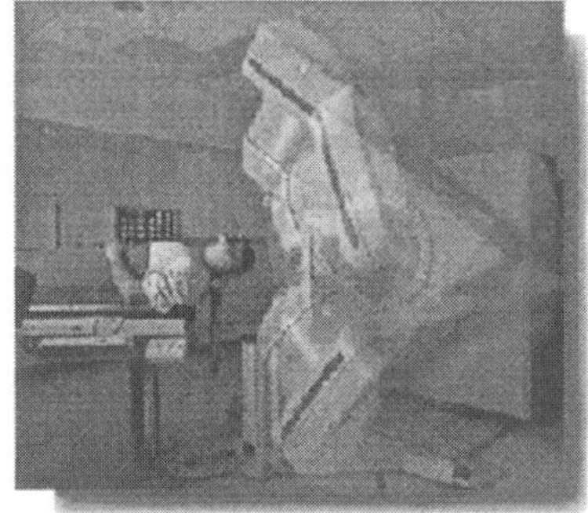

Figure 4.2. A Linear accelerator rotating through various angles. Note that the treatment couch is rotated.

of rotating about a single axis of rotation so that beams may be delivered from essentially 360 degrees about the patient. Additionally, the treatment couch, on which the patient lies, can also be rotated through, typically, 180 degrees. The combination of gantry and couch rotation can facilitate the delivery of radiation beams from almost any feasible angle. The point defined by the physical intersection of the axis of rotation of the linac gantry with the central axis of the beam which emerges from the "head" of the linac is referred to as *isocenter.* Isocenter is, essentially, a geometric reference point associated with the beam of radiation, which is strategically placed inside of the patient to cause the tumor to be intersected by the treatment beam.

External beam radiation therapy can be loosely subdivided into the general categories of *conventional radiation therapy* and, more recently, *conformal radiation therapy* techniques. Generally speaking, conventional RT differs from conformal RT in two regards; complexity and intent. The goal of conformal techniques is to achieve a high degree of conformity of the delivered distribution of dose to the shape of the target. This means that if the target surface is convex in shape at some location, then the delivered dose distribution will also be convex at that same location. Such distributions of dose are typically represented in graphical form by what are referred to as *isodose distributions*. Much like the isobar lines on a weather map, such representations depict iso-levels of absorbed dose, wherein all tissue enclosed by a particular isodose level is understood to see that dose, or higher. An isodose line is defined as a percentage of the target dose, and an isodose volume is that amount of anatomy receiving at least that much radiation dose. Figure 4.3 depicts a conformal isodose distribution used for treatment of a tumor. The high dose region is represented by the 60 Gy line(dark line), which can be seen to follow the shape of the convex shaped

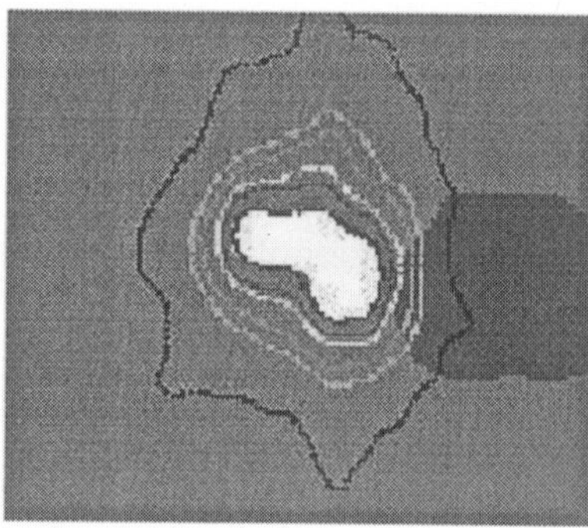

Figure 4.3. Conformal dose distribution. The target is shaded white and the brain stem dark grey. Isodose lines shown are 100%, 90%, 70%, 50%, 30% and 20%.

tumor nicely. The outer most curve is the 20 percent isodose curve, and the tissue inside of this curve receives at least 20 percent of the tumorcidal dose. By conforming the high dose level to the tumor, nearby healthy tissues are spared from the high dose levels. The ability to deliver a conformal distribution of dose to a tumor does not come without a price, and the price is complexity. Interestingly, the physical ability to deliver such convex-shaped distributions of dose has only recently been made possible by the advent of Intensity Modulating Technology, which will be discussed in a later section.

In conventional external beam radiation therapy, radiation dose is delivered to a target by the aiming of high-energy beams of radiation at the target from an origin point outside of the patient. In a manner similar to the way one might shine a diverging flashlight beam at an object to illuminate it, beams of radiation which are capable of penetrating human tissue are shined at the targeted tumor. Typically, such beams are made large enough to irradiate the entire target from each particular delivery angle that a beam might be delivered from. This is in contrast to IMRT approaches, which will be discussed in a later section, wherein each beam may treat only a small portion of the target. A fairly standard conventional delivery scheme is a so-called 2 field parallel-opposed arrangement (Figure 4.4). The figure depicts the treatment of a lesion of the liver created by use of an anterior to posterior-AP (i.e., from patient front to patient back) and posterior to anterior field-PA (i.e., from patient back to patient front). The isodose lines are depicted on computed tomography (CT) images of the patient's internal anatomy. The intersection of two different divergent fields delivered from two opposing angles results in a roughly rectangular shaped region of high dose (depicted by the resulting isodose lines for this plane). Note that the resulting high dose region encompasses almost the entire front to back dimension of the patient, and that this region includes the spinal cord critical structure. The addition of a third field, which is perpendicular to the

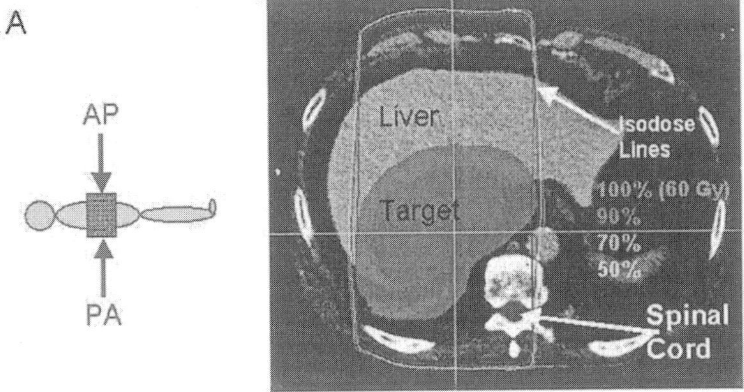

Figure 4.4. Two field, parallel opposed treatment of liver lesion.

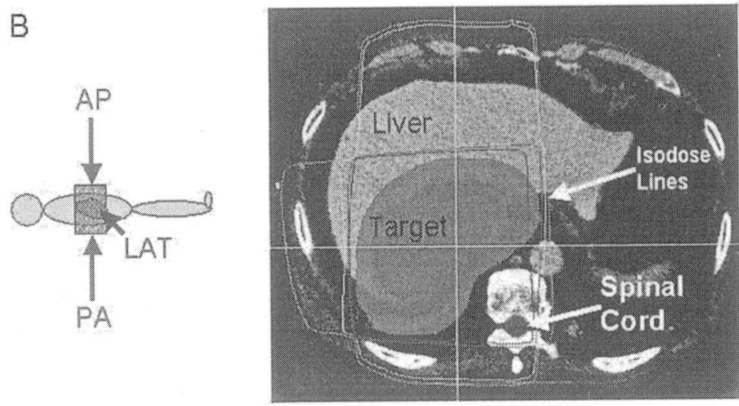

Figure 4.5. Three field treatment of liver lesion.

opposing fields, results in a box or square shaped distribution of dose, as seen in Figure 4.5. Note that the high dose region has been significantly reduced in size, but still includes the spinal cord. For either of these treatments to be viable, the dose prescribed by the physician to the high dose region would have to be maintained below the tolerance dose for the spinal cord (typically 44 Gy in 2 Gy fractions, to keep the probability of paralysis acceptably low) or a higher probability of paralysis would have to be accepted as a risk necessary to the survival of the patient. Such conventional approaches, which typically use 2-4 intersecting beams of radiation to treat a tumor, have been the cornerstone of radiation therapy delivery for years. By using customized beam blocking

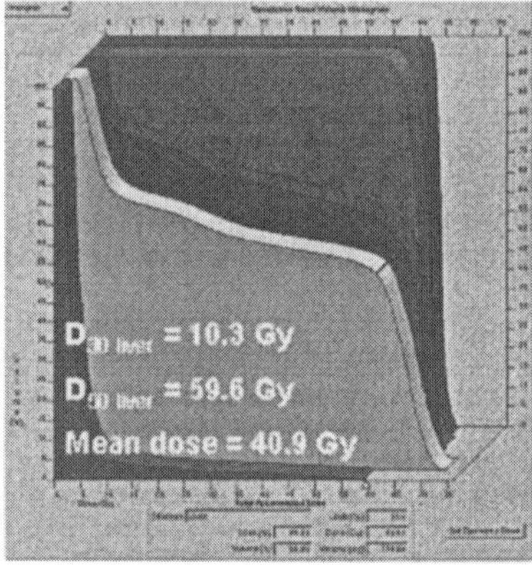

Figure 4.6. CDVH of two field treatment depicted in Figure 4.4

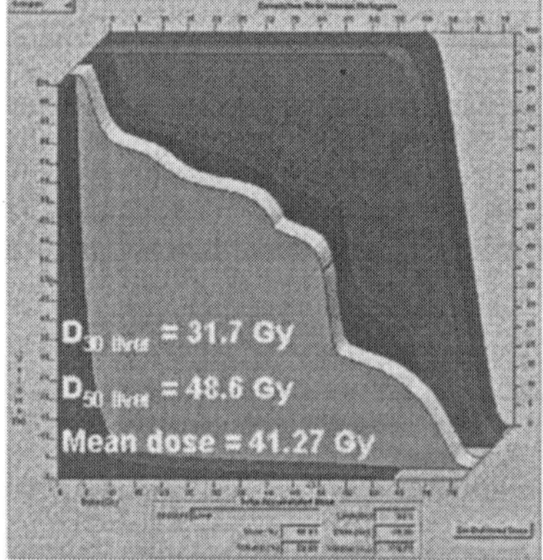

Figure 4.7. CDVH of three field treatment depicted in Figure 4.5

devices called "blocks" the shape of each beam can be matched to the shape of the projection of the target from each individual gantry angle, thus causing the total delivered dose distribution to match the shape of the target more closely.

The quality of a treatment delivery approach is characterized by several methods. Figures 4.6 and 4.7 show what is usually referred to as a "dose volume histogram" (DVH). More accurately, it is a cumulative DVH (CDVH). The curves describes the volume of tissue for a particular structure that is receiving a certain dose, or higher, and as such represents a plot of percentage of a particular structure versus Dose. The two CDVH's shown in Figures 4.6 and 4.7 are for the two conventional treatments shown in Figure 4.4 and 4.5, respectively. Five structures are represented in the figures from back to front, the Planning Target Volume (PTV) — a representation of the tumor that has been enlarged to account for targeting errors, such as patient motion; Clinical Target Volume (CTV) — The targeted tumor volume as defined by the physician on the 3-dimensional imaging set; The spinal Cord; the healthy, or non-targeted, Liver; all non-specific Healthy Tissue not specified as a critical structure. An ideal tumor CDVH would be a step function, with 100% of the target receiving exactly the prescribed dose (i.e., the 100% of prescribed level). Both treatments (i.e., Two Field and Three Field) produce near-step-function-like tumor DVH's. An ideal healthy tissue or critical structure DVH would be similar to that shown in Figures 4.6 and 4.7 for the Healthy Tissue, with 100% of the volume of the structure seeing 0% of the prescribed dose. The three field treatment in Figure 4.7 delivers less dose to the liver (second curve from front) and spinal cord (third curve from front) in that the CDVH's for these structures are pushed to the left, towards lower delivered doses. With regard to volumetric sparing of the liver and spinal cord, the three field treatment can be seen to represent a superior treatment. Dose volume histograms capture the volumetric information that is difficult to ascertain from the isodose distributions, but they do not provide information about the location of high or low dose regions. Both the isodose lines and the DVH information are needed to adequately judge the quality of a treatment plan.

Thus far, the general concept of cancer and its treatment by delivery of tumorcidal doses of radiation have been outlined. The concepts underlying the various delivery strategies which have historically been employed were summarized, and general terminology has been presented. What has not yet been discussed is the method by which a treatment "plan" is developed. The treatment plan is the strategy by which beams of radiation will be delivered, with the intent of killing the tumor and sparing from collateral damage the surrounding healthy tissues. It is, quite literally, a plan of attack on the tumor. The process by which a particular patient is taken from initial imaging visit, through the treatment planning phase and, ultimately, to treatment delivery will now be outlined.

4.1.2 The Clinical Process

A patient is often diagnosed with cancer following the observation of symptoms related to the disease. The patient is then typically referred for imaging studies and/or biopsy of a suspected lesion. The imaging may include CT scans, magnetic resonance imaging (MRI) or positron emission tomography (PET). Each imaging modality provides different information about the patient, from bony anatomy and tissue density information provided by the CT scan, to excellent soft tissue information from the MRI, to functional information on metabolic activity of the tumor from the PET scan. Each of these sets of three dimensional imaging information may be used by the physician both for determining what treatment approach is best for the patient, and what tissues should be identified for treatment and/or sparing. If external beam radiotherapy is selected as the treatment option of choice, the patient will be directed to a radiation therapy clinic where they will ultimately receive radiation treatment(s) for a period of time ranging from a single day, to several weeks.

Before treatment planning begins, a 3-dimensional representation of the internal anatomy of the patient must be obtained. For treatment planning purposes such images are typically created by CT scan of the patient, because of CT's accurate rendering of the attenuation coefficients of each voxel of the patient, as will be discussed in the section on Dose Calculation. The 3-dimensional CT representation of the patient is built by a series of 2-dimensional images (or slices), and the process of acquiring the images is often referred to as the Simulation phase. Patient alignment and immobilization is critical to this phase.

The treatment that will ultimately be delivered will be based on these images, and if the patient's position and orientation at the time of treatment do not agree with this "treatment planning position", then the treatment will not be delivered as planned. In order to ensure that the patient's position can be reproduced at treatment time, an immobilization device may be constructed. Such devices may be as invasive as placing screws into the skull of the patient's head to ensure precise delivery of a radiosurgical treatment to the brain, to as simple as placing a rubber band around the feet of the patient to help them hold still for treatment of a lesion of the prostate. Negative molds of the patient's posterior can be made in the form of a cradle to assist in immobilization, and pediatric patient's may need to be sedated for treatment. In all cases, alignment marks are placed on the patient to facilitate alignment to the linac beam via lasers in the treatment vault.

Once the images and re-positioning device(s) are constructed, the treatment plan must be devised. Treatment plans are designed by a medical physicist, or a dosimetrist working under the direction of a medical physicist, all according to the prescription of a radiation oncologist. The planning process depends heavily on the treatment machine and software, and without discussing the nuances of different facilities, we explain the important distinction between forward and inverse planning. During treatment, a patient is exposed to the beams of radiation created by a high-energy radioactive source, and these beams deposit their energy as they travel through the anatomy (see Subsection 4.2). Treatment design is the process of selecting how these beams will pass through the patient so that maximum damage accumulates in the target and minimal damage in healthy tissues. Forward treatment design means that a physicist or dosimetrist manually selects beam angles and fluences (the **amount** of radiation delivered by a beam, controlled by the amount of time that a beam is "turned on"), and calculates how radiation dose accumulates in the anatomy as a result of these choices. If the beams and exposure times result in an unacceptable dose distribution, different beams and fluences are selected. The process repeats until a satisfactory treatment is found.

The success of the trial-and-error technique of forward planning depends on the difficulty of the treatment and the expertise of the planner. Modern technology is capable of delivering complicated treatments, and optimally designing a treatment that considers the numerous options is beyond the scope of human ability. As its name suggests, inverse planning reverses the forward paradigm. Instead of selecting beams and fluences, the idea is to prescribe absorbed dose in the anatomy, and then algorithmically find a collection of beams and fluences that satisfy the anatomical restrictions. This means that inverse planning relies on optimization software, and the models that make this possible are the primary focus of this work.

Commercial software products blend forward and inverse planning, with most packages requiring the user to select the beam directions but not the fluences. The anatomical restrictions are defined on the patient images by delineating the target volume and any surrounding sensitive regions. A target dose is prescribed and bounds on the sensitive tissues are defined as percentages of this dose. For example, the tumor in Figure 4.4 is embedded in healthy surrounding liver, and located near the spinal cord. After manually identifying the tumor, the healthy liver, and the spinal cord on each 2-dimensional image, the dosimetrist enters a physician prescribed target dose, and then bounds how much radiation is delivered to the remaining structures as a percentage of the target dose. The dosimetrist continues by selecting a collection of beam angles and then uses inverse planning software to determine optimal beam fluences. The optimization problems are nontrivial, and modern computing power can calculate optimal fluence maps in about 20 minutes. We mention that commercial software varies substantially, with some using linear and quadratic models and others using complex, global optimization models solved by simulated annealing. Input parameters to the optimization software are often adjusted several times before developing a satisfactory treatment plan. Once an acceptable treatment plan has been devised, treatment of the patient, according to the radiation oncologist's dose and fractionation directive can begin.

In the following sections we investigate the underpinnings of the physics describing how radiation deposits energy in tissue, as well as many of the optimization models suggested in the literature. This discussion requires a more detailed description of a clinic's technology, and different clinical applications are explained as needed. We want to again stress that a continued dialog with a treatment facility is needed for OR techniques to impact clinical practice. In the author's experience, medical physicists are very receptive to collaboration. The OR & Oncology web site (`http://www.trinity.edu/aholder/HealthApp/oncology/`) lists several interested researchers, and we encourage interested readers to contact people on this list.

4.2 Dose Calculations

Treatment design hinges on the fact that we can accurately model how beams of high-energy radiation interact with the human anatomy. While an entire tutorial could be written on this topic alone, our objective is to provide the basics of how these models work. An academic dose model does not need to precisely replicate clinical dose calculations but does need to approximate how radiation is deposited into the anatomy. We develop a simple, 2-dimensional, continuous dose model and its discrete counterpart. The 3-dimensional model is a natural extension but is more complicated to describe.

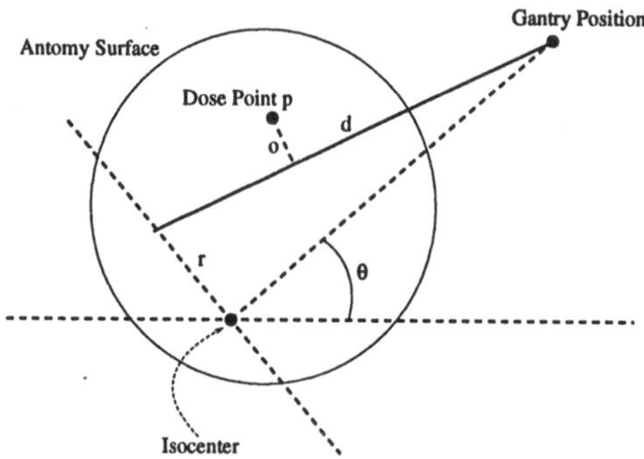

Figure 4.8. The geometry involved in calculating the contribution of sub-beam (θ, r) to the Dose Point.

Consider the diagram in Figure 4.8. The isocenter is in the lower part of the diagram, and the gantry is rotated to angle θ. Patients are often shielded from parts of the beam by devices such as a *multileaf collimator*, which are discussed in detail in Section 4.3. The sub-beam considered in Figure 4.8 is (θ, r), and we calculate this sub-beam's contribution to the dose point p. A simple but effective model uses the depth of the dose point along sub-beam (θ, r), labeled *d*, and the distance from the dose point to the sub-beam, denoted o (o is used because this is often referred to as the 'off axis' distance). The radiation being delivered along sub-beam (θ, r) attenuates and scatters as it travels through the anatomy. Attenuation means that photons of the beam are removed by scattering and absorption interactions as depth increases. So, if the dose point was directly in the path of sub-beam (θ, r), it would receive more radiation the closer it is to the gantry. While the dose point is not directly in the path of sub-beam (θ, r), it still receives radiation from this sub-beam because of scatter. A common model assumes that the percentage of deposited dose falls exponentially as d and o increase. So, if $g(\theta, r)$ is the amount of energy being delivered along sub-beam (θ, r) (or equivalently, the amount of time this sub-beam is not blocked), the dose point receives

$$g(\theta, r)e^{\eta o}e^{\mu d}$$

units of radiation from sub-beam (θ, r), where μ and η are parameters decided by the beam's energy. If $L_\theta = \{r : (\theta, r) \text{ is a sub-beam of angle } \theta\}$, we have that the total (or integral) amount of radiation delivered to the dose point from all gantry positions is

$$D_p = \int_L g(\theta, r)e^{\eta o}e^{\mu d} d\theta. \tag{2.1}$$

Calculating the amount of radiation deposited into the anatomy is a *forward* problem, meaning that the amount of radiation leaving the gantry is known and the radiation deposited into the patient is calculated. An *inverse* problem is one in which we know the radiation levels in the anatomy and then find a way to control the beams at the gantry to achieve these levels. Treatment design problems are inverse problems, as our goal is to specify the distribution of dose being delivered and then calculate a 'best' way to satisfy these limits. As an example, if the dose point p' is inside a tumor, we may desire that $D_{p'}$ be at least 60Gy. Similarly, if the dose point p'' was in a nearby, sensitive organ, we may want $D_{p''}$ to be no greater than 20Gy. So, our goal is to calculate $g(\theta, r)$ for each sub-beam so that

$$\begin{aligned} D_{p'} &= \int_L g(\theta, r) e^{\eta o} e^{\mu d} d\theta \geq 60, & (2.2) \\ D_{p''} &= \int_L g(\theta, r) e^{\eta o} e^{\mu d} d\theta \leq 20, \text{ and} & (2.3) \\ g(\theta, r) &\geq 0 \text{ for all } (\theta, r). & (2.4) \end{aligned}$$

From these constraints it is obvious that we need to invert the integral transformation that calculates dose, and while there are several numerical techniques to do so, such techniques do not guarantee the non-negativity of g. Moreover, the system may be inconsistent, which means the physician's restrictions are not possible. However, the typical case is that there are many choices of $g(\theta, r)$ that satisfy the physician's requirements, and in such a situation, the optimization question is which collection of $g(\theta, r)$'s is best?

The discrete approximation to (2.1) depends on a finite collection of angles and sub-beams. Instead of the continuous variables θ and r, we assume that there are q gantry positions, indexed by a, and that each of the gantry positions is comprised of τ sub-beams, indexed by s. The amount of radiation to deliver along sub-beam (a, s), which is equivalent to deciding how long to leave this sub-beam unblocked, is denoted by $x_{(a,s)}$. For the dose point p, we let $a_{(p,a,s)}$ be $e^{\eta o} e^{\mu d}$. The discrete counterpart of (2.1) is

$$\sum_{(a,s)} a_{(p,a,s)} x_{(a,s)} \approx D_p = \int_L g(\theta, r) e^{\eta o} e^{\mu d} d\theta.$$

We construct the *dose matrix*, A, from the collection of $a_{(p,a,s)}$'s by indexing the rows and columns of A by p and (a, s), respectively.

The dose matrix A adequately models how radiation is deposited into the anatomy as the gantry rotates around a single isocenter, which can be located at any position within the patient. Moreover, modern linear accelerators are capable of producing beams with different energies, and these energies correspond to different values of μ and η. So, for each isocenter i and beam energy e, we construct the dose matrix $A_{(i,e)}$. The entire dose matrix is then

$$\left[A_{(1,1)}|A_{(1,2)}|\cdots|A_{(1,E)}|A_{(2,1)}|\cdots|A_{(2,E)}|\cdots|A_{(I,1)}|\cdots|A_{(I,E)}\right],$$

where there are I different isocenters and E different energies. The index on x is adjusted accordingly to (i, e, a, s) so that $x_{(i,e,a,s)}$ is the radiation leaving the gantry along sub-beam (a, s) while the gantry is rotating around isocenter i and the linear accelerator is producing energy e. Many of the examples in this chapter use a single isocenter, and all use a single energy, but the reader should be aware that clinical applications are complicated by the possibility of having multiple isocenters and energies.

The cumulative dose at point p is the pth component of the vector Ax, denoted by $(Ax)_p$. We now see that the discrete approximations to (2.2)–(2.4) are

$$D_{p'} \approx (Ax)_{p'} \geq 60, \quad D_{p''} \approx (Ax)_{p''} \leq 20 \quad \text{and} \quad x \geq 0.$$

As before, there may not be an x that satisfies the system. In this case, we know that the physician's bounds are not possible with the discretization described by A. However, there may be a different collection of angles, sub-beams, and isocenters, and hence a different dose matrix, that allows the physician's bounds to be satisfied. Selecting the initial discretization is an important and challenging problem that we address in Section 4.3.

The vector x is called a *treatment plan* (or more succinctly a plan) because it indicates how radiation leaves the gantry as it rotates around the patient. The linear transformation $x \mapsto Ax$ takes the radiation at the gantry and deposits it into the anatomy. Both the continuous model and the discrete model are linear — i.e., the continuous model is linear in g and the discrete model is linear in x. The linearity is not just an approximation, as experiments have shown that the dose received in the anatomy scales linearly with the time a sub-beam is left unblocked. So, linearity is not just a modeling assumption but is instead natural and appropriate.

The treatment area and geometry are different from patient to patient, and the clinical dose calculations are patient specific. Also, depending on the region being treated, we may modify the attenuation to reflect different tissue densities, with the modified distances being called the *effective* depth and off-axis distance. As an example, if the sub-beam (a, s) is passing through bone, the effective depth is increased so that the attenuation (exponential decay) of the beam is greater as it travels through the bone. Similarly, if the sub-beam is passing through air, the effective depth is shortened so that less attenuation occurs.

We reiterate that there are numerous models of widely varying complexity that calculate how radiation is deposited into the anatomy. Our goal here was to introduce the basic concepts of a realistic model. Again, it is important to remember that for academic purposes, the dose calculations need only be reasonably close to those used in a clinic.

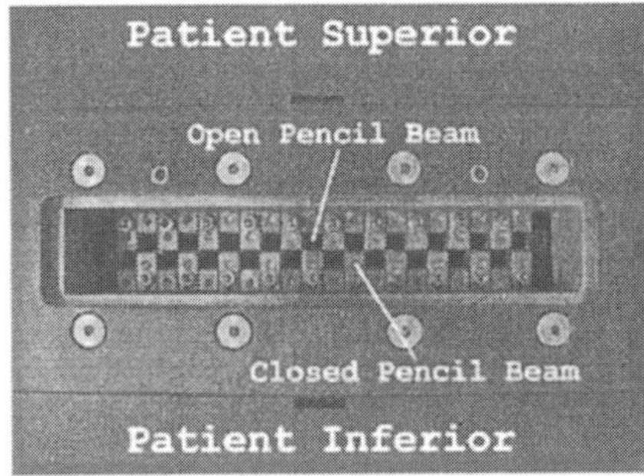

Figure 4.9. A tomotherapy multileaf collimator. The leaves are either open or closed.

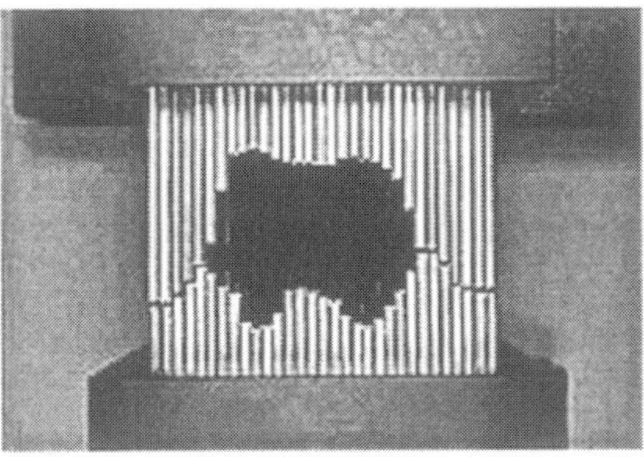

Figure 4.10. A multileaf collimator for static gantry IMRT.

4.3 Intensity Modulated Radiotherapy (IMRT)

A recent and important development in the field of RT is that of Intensity Modulated Radiotherapy (IMRT). Regarded by many in the field as a quantum leap forward in treatment delivery capability, IMRT allows for the creation of dose distributions that were previously not possible. As a result, IMRT has allowed for the treatment of patients that previously had no viable treatment options.

The distinguishing feature of IMRT is that the normally large, rectangular beam of radiation produced by a linear accelerator is shaped by a multileaf collimator into smaller so-called pencil beams of radiation, each of which can be varied, or modulated, in intensity (or fluence). Figures 4.9 and 4.10 show images of two multileaf collimators used for delivery of IMRT treatments. The leaves in Figure 4.9 are pneumatically controlled by individual air valves that cause the leaves to open or close in about 30 to 40 milliseconds. By varying the amount of time that a given leaf is opened from a particular gantry angle the intensity, or fluence, of the corresponding pencil beam is varied, or modulated. This collimator is used in *tomotherapy*, which treats the 3-dimensional problem as a series of 2-dimensional sub-problems. In tomotherapy a treatment is delivered as a summation of individually delivered "slices" of dose, each of which is optimized to the specific patient anatomy that is unique to the treatment slice. Tomotherapy treatments are delivered by rapidly opening and closing the leaves as the gantry swings continuously about the patient.

The collimator in Figure 4.10 is used for static gantry IMRT. This is a process where the gantry moves to several static locations, and at each position the patient is repeatedly exposed to radiation using different leaf configurations. Adjusting the leaves allows for the modulation of the fluence that is delivered along each of the many sub-beams. This allows the treatment of different parts of the tumor with different amounts of radiation from a single angle. Similar to tomotherapy, the idea is to accumulate damage from many angles so that the target is suitably irradiated.

4.3.1 Clinically Relevant IMRT Treatments

For an optimized IMRT treatment to be clinical useful, the problem must be modeled assuming clinically reasonable values for the relevant input variables.The clinical restrictions of IMRT depend on the type of delivery used. Tomotherapy has fewer restrictions with regard to gantry angles, in that any and all of the possible pencil beams may be utilized for treatment delivery. The linac gantry performs a continuous arc about the patient regardless of whether or not pencil beams from each gantry angle are utilized by the optimized delivery scheme. This is in contrast to the static gantry model where clinical time limitations make it impractical to deliver treatments comprised of, typically, more than 7-9 gantry angles. This means that the optimization process must necessarily select the optimal set of 7 to 9 gantry angles of approach from which to deliver pencil beams, from the much larger set of *possible* gantry angles of delivery, which leads to mixed integer problems. For either delivery approach, the gantry angles considered must, of course, be limited to those angles that do not lead to collisions of the gantry and treatment couch or patient. Clinical optimization software for static gantry approaches typically requires that the user pre-select the static gantry angles to be used. Such software provides visualization tools that help the user intelligently select gantry angles that can be visually recognized to provide unobstructed angles. This technique serves to reduce the complexity of the problem to manageable levels but does not, of course, guarantee a truly optimal solution. The continuous gantry movement of a tomotherapy treatment is approximated by modeling the variation of leaf positions every 5°, and the large number of potential angles coupled with a typical fluence variation of 0 to 100% in steps of 10% causes tomotherapy to possess an extremely large solution space.

4.3.2 Optimization Models

Before we begin describing the array of optimization models that are used to design treatments, we point out that several reviews are already in the literature. Shepard, Ferris, Olivera, and Mackie have a particularly good article in *SIAM Review* [56]. Other OR reviews include the exposition by Bartolozzi, et. al. in the *European Journal of Operations Research* [2] and the introductory material by Holder in the *Handbook of Operations Research/Management Science Applications in Health Care* [24]. In the medical physics literature, Rosen has a nice review in *Medical Physics* [55]. We also mention two web resources: the *OR & Oncology Web Site* at `www.trinity.edu/aholder/HealthApp/oncology/`

and *Pub Med* at `www.ncbi.nlm.nih.gov/`. The medical literature can be overwhelming, with a recent search at Pub Med on "optimization" and "oncology" returning 652 articles.

We begin our review of optimization models by studying linear programs. This is appropriate because dose deposition is linear and because linear programming is common to all OR experts. Also, many of the models in the literature are linear [1, 22, 24, 33, 36]. Let A be the dose deposition matrix described in Section 4.2, and partition the rows of A so that

$$A = \begin{bmatrix} A_T \\ A_C \\ A_N \end{bmatrix} \begin{array}{l} \leftarrow \text{ Target Volume} \\ \leftarrow \text{ Critical Structures} \\ \leftarrow \text{ Unrestricted, Normal Tissue,} \end{array}$$

where A_T is $m_T \times n$, A_C is $m_C \times n$, and A_N is $m_N \times n$. The sets T, C, and N partition the dose points in the anatomy, with T containing the dose points in the target volume, C containing the dose points in the critical structures, and N contains the remaining dose points. We point out that A is typically large. For example, if we have a 512×512 patient image with each pixel having its own dose point, then A has $262,144$ rows. Moreover, A has $360,000$ columns if we design a treatment using 4 energies, 5 isocenters, 360 angles per isocenter, and 50 sub-beams per angle. So, for a single image we would need to apriori make 9.44×10^{10} dose calculations. Since there are usually several images involved, it is easy to see that generating the data for a model instance is time consuming. Romeijn, Ahuja, Dempsey and Kumar [54] have developed a column generation technique to address this computational issue.

The information provided by a physician to build a model is called a *prescription.* This clinical information varies from clinic to clinic depending on the design software. A prescription is initially the triple (TG, CUB, NUB), where TG is a m_T vector containing the goal dose for the target volume, CUB is a m_C vector listing the upper bounds on the critical structures, and NUB is a m_N vector indicating the highest amount of radiation that is allowed in the remaining anatomy. In many clinical settings, NUB is not decided before the treatment is designed. However, clinics do not routinely allow any part of the anatomy to receive doses above 10% of the target dose, and one can assume that $NUB = 1.1 \times TG$.

The simplest linear models are feasibility problems [5, 48]. In these models the goal is to satisfy

$$A_T x \geq TG, \quad A_C x \leq CUB, \quad A_N x \leq NUB, \text{ and } \; x \geq 0.$$

The consistency of this system is not guaranteed because physicians are often overly demanding, and many authors have complained that infeasibility is a shortcoming of linearly constrained models [22, 33, 43, 55]. In fact, the argument that feasibility alone correctly addresses treatment design is that the region defined by these constraints is relatively small, and hence, optimizing over this region does not provide significant improvements in treatment quality.

If a treatment plan that satisfies the prescription exists, the natural question is which plan is best. The immediate, but naive, ideas are to maximize the tumor dose or minimize the critical structure dose. Allowing e to be the vector of ones, where length is decided by the context of its use, these models are variants of

$$\max\{e^T A_T x : A_T x \geq TG,\ A_C x \leq CUB,\ A_N x \leq NUB,\ x \geq 0\}, \quad (3.1)$$

$$\min\{e^T A_C x : A_T x \geq TG,\ A_C x \leq CUB,\ A_N x \leq NUB,\ x \geq 0\}, \quad (3.2)$$

$$\max\{z : A_T x \geq TG + ze,\ A_C x \leq CUB,\ A_N x \leq NUB,\ x \geq 0,\ z \geq 0\}, \text{ or} \quad (3.3)$$

$$\min\{z : A_T x \geq TG,\ A_C x \leq CUB - ze,\ A_N x \leq NUB,\ x \geq 0,\ z \geq 0\}. \quad (3.4)$$

Models (3.1) and (3.2) maximize and minimize the cumulative dose to the tumor and critical structures, respectively. Model (3.3) maximizes the minimum dose received by the target volume and (3.4) minimizes the maximum dose received by a critical structure.

The linear models in (3.1)–(3.4) are inadequate for several reasons. As already mentioned, if the feasibility region is empty, most solvers terminate by indicating that infeasibility has been detected. While there is a substantial literature on analyzing infeasibility (see for example [8, 9, 7, 20, 21]), discovering the source of infeasibility is an advanced skill, one that we can not expect physicians to acquire. Model (3.1) further suffers from the fact that it is often unbounded. This follows because it is possible to have sub-beams that intersect the tumor but that do not deliver numerically significant amounts of radiation to the critical structures. In this situation, it is obvious that we can make the cumulative dose to the tumor as large as possible. Lastly, these linear models have the unintended consequence of achieving the physician's bounds. For example, as model (3.3) increases the dose to the target volume, it is also increasing the dose to the critical structures. So, an optimal solution is likely to achieve the upper bounds placed on the critical structures, which is not desired. We also point out that because simplex based optimizers terminate with an extreme point solution, we are **guaranteed** that several of the inequalities hold with equality when the algorithm terminates [22]. So, the choice of algorithm plays a role as well, a topic that we address later in this section.

An improved linear objective was suggested by Morrill [44]. This objective maximizes the difference between the dose delivered to the tumor and the dose received by the critical structures. For example, consider the following models,

$$\max\{e^T A_T x - e^T A_C x : A_T x \geq TG,\\ A_C x \leq CUB, A_N x \leq NUB, x \geq 0\} \quad \text{and} \tag{3.5}$$

$$\max\{z - q : A_T x \geq TG + ze, A_C x \leq CUB - qe,\\ A_N x \leq NUB, x \geq 0, z \geq 0, q \geq 0\}. \tag{3.6}$$

These models attempt to overcome the difficulty of attaining the prescribed limits on the target volume and the critical structures. However, model (3.5) is often unbounded for the same reason that model (3.1) is. Also, both of these models are infeasible if the physician's goals are overly restrictive.

Many of the limitations of models (3.1)–(3.6) are addressed by parameterizing the constraints. This is similar to goal programming, where we think of the prescription as a goal instead of an absolute bound. Constraints that use parameters to adjust bounds are called *elastic*, and Holder [25] used these constraints to build a linear model that overcame the previous criticisms. Before presenting this model, we discuss another pitfall that new researchers often fall into. The target volume is not exclusively comprised of tumorous cells, but rather normal and cancerous cells are interspersed throughout the region. Recall that external beam radiotherapy is successful because cancerous cells are slightly more susceptible to radiation damage than are normal tissues. The goal is to deliver enough dose to the target volume so that the cancerous cells die but not enough to kill the healthy cells. So, one of the goals of treatment planning is to find a plan that delivers a uniform dose to the tumor. The model suggested in [25] uses a uniformity index, ρ, and sets the tumor lower bound to be $TLB = TG - \rho e$ and the tumor upper bound to be $TUB = TG + \rho e$ (typical values of ρ in the literature range from 0.02 to 0.15). Of course, there is no reason why the upper and lower bounds on the target volume need to be a fixed percentage of TG, and we extended a prescription to be the 4-tuple (TUB, TLB, CUB, NUB), where TUB and TLB are arbitrary positive vectors such that $TUB \geq TLB$. Consider the model below.

$$\min\{\omega \cdot l^T \alpha + u_C^T \beta + u_N^T \gamma : TLB - L\alpha \leq A_T x \leq TUB,\\ A_C x \leq CUB + U_C\beta, A_N x \leq NUB + U_N\gamma, -CUB \geq U_C\beta,\\ 0 \leq U_N\gamma, 0 \leq x\} \tag{3.7}$$

In this model, the matrices L, U_C, and U_N are assumed to be non-negative, semimonotone matrices with no row sum being zero. The term La measures the target volume's under dose, and the properties of L ensure that the target volume receives the minimum dose if and only if α is zero. Similarly, $U_C\beta$ and $U_N\gamma$ measure the amount the non-cancerous tissues are over their prescribed bounds. The difference between β and γ is that they have different lower bounds. If $U_C\beta$ attains its lower bound of $-CUB$, we have found a treatment plan that delivers no radiation to the critical structures. The lower bound on $U_N\gamma$ is 0, which indicates that we are willing to accept any plan where the dose to the non-critical tissue is below its prescribed limit.

The objective function in (3.7) penalizes adverse deviations and rewards desirable deviations. The term $l^T\alpha$ penalizes under dosing the target volume and $u_N^T\gamma$ penalizes overdosing the normal tissue. The role of $u_C^T\beta$ is twofold. If β is positive, it penalizes overdosing the critical structures, and if β is negative, it rewards under dosing the critical structures. The parameter ω weights the importance placed on attaining tumor uniformity.

One may ask why model 3.7 is stated in such general terms of measure and penalty. The reason is that there are two standard ways to measure and penalize discrepancies. If we want the sum of the discrepancies to be the penalty, then we let l, u_C, and u_N be vectors of ones and L, U_C, and U_N be the identity matrices. Alternatively, if we want to penalize the largest deviation, we let l, u_c, and u_N each be the scalar 1 and L, U_C and U_N be vectors of ones. So, this one model allows deviations to be measured and penalized in many ways but has a single mathematical analysis that applies to all of these situations.

The model in (3.7) has two important theoretical advantages to the previous models. The first result states that the elastic constraints of the model guarantee that both the primal and dual problems are feasible.

THEOREM 4.1 (HOLDER [25]) *The linear model in 3.7 and its dual are strictly feasible, meaning that each of the constraints can simultaneously hold without equality.*

The conclusion of Theorem 4.1 is not surprising from the primal perspective, but the dual statement requires all of the assumptions placed on l, u_C, u_N, L, U_C and U_N. The feasibility guaranteed by this result is important for two reasons. First, if the physician's goals are not possible, this model minimally adjusts the prescription to attain feasibility. Hence, this model returns a treatment plan that matches the physician's goals as closely as possible even if the original desires were not achievable. Second, Theorem 4.1 assures us that interior-point algorithms can be used, and we later discuss why these techniques are preferred over simplex based approaches.

The second theoretical guarantee about model (3.7) is that it provides an analysis certificate. Notice that the objective function is a weighted sum of the competing goals of delivering a large amount of radiation to the target volume and a small amount of radiation to the remaining anatomy. The next result shows that the penalty assigned to under dosing the target volume is uniformly bounded by the inverse of ω.

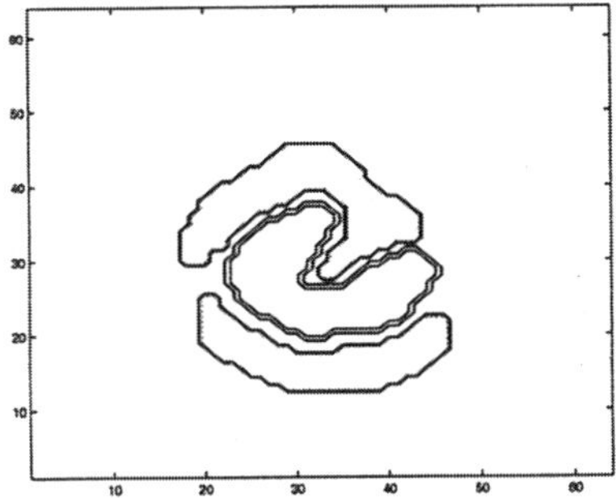

Figure 4.11. A tumor surrounded by two critical structures. The desired tumor dose is 80Gy±3%, and the critical structures are to receive less than 40Gy.

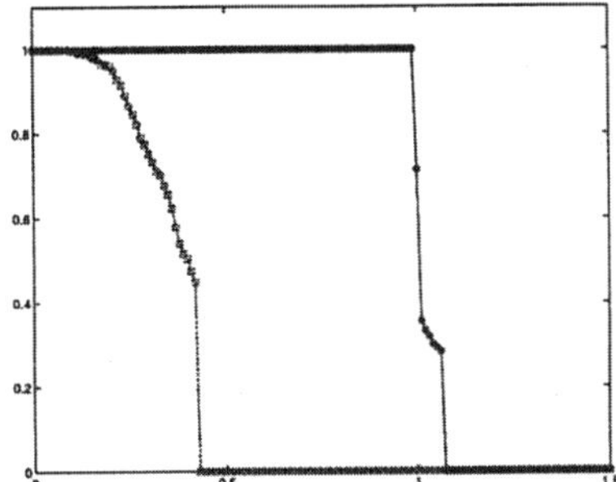

Figure 4.12. The dose-volume histogram indicates that 100% of the tumor receives its goal dose and that about 60% of the critical structures is below its bound of 40Gy.

THEOREM 4.2 (HOLDER [25]) *Allowing* $(x^*(\omega), \alpha^*(\omega), \beta^*(\omega), \gamma^*(\omega))$ *to be an optimal solution for a particular* ω, *we have that* $l^T\alpha^*(\omega) = O(1/\omega)$.

A consequence of Theorem 4.2 is that there is a positive scalar κ such that for any positive ω, we have that $l^T\alpha^*(\omega) \leq \kappa/\omega$. This is significant because we can apriori calculate an upper bound on κ that depends on the dose matrix A. If κ' is this upper bound, we have that $l^T\alpha^*(\omega) \leq \kappa/\omega \leq \kappa'/\omega$. So, we can make $l^T\alpha^*(\omega)$ as small as we want by selecting a sufficiently large ω. If we use this ω and $l^T\alpha^*(\omega)$ is larger that κ'/ω, then we know with certainty that we can not achieve the desired tumor uniformity. Moreover, we know that if $l^T\alpha^*(\omega)$ is less than κ'/ω and the remaining terms of the objective function are positive, then we can attain the tumor uniformity only at the expense of the critical structures. So, the importance of Theorem 4.2 is that it provides a guaranteed analysis.

Consider the geometry in Figure 4.11, where a tumor is surrounded by two critical structures. The goal dose for the tumor is 80Gy±3%, and the upper bound on the critical structures is 40Gy. Figure 4.12 is a dose-volume histogram for the treatment designed by Model (3.7), and from this figure we see that 100% of the tumor receives it's goal dose. Moreover, we see that about 60% of the critical structure is below its upper bound of 40Gy.

Outside of linear models, the most prevalent models are quadratic [36, 42, 60]. A popular quadratic model is

$$\min\{\|A_T x - TG\|_2 : A_C x \leq CUB, A_N x \leq NUB, x \geq 0\}. \qquad (3.8)$$

This model attempts to exactly attain the goal dose over the target volume while satisfying the non-cancerous constraints. This is an attractive model because the objective function is convex, and hence, local search methods like gradient descent and Newton's method work well. However, the non-elastic, linear constraints may be inconsistent, and this model suffers from the same infeasibility complaints of previous linear models. Some medical papers have suggested that we instead solve

$$\begin{aligned}\min\{\|A_T x - TG\|_2 + \|A_C x - CUB\|_2 + \\ \|A_N x - NUB\|_2 : x \geq 0\}.\end{aligned} \tag{3.9}$$

While this model is never infeasible, it is inappropriate for several reasons. Most importantly, this model attempts to **attain** the bounds placed on the non-cancerous tissue, something that is clearly not desirable. Second, this model could easily provide a treatment plan that under doses the target volume and over doses the critical structures, even when there are plans that sufficiently irradiate the tumor and under irradiate the critical structures. A more appropriate version of (3.9) is

$$\min\{\|A_T x - TG\|_2 + \|A_C x\|_2 + \|A_N x\|_2 : x \geq 0\}, \tag{3.10}$$

but again, without constraints on the non-cancerous tissues, there is no guarantee that the prescription is (optimally) satisfied.

The only real difference between the quadratic and linear models is the manner in which deviations from the prescription are measured. Since there is no clinically relevant reason to believe that one measure is more appropriate than another, the choice is a personal preference. In fact, all of the models discussed so far have a linear and a quadratic counterpart. For example, the quadratic manifestation of (3.7) is

$$\begin{aligned}\min\{\omega \cdot \|l^T \alpha\|_2 + \|u_C^T \beta\|_2 + \|u_N^T \gamma\|_2 : TLB - L\alpha \leq A_T x \leq TUB, \\ A_C x \leq CUB + U_C \beta, A_N x \leq NUB + U_N \gamma, -CUB \geq U_C \beta, \\ 0 \leq U_N \gamma, 0 \leq x\}\end{aligned} \tag{3.11}$$

and the linear counterparts of (3.10) are

$$\min\{\|A_T x - TG\|_1 + \|A_C x\|_1 + \|A_N x\|_1 : x \geq 0\}, \text{ and} \tag{3.12}$$
$$\min\{\|A_T x - TG\|_\infty + \|A_C x\|_\infty + \|A_N x\|_\infty : x \geq 0\}. \tag{3.13}$$

We point out that Theorems 4.1 and 4.2 apply to model (3.11), and in fact, these results hold for any of the p-norms.

Each of the above linear and quadratic models attempts to 'optimally' satisfy the prescription, but the previous prescriptions of $(TG, CUB,$

NUB) and (TLB, TUB, CUB, NUB) do not adequately address the physician's goals. The use of dose-volume histograms to judge treatments enables physicians to express their goals in terms of tissue percentages that are allowed to receive specified doses. For example, we could say that we want less than 80% of the lung to receive more than 60% of the target dose, and further, that less than 20% of the lung receives more than 75% of the target dose.

Constraints that model the physician's goals in terms of percent tissue receiving a fixed dose are called *dose-volume constraints.* These restrictions are biologically natural because different organs react to radiation differently. For example, the liver and lung are modular, and these organs are capable of functioning with substantial portions of their tissue destroyed. Other organs, like the spinal cord and bowel, lose functionality as soon as a relatively small region is destroyed. Organs are often classified as *rope* or *chain* organs [19, 53, 63, 64], with the difference being that rope organs remain functional even with large amounts of inactive tissue and that chain organs fail if a small region is rendered useless. Rope organs typically fail if the entire organ receives a relatively low, uniform dose, and the radiation passing through these organs should be accumulated over a contiguous portion of the tissue. Alternatively, chain organs are usually capable of handling larger, uniform doses over the entire organ, and it is desirable to disperse the radiation over the entire region. So, there are biological differences between organs that need to be considered. Dose-volume constraints capture a physician's goals for these organs.

We need to alter the definition of a prescription to incorporate dose-volume constraints. First, we partition C into $C^1, C^2, \ldots, C^K$, where C^k contains the dose points within the kth critical structure. We know have that

$$A_C = \begin{bmatrix} A_{C^1} \\ A_{C^2} \\ \vdots \\ A_{C^K} \end{bmatrix} \begin{matrix} \leftarrow \text{ Critical Structure 1} \\ \leftarrow \text{ Critical Structure 2} \\ \\ \leftarrow \text{ Critical Structure K.} \end{matrix} \tag{3.14}$$

The vector of upper bounds, CUB, no longer has the same meaning since we instead want to calculate the volume of tissue that is above the physician defined thresholds. For each k, let $T^{k_1}, T^{k_2}, \ldots, T^{k_{\Lambda_k}}$ be the thresholds for critical structure k. We let $\alpha_p^{k_\lambda}$ be a binary variable that indicates whether or not dose point p, which is in critical structure k, is below or above threshold T^{k_λ}. The percentage of critical structure k that is desired to be under threshold T^{k_λ} is $1 - \rho^{k_\lambda}$, or equivalently, ρ^{k_λ} is the percent of critical structure k that is allowed to violate threshold T^{k_λ}. Allowing M to be an upper bound on the amount of radiation deposited in the anatomy, we have that any x satisfying the following constraints also satisfies the physician's dose-volume and tumor uniformity goals,

$$\left.\begin{array}{rcll} TLB \leq & A_T x & \leq & TUB & \\ & A_{C^k} x & \leq & T^{k_\lambda} e + \alpha^{k_\lambda} M, & \text{for each } k_\lambda \\ & e^T \alpha^{k_\lambda} & \leq & \rho^{k_\lambda} |C^k|, & \text{for each } k_\lambda \\ & A_N x & \leq & NUB & \\ & x & \geq & 0 & \\ & \alpha_p^{k_\lambda} & \in & \{0,1\} & p \in C^k. \end{array}\right\} \quad (3.15)$$

The binary dose-volume constraints on the critical structures have replaced the previous linear constraints. In a similar fashion, we can add binary variables β_p, for $p \in T$, to measure the amount of target volume that is under dosed. If γ is the percentage of tumor that is allowed to be under its prescribed lower bound, we change the first set of inequalities in (3.15) to obtain,

$$\left.\begin{array}{rcll} A_T x & \leq & TUB, & \\ A_T x & \geq & TLB - \mathrm{diag}(TLB)\beta, & \\ e^T \beta & \leq & \gamma |T|, & \\ A_T x & \leq & TUB, & \\ A_{C^k} x & \leq & T^{k_\lambda} e + \alpha^{k_\lambda} M, & \text{for each } k_\lambda, \\ e^T \alpha^{k_\lambda} & \leq & \rho^{k_\lambda} |C^k|, & \text{for each } k_\lambda, \\ A_N x & \leq & NUB, & \\ x & \geq & 0, & \\ \alpha_p^{k_\lambda} & \in & \{0,1\}, & p \in C^k, \\ \beta_p & \in & \{0,1\}, & p \in T. \end{array}\right\} \quad (3.16)$$

Of course we could add several threshold levels for the target volume, but the constraints in (3.16) describe how a physician prescribes dose in common commercial systems. Notice that a prescription now takes the form

$$(TLB, TUB, NUB, T^{1_1}, \ldots, T^{1_{\Lambda_1}}, T^{2_1}, \ldots, T^{2_{\Lambda_2}}, \ldots,$$
$$T^{K_1}, \ldots, T^{2_{\Lambda_K}}, \gamma, \rho^{1_1}, \ldots, \rho^{1_{\Lambda_1}}, \rho^{2_1}, \ldots, \rho^{2_{\Lambda_2}}, \ldots, \rho^{K_1}, \ldots, \rho^{K_{\Lambda_K}}).$$

For convenience, we let $\mathcal{P}$ be the collection of

$$u = (x, \alpha^{1_1}, \ldots, \alpha^{1_{\Lambda_1}}, \alpha^{2_1}, \ldots, \alpha^{2_{\Lambda_2}}, \ldots, \alpha^{K_1}, \ldots, \alpha^{K_{\Lambda_K}}, \beta)$$

that satisfy the constraints in (3.16).

From an optimization perspective, the difficulty of the problem has significantly increased from the earlier linear and quadratic models. Common objective functions are those that improve the under and over dosing. A linear objective is

$$\min \left\{ w^1 \cdot e^T \beta + \sum_{k_\lambda} w^{k_\lambda} \cdot e^T \alpha^{k_\lambda} : u \in \mathcal{P} \right\}, \quad (3.17)$$

where the w's weight the importance of the respective under and over dosing. For more information on similar models, we point to Lee [45, 29, 32, 30, 31].

A different modeling approach is to take a biological perspective [43, 53]. The concept behind these models is to use biological probabilities to find desirable treatments. In [53], Raphael presents a stochastic model that maximizes the probability of a successful treatment. The assumption is that tumorous cells are uniformly distributed throughout the target volume and that cells are randomly killed as they are irradiated. Allowing d_p to be the dose delivered at point p, we let $S(d_p)$ be the probability that any particular cell survives in the region represented by p under dose d_p. So, if there are $C(p)$ cancerous cells near dose point p, the expected number of survivors in this region under dose d_p is $\bar{N}(p) = C(p)S(d_p)$. If TV is the set of dose points within the target volume, then the expected number of surviving cancer cells is $\sum_{p \in TV} C(p)S(d_p)$. The actual number of survivors is the sum of many independent Bernoulli trials, whose distribution is assumed to be Poisson. This means that the probability of tumor control — i.e., when the expected number of survivors is zero, is

$$e^{-\sum_{p \in TV} C(p)S(d_p)}.$$

We want to maximize this probability, and the corresponding optimization problem is

$$\begin{aligned} \max \Big\{ & e^{-\sum_{p \in TV} C(p)S(d(p))} : \\ & A_T x = d,\ A_C x \leq CUB,\ A_N x \leq NUB,\ x \geq 0 \Big\}. \end{aligned} \tag{3.18}$$

This model has the favorable quality that it attempts to measure the overriding goal of treatment, that of killing the cancerous cells. However, this model simply introduces an exponential measure that increases the dose to the target volume as much as possible. As such, this model is similar to models (3.1) and (3.3), and it suffers from the same inadequacies.

Morrill develops another biologically based model in [43], where the goal is to maximize the probability of a complication free treatment. The idea behind this model is that different organs react to radiation differently, and that there are probabilistic ways to measure whether or not an organ will remain *complication free* [28, 41, 39, 40, 46]. To represent this model, we divide the rows of A_C as in (3.14). If $f(d^k)$ is the probability of critical structure k remaining complication free, where d^k is a vector of dose values in critical structure k, the optimization model is

$$\begin{aligned} \max \Big\{ & \prod_{k=1}^{K} f(d^k) : TLB \leq A_T x \leq TUB, \\ & A_{C_k} x = d^k,\ k = 1, 2, \ldots, K,\ A_N x \leq NUB \Big\}. \end{aligned} \tag{3.19}$$

As one can see, the scope of designing radiotherapy treatments intersects many areas of optimization. In addition to the models just discussed, several others have been suggested, and in particular, we refer to [14, 15, 33, 43, 56] for further discussions of nonlinear, nonquadratic models. All of the models in this subsection measure an aspect of treatment design, but they each fall short of mimicking the design process faced by a physician. Hence, it is crucial to continue the investigation of new models. While it may be impossible to include all of the patient specific information, the goal is to consistently improve the models so that they become flexible enough to work in a variety of situations.

4.3.3 New Directions

The models presented in Subsection 4.3.2 are concerned with the difficult task of optimally satisfying a physician's goals. In this section, we address some related treatment design questions that are beginning to benefit from optimization. The models in Subsection 4.3.2 have made significant inroads into the design of radiotherapy treatments, and the popular commercial systems use variants of these models in their design process. While this is a success for the field of OR, this is not the end of the story, and there are many, many clinically related problems that can benefit from optimization. This subsection focuses on the design questions that need to be made before a treatment is developed.

Several questions need to be answered before building any of the models in Subsection 4.3.2. These include deciding: 1) the distribution of the dose-points, 2) the number and location of the isocenters, and 3) the number and location of the beams. Each of these decisions is currently made by trial-and-error, and once these decisions are made, the previous models optimize the treatment. However, the fact that a model's representation depends on these decisions means that a treatment's quality depends on a physician's experience. Replacing the trial-and-error process with an optimization technique that provides consistently favorable answers to these questions is an important and relatively untapped area of research.

To formally study how these three questions affect a treatment plan, we let Opt be a function with the following arguments: $\mathbb{B}$ is a collection of isocenters and beams, $\mathbb{D}$ is a vector of dose points, and $\mathbb{P}$ is a prescription. Opt$(\mathbb{B}, \mathbb{D}, \mathbb{P})$ returns the optimal value of the optimization routine, denoted by optval, and an optimal treatment, x. The argument $\mathbb{B}$ has the following form,

$$\mathbb{B} = \begin{pmatrix} ((x_1, y_1, z_2), (\theta_{(1,1)}, \psi_{(1,1)}, \rho_{(1,1)}), \ldots, (\theta_{(1,B_1)}, \psi_{(1,B_1)}, \rho_{(1,B_1)}) \\ ((x_2, y_2, z_2), (\theta_{(2,2)}, \psi_{(2,2)}, \rho_{(2,2)}), \ldots, (\theta_{(2,B_2)}, \psi_{(2,B_2)}, \rho_{(2,B_2)}) \\ \vdots \\ ((x_I, y_I, z_I), (\theta_{(I,I)}, \psi_{(I,I)}, \rho_{(I,I)}), \ldots, (\theta_{(I,B_I)}, \psi_{(I,B_I)}, \rho_{(I,B_I)}) \end{pmatrix},$$

where (x_i, y_i, z_i) is an isocenter and $\{(\theta_{(i,j)}, \psi_{(i,j)}, \rho_{(i,j)}) : j = 1, 2, \ldots, B_i\}$ is the collection of spherical coordinates for the beams around isocenter i. Assuming that there are m dose points in the anatomy (so the dose matrix has m rows), the vector of dose points looks like

$$\mathbb{D} = ((x_1, y_1, z_1), (x_2, y_2, z_2), \ldots, (x_m, y_m, z_m)).$$

The form of the prescription $\mathbb{P}$ depends on the optimization problem. The functional dependence a treatment has on the isocenters, the beam positions, the dose points, and the prescription is represented by $\text{Opt}(\mathbb{B}, \mathbb{D}, \mathbb{P}) = (\text{optval}, x)$. The form of Opt is defined by the optimization model, and most of the research has been directed toward having a useful representation of Opt. However, the optimization models and subsequent treatments depend on $\mathbb{B}$, $\mathbb{D}$, and $\mathbb{P}$, and this dependence is not clearly understood.

The question of deciding how the dose points are distributed has received some attention (see for example [44]) and is often discussed as authors describe their implementation. However, a model that 'optimally' positions dose points within an anatomy has not been considered, and deciding what optimal means is open for debate. Most researchers use either a simple grid or the more complicated techniques of skeletization or octree, but none of these processes are supported by rigorous mathematics.

The question of deciding the number and position of the isocenter(s) is the least investigated of the three problems. Most treatment systems place the isocenter at the center-of-mass of a user defined volume, typically the target volume. This placement is intuitive, but there is no reason to believe that this is the 'best' isocenter. In fact, the clinics with which the authors are familiar have developed techniques for special geometries that place the isocenter at different positions. In addition to the location question, there has been no mathematical work on deciding the number of isocenters. Investigating these questions promises to be fruitful research.

The question of pre-selecting a candidate set of beams has witnessed some work [4, 12, 17, 51, 52, 62, 61]. The breadth of the research exhibits that the problem is complicated enough so that there is no clearly defined manner to address the problem. Indeed, the first author of this tutorial spent several years working on this problem to no avail. Much of the current research is structural, meaning that the set of beams is constructed by adding beams under a decision rule. The other work selects a candidate set by solving a large, mixed-integer problem. For the sake of brevity, we omit a detailed discussion of these techniques and instead investigate a promising new process.

We suggest that rather than constructing a candidate set of beams, we instead begin with an unusually large collection of beams and then prune them to the desired number. The premise of the idea is that a treatment based on many beams indicates which beams are most useful. Ehrgott [12] uses this idea to select a candidate set of beams from a larger collection by solving a large, mixed-integer problem. Our technique is different and is based on the data compression technique of vector quantization.

A *quantizer* is a mapping that has a continuous, random variable as its argument and maps into a discrete set, called the code book. Each quantizer is the composition of an *encoder* and a *decoder.* If u is a random variable with possible values in V, an encoder takes the form $f : V \to \{1, 2, \ldots, n\}$, and a decoder looks like $g : \{1, 2, \ldots, n\} \to V$. The quantizer defined by f and g is $Q(u) = g(f(u))$. The encoder maps the realizations of u into the index set $\{1, 2, \ldots, n\}$ and partitions V into n sets. The decoder completes the quantization by assigning the possible realizations of u to a discrete subset of V, and the elements in this subset are called codewords. As an example, let u be uniformly distributed on $[0, 1]$. The process of rounding is a quantizer, and in this case we have that

$$f : [0,1] \to \{1,2\} : u \mapsto \begin{cases} 1, & 0 \leq u < 0.5 \\ 2, & 0.5 \leq u \leq 1 \end{cases}$$

$$g : \{1,2\} \to [0,1] : u \mapsto \begin{cases} 0, & u = 1 \\ 1, & u = 2. \end{cases}$$

In this example, the interval $[0, 1]$ is partitioned by $\{[0, 0.5), [0.5, 1]\}$, with the first interval mapping to the codeword 0 and the second interval mapping to the codeword 1.

In the previous example, the interval $[0, 1]$ is quantized to the discrete set $\{0, 1\}$. The quantization error for any realization of u is $d(u, Q(u))$, where d is a metric on V (the most common error is $\|u - Q(u)\|_2$). The quantizer's distortion is the average error,

$$D_Q = E\, d(u, Q(u)) = \int_V P(u) \cdot d(u, Q(u))du,$$

where $E\, d(u, Q(u))$ is the expected value of d and $P(u)$ is the probability distribution of u. A quantizer is *uniform* if the partitioning sets have the same measure and is *regular* if it satisfies the nearest neighbor condition — i.e.,

$$d(u, Q(u)) = \min\{d(u, y) : y \text{ is a codeword}\}.$$

The quantizer design problem is to build a quantizer that minimizes the distortion, and the following necessary conditions [18] guide the design process:

- For any codebook, the partition must satisfy the nearest neighbor condition, or equivalently, the quantizer must be regular.
- For any Partition, the codevectors must be the centers of mass of the probability density function.

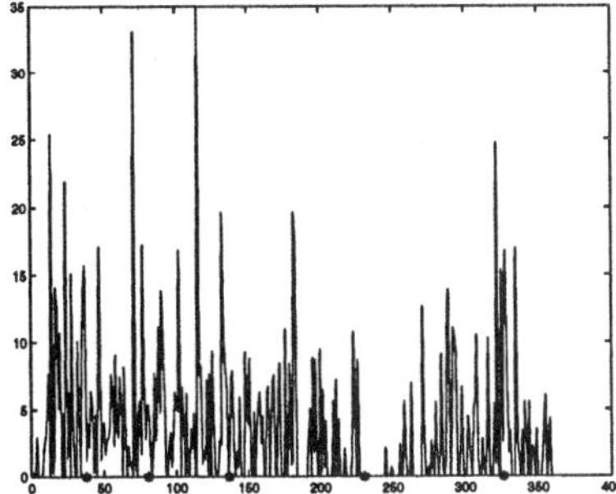

Figure 4.13. The dose profile of a treatment with one isocenter and 360 beams.

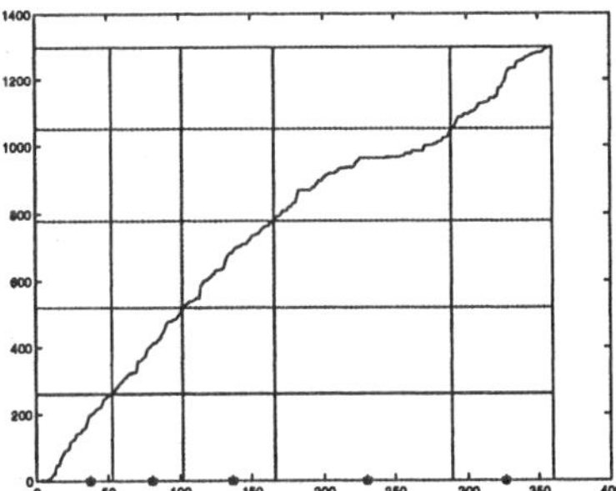

Figure 4.14. The cumulative dose distribution.

Since these are only necessary conditions, a quantizer satisfying these conditions may not minimize distortion. However, there are cases where these are necessary and sufficient, such as when the logarithm of the probability density function is convex [18].

We address the 2-dimensional beam selection problem by designing a quantizer from $[0, 2\pi)$ into $[0, 2\pi)$. The probability density function is patient specific and is calculated by approximating the continuous planning problem. For each isocenter, assume that there is a large number of beams, something on the order of one every degree. Solve $\text{Opt}(\mathbb{B}, \mathbb{D}, \mathbb{P}) = (\text{optval}, x)$, and from the treatment x calculate the amount of radiation delivered along each beam — i.e., aggregate each beam's sub-beams to attain the total radiation for the beam. As an example, the dose profile for the problem in Figure 4.11 is in Figure 4.13, where Opt is defined by model (3.7), there is a single isocenter in the middle of the 64×64 image, there are 360 beams, and dose points are centered within each pixel. The normalized dose profile is the probability density function, and the idea is that this function estimates the likelihood of using an angle. This assumption is reasonable because beams that deliver large amounts of radiation to the tumor are often the ones that intersect the tumor but not the critical structures.

Figure 4.14 is the cumulative dose distribution, and this function is used to define the encoder. Let h be the function that calculates the cumulative dose distribution from a treatment plan x, and let the range of h be the interval $[0, \gamma]$. So, $h_{(\text{Opt}(\mathbb{B},\mathbb{D},\mathbb{P}))}(u)$ is a bijective mapping from $[0, 2\pi)$ onto $[0, \gamma)$ that depends on the isocenters, the beams, the dose points, the prescription, and the optimization model. If the physician desires an N beam plan, the encoder is

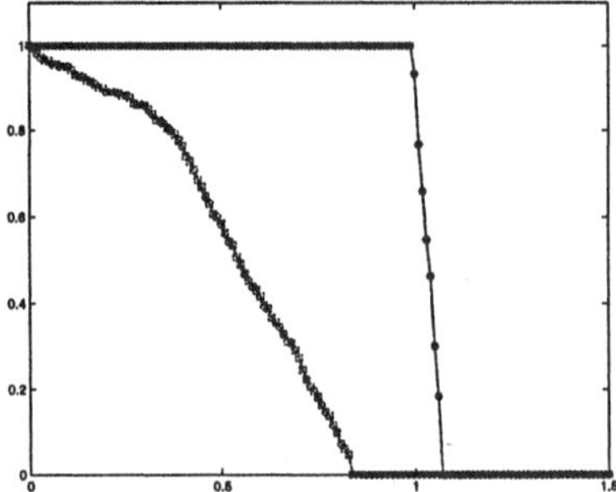

Figure 4.15. The dose-volume histogram for a treatment pruned from 360 to 5 angles for the problem in Figure 4.11.

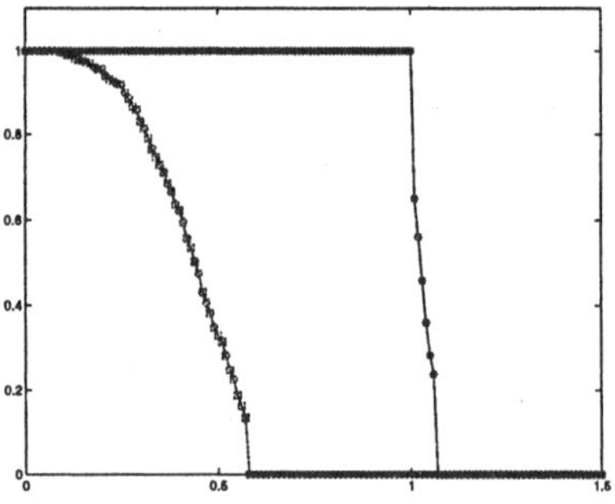

Figure 4.16. The dose-volume histogram for a treatment pruned from 360 to 9 angles for the problem in Figure 4.11.

defined by

$$f : [0, 2\pi) \to \{1, 2, \ldots, N\} :$$
$$h_{(\text{Opt}(\mathbb{B},\mathbb{D},\mathbb{P}))}(u) \to i, \; u \in [(i-1)\gamma/N, i\gamma/N).$$

The encoder partitions the interval $[0, 2\pi)$ into code regions, and because h is monotonic, we are guaranteed that the quantizer is regular. For the example in Figure 4.14, the partition of $[0, 2\pi)$ for a 5 beam treatment is depicted along the horizontal axis (which is in degrees instead of radians).

The decoder assigns a codeword to each partition, and from the necessary conditions of optimality we have that the codewords of an optimal quantizer must be the centers of mass of the normalized dose profile. These codewords are highlighted on the horizontal axes of Figures 4.13 and 4.14 and are beams 32, 78, 140, 232, and 327. The quantizer for this example has the final form,

$$[0, 52) \mapsto 32, \quad [52, 101) \mapsto 78,$$
$$[101, 164) \mapsto 140, \quad [164, 290) \mapsto 232, \quad [290, 360) \mapsto 327.$$

Dose-volume histograms of two pruned plans are in Figures 4.15 and 4.16. These images indicate that the critical structures fare better as more angles are used (compare to Figure 4.12).

Our initial investigations into selecting beams with vector quantization are promising, but there are many questions. The partition depends on the cumulative dose distribution, and this function depends on where accumulation begins. We currently start accumulating dose at angle 0, but this choice is arbitrary and not substantiated. A more serious challenge is to define a clinically relevant error measure that allows us to analyze distortion. We point out that any meaningful error measure relies on Opt and moreover that the dose matrices are different for the quantized and unquantized beams. Lastly, for this technique to have clinical meaning, we need to extend it to 3-dimensions. Any advancement in these areas will lead to immediate improvement in patient care.

We conclude this section by mentioning that the quality of a treatment not only depends on the model, which we have discussed in detail, but also on the algorithm, which we have ignored. As an example, many of the nonlinear models, including the least squares problems, are often solved by simulated annealing. Because this is a stochastic algorithm, it is possible to design different treatments with the same model. An interesting numerical paper for the nonlinear models would be to solve the same model with several algorithms to find if some of them naturally design better treatments. The linear models are likely to have multiple optimal solutions, and in this case, the solutions from a simplex algorithm and an interior-point algorithm are different. Both solutions have favorable and unfavorable characteristics. The basic solutions have the favorable property that the number of sub-beams is restricted by the number of constraints, and if constraints are aggregated, we can control the number of sub-beams [36]. However, the simplex solutions have the unfavorable quality that they guarantee that some of the prescribed bounds are attained [22]. The interior-point solutions have the reverse qualities, as they favorably ensure the prescribed bounds are strictly satisfied and they unfavorably use as many sub-beams as possible [23–25]. The fact that interior-point algorithms inherently produce treatments with many beams makes them well suited to tomotherapy.

4.4 The Gamma Knife

The Gamma Knife treatment system was specifically designed to treat brain, or intracranial, lesions. The first Gamma Knife was built in 1968 and is *radiosurgical* in its intent (the term radiosurgery was first used in [35]). The difference between radiosurgical and radiotherapy approaches has been previously described. The high dose delivered during a radiosurgery makes accuracy in both treatment planning and delivery crucial to a treatment's success.

The Gamma Knife uses 201 radioactive cobalt-60 sources to generate its treatment pencil beams instead of a linear accelerator. These sources are spherically distributed around the patient, and their width is controlled by a series of collimators. These collimators are different than those used in IMRT, with the Gamma Knife collimators being located in a helmet that fits the patient's head. Each helmet consists of 201 cylindrical holes of either 4, 8, 14, or 18mm. The 201 radiation beams thus produced intersect at a common focal point and form a spherically shaped high dose region whose diameter is roughly equal to the collimator size. These spheres are called *shots* and have the favorable property that radiation dose outside these regions falls off very quickly (i.e., a high dose gradient). It is this fact that makes the Gamma Knife well suited to deliver

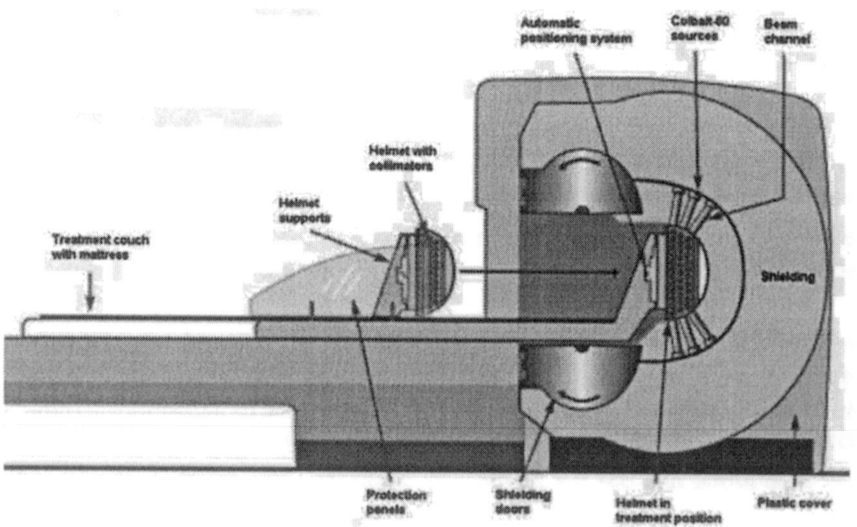

Figure 4.17. A Gamma Knife treatment machine.

radiosurgeries, in that very high doses of radiation may be delivered to a target which is immediately adjacent to a critical structure, with relatively little dose delivered to the structure.

4.4.1 Clinically Relevant Gamma Knife Treatments

The primary clinical restriction on Gamma Knife treatments is that the number of shots must be controlled. Between each shot the patient is removed from the treatment area, re-aligned, possibly re-fitted with a different collimator helmet, and returned to the treatment area. This is a time consuming process, and most treatment facilities attempt to treat a patient in under 10 to 15 shots. We mention that it is possible to 'plug' some of the 201 collimator holes, which can produce an ellipsoidally shaped distribution of dose. While this is clinically possible, this is rarely undertaken because of time restrictions and the possibility of errors related to the manual process. In this tutorial we do not consider plugged collimators, and we therefore assume that the dose is delivered in spherical packets.

4.4.2 Optimization Models

From a modeling perspective, the Gamma Knife's sub-beams are different than the sub-beams of IMRT. The difference is that in IMRT the amount of radiation delivered along each sub-beam is controlled by a multileaf collimator, but in the Gamma Knife each sub-beam delivers the same amount of radiation. So, Gamma Knife treatments do not depend on the same decision variables as IMRT, and consequently, the structure of the dose matrix for the Gamma Knife is different. The basic dose model discussed in Section 4.2 is still appropriate, but we need to alter the indices of $a_{(p,a,s,i,e)}$, which recall is the rate at which radiation accumulates at dose point p from sub-beam (a, s) when the gantry is focused on the ith isocenter and energy e is used. The Gamma Knife delivers dose in spherical packets called *shots*, which are defined by their centers and radii. A shot's center is the point at which the sources are focused and is the same as the isocenter. The radius of a shot is controlled by the collimators that are placed on each source. As mentioned in the previous subsection, the same collimator size is used for every source per shot, and hence, a shot is defined by its isocenter i and its collimator c. Moreover, unlike the linear accelerators used in IMRT, the cobalt sources of the Gamma Knife produce a single energy, and hence, there is no functional dependence on the energy e. We alter the indices of $a_{(p,a,s,i,e)}$ by letting $a_{(p,c,i)}$ be the rate at which radiation accumulates at dose point p from shot (c, i). These values form the dose matrix A, where the rows are indexed by p and the columns by (c, i).

Since the Gamma Knife delivers dose in spherical shots, the geometry of treatment design is different than that of IMRT. The basic premise of irradiating cancerous tissue without harming surrounding structures remains, but instead of placing beams of radiation so that they avoid critical areas, we rather attempt to cover (or fill) the target volume with spheres. Since the cumulative dose is additive, regions where shots overlap are significantly over irradiated. These hot spots do not necessarily degrade the treatment because of its radiosurgical intention. However, it is generally believed that the best treatments are those that sufficiently irradiate the target and at the same time reduce the number and size of hot spots. This means that favorable Gamma Knife treatments fill the target with spheres of radiation so that 1) shots intersections are small and 2) shots do not intersect non-target tissue. Outside the fact that designing Gamma Knife treatments is clinically important, the problem is mathematically interesting because of its relationship to the sphere packing problem. While there is a wealth of mathematical literature on sphere packing, this connection has not been exploited, and this promises to be a fruitful research direction.

The problem of designing Gamma Knife treatments received significant exposure when it was one of the modeling problems for the 2003 COMAP competition in mathematical modeling [10], and there are many optimization models that aid in the design of treatments [15, 34, 38, 57–59, 65, 66]. We focus on the recent models by Ferris, Lim, and Shepard [14] (winner of the 2002 Pierskalla award) and Cheek, Holder, Fuss, and Salter [6]. Both of these models use dose-volume constraints and segment the anatomy into target and non-target, making A_N vacuous. The model proposed in [14] is

$$\begin{aligned}\min\{e^T u_T : d_T = A_T x, d_C = A_C x, \theta \leq u_T + d_T,\\ 0 \leq x \leq sM, \rho(e^T d_T + e^T d_C) \leq e^T d_T, e^T s \leq n,\\ s_i \in \{0,1\}, 0 \leq u_T, 0 \leq u_C\}.\end{aligned} \tag{4.1}$$

The dose to the target and non-target tissues is contained in the vectors d_T and d_C, and u_T measures how much the target volume is under the goal dose θ. The objective is to minimize the total amount the target volume is under irradiated. The binary variables s_i indicate whether or not a shot is used or not, and the constraint $e^T s \leq n$ limits the treatment to n shots (M is an arbitrarily large number). The parameter ρ is a measure of desired conformality, and the constraint $\rho(e^T d_T + e^T d_C) \leq e^T d_T$ ensures that the target dose is at least ρ of the total dose. If ρ is 1, then we are attempting to design a treatment in which the entire dose is within the target.

Model (4.1) is a binary, linear optimization problem. The authors of [14] recognize that the size of the problem makes it impossible for modern optimization routines to solve the problem to optimality (small Gamma Knife treatments often require more than 500 Gigabytes of data storage). The authors of [14] replace the binary variables with a $\tan^{-1}$ constraint that transforms the problem into a continuous, nonlinear program. Specifically, they replace the constraints

$$\left.\begin{array}{r} 0 \leq x \leq sM, \\ e^T s \leq n, \\ s \in \{0,1\} \end{array}\right\} \quad \text{with} \quad \left\{ \sum_{(c,i)} \tan^{-1}(\alpha x_{(c,i)}) \leq n, \right.$$

where larger α values more accurately resemble the binary constraints. Together with other reductions and assumptions, this permits the authors to use CONOPT [11] to design clinically acceptable treatments.

Model (4.1) is similar to the IMRT models that use dose-volume constraints because it's objective function measures dose and ρ describes the volume of non-target tissue that we are allowed to irradiate. While this model successfully designed clinically relevant treatments, physicians often judge Gamma Knife treatments with a conformality index. These indices are scoring functions that quantify a treatment's quality by measuring how closely the irradiated tissue resembles the target volume. So, in addition to the dose-volume histograms and the 2-dimensional isodose lines, Gamma Knife treatments are often judged by a single number. Collapsing large amounts of information into a single score is not always appropriate, but the radiosurgical intent of a Gamma Knife treatment lends itself well to such a measure—i.e., the primary goal is to destroy the target with an extremely high level of radiation and essentially deliver no radiation to the remaining anatomy. This means that conforming the high-dose region to the target is crucial, and hence, judging treatments on their conformity is appropriate.

Several conformality indices are suggested in the literature (see [37] for a review). Let D be the suggested target dose (meaning the physician desires $A_T x \geq De$) and define

$$\begin{aligned} TV &= \{p: \text{ dose point } p \text{ is in the target volume }\} \text{ and} \\ IV_T &= \{p: \text{ the dose at point } p \text{ is at least } T \cdot D\}, \end{aligned}$$

where T is between 0 and 1. If we assume that each dose point represents a volume V, the target volume is $V \cdot |TV|$ and the Tth isodose line encloses a volume of $V \cdot |IV_T|$. The standard indices are expressed in terms of the $\%100$ isodose line and are

$$\begin{aligned} PITV &= |IV_1|/|TV|, \\ CI &= |TV \cap IV_1|/|IV_1|, \text{ and} \\ IPCI &= (|TV \cap IV_1|/|TV|) \cdot (|TV \cap IV_1|/|IV_1|). \end{aligned}$$

The last index is called Ian Paddick's conformality index [47] and is the product of the *over treatment ratio* and the *under treatment ratio*. These are defined for any isodose value by

$$OTR_T = |TV \cap IV_T|/|IV_T| \quad \text{and} \quad UTR_T = |TV \cap IV_T|/|TV|.$$

The over treatment ratio is at most 1 if the target volume contains the Tth isodose volume. Otherwise, OTR_T is between 0 and 1, and $1 - OTR_T$ is the volume of non-target tissue receiving a dose of at least $T \cdot D$. Similarly, the under treatment ratio is 1 if the target volume is contained in the Tth isodose line, and $1 - UTR_T$ is the percentage of target volume receiving less than $T \cdot D$Gy. The over and under treatment ratios are 1 only if the Tth isodose volume matches the target volume, and the conformality objective is to design plans that have $OTR_T = UTR_T = 1$. For any T, the Ian Paddick conformality index is $IPCI_T = UTR_T \cdot OTR_T$.

The authors of [6] suggest a model whose objective function is based on Ian Paddick's conformality index. Assume that there are I isodose lines and that Θ_i is the ith column of the matrix Θ. The model in [6] is

$$\begin{aligned} \min \Bigg\{ & \sum_{i=1}^{I} w_i(1 - e_{TV}^T \Theta_i / e^T \Theta_i) + u_i(1 - (V/K) e_{TV}^T \Theta_i) : \\ & Ax = d, \ \text{diag}(d) ee^T \leq ee^T HD + M\Theta, \ 0 \leq x \leq M\beta, \ e^T \beta \leq L, \\ & \beta_i \in \{0,1\}, \ \Theta_{(p,i)} \in \{0,1\} \Bigg\}. \end{aligned} \tag{4.2}$$

The parameters V and K are the voxel and target volumes, and e_{TV} is the binary vector with ones where an index corresponds to a targeted dose point.

The number of shots is measured by the binary vector β and is restricted by L (M is an arbitrarily large value). The vector d is the delivered dose, and diag(d) is the diagonal matrix formed by d. The diagonal matrix H contains the isodose values that we are using, and $ee^T HD$ is a matrix with each column being $T_i De$. The matrix constraint diag$(d)ee^T \leq ee^T H + M\Theta$ guarantees that if the dose at point p is above the isodose value $T_i D$, then $\Theta_{(p,i)}$ is 1. From this we see that $OTR_{T_i} = e^T_{TV}\Theta_i / e^T\Theta_i$ and that $UTR_{T_i} = (V/K)e^T_{TV}\Theta_i$. The weights w_i and u_i express the importance of having the T_i^{th} isodose line conform to the target volume. We point out that the objective function of Model (4.2) is not *IPCI* but is rather a weighted sum of the over and under treatment ratios.

Neither model (4.1) or (4.2) penalizes over irradiating portions of the target, and controlling hot spots complicates the problem. This follows because measuring hot spots is often accomplished by adding a variable for each dose point that increases as the delivered dose grows beyond an acceptable amount. The problem is not with the fact that there are an increased number of variables, but rather that physicians are not concerned with high doses over small regions. A more appropriate technique is to partition the dose points into subsets, say H_r, and then aggregate dose over these regions to control hot spots. If a hot spot is defined by the average dose of a region exceeding τ, then adding the constraints,

$$\sum_{p \in H_r} d_p \leq |H_r|(\tau + q) \quad \text{and} \quad q \geq 0$$

to model (4.1) or (4.2) enables us to calculate the largest hot spot. Such a tactic was used in [6] for model (4.2), where each H_r contained 4 dose points in a contiguous, rectangular pattern. The objective function was altered to

$$\sum_{i=1}^{I} \left(w_i(1 - e^T_{TV}\Theta_i / e^T\Theta_i) + u_i(1 - (V/K)e^T_{TV}\Theta_i) \right) + 0.5q.$$

Model (4.2) is easily transformed into a binary, quadratic problem, but again, it's size makes standard optimization routines impractical. As an alternative, fast simulated annealing is used in [6], where the research goal was to explore how treatment quality depends on the number of shots — i.e., how the standard indices depend on L. Treatments designed with this model are shown in Figure 4.18, and the *CI* and *IPCI* values for different choices of L are in Table 4.1. Figure 4.19 shows how the dose-volume histograms improve as more shots are allowed.

4.5 Treatment Delivery

The previous sections focused on treatment design, and while these problems are interesting and important, the optimization community is now poised

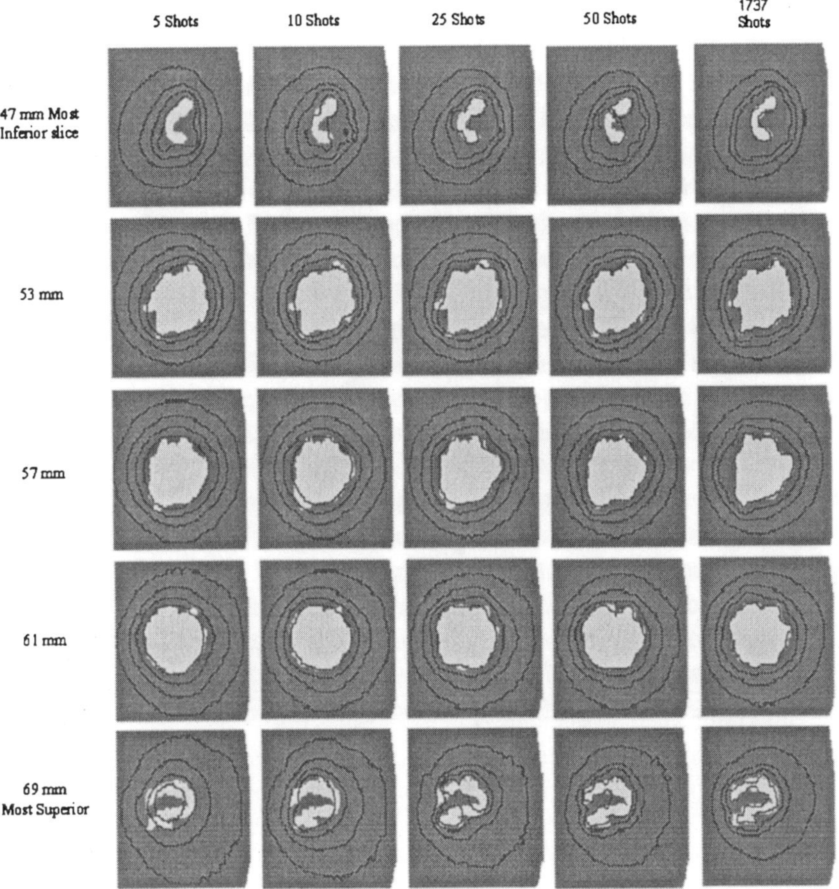

Figure 4.18. Isodose curves from treatments designed with Model (4.2). The value of L is listed across the top of each treatment, and the millimeter value on the left indicates the depth of the image.

Table 4.1. How the *PITV*, *CI* and *IPCI* indices react as the number of possible shots increases.

	5 Shots	10 Shots	25 Shots	50 Shots	Unlimited	Ideal
PITV	0.934	0.996	0.992	0.990	0.999	1
CI	0.846	0.897	0.925	0.954	0.997	1
IPCI	0.767	0.808	0.863	0.919	0.995	1

to significantly improve patient care with respect to the design process. So, even though it is important to continue the study of treatment design, there

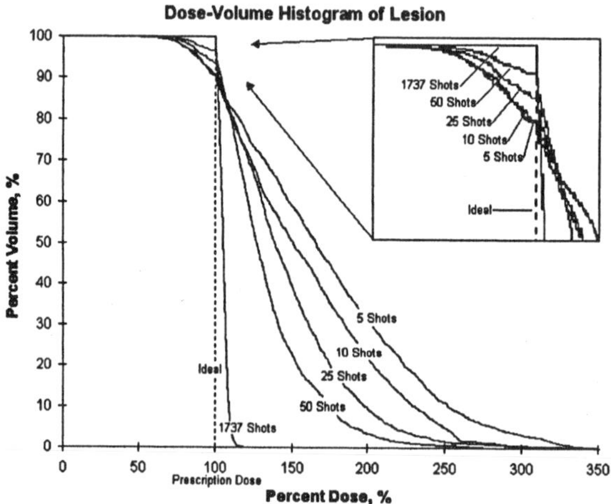

Figure 4.19. The dose-volume histograms for treatment plans with differing numbers of shots.

are related clinical questions where beginning researchers can make substantial contributions. In this section we focus on two treatment delivery questions that are beginning to receive attention.

As mentioned earlier, the difference between radiotherapy and radiosurgery is that radiotherapy is delivered in fractional units over several days. Current practice is to divide the total dose into N equal parts and deliver the overall treatment in uniform, daily treatments. The value of N is based on studies that indicate how healthy tissue regenerates after being irradiated, and the overall treatment is fractionated to make sure that healthy tissue survives. Dividing the total dose was particularly important when technology was not capable of conforming the high-dose region to the target, as this meant that surrounding tissues were being irradiated along with the tumor. However, modern technology permits us to magnify the difference between the dose delivered to the target and the dose delivered to surrounding tissues. The support for a uniform division does not make sense with our improved technology, and Ferris and Voelker [16] have investigated different approaches.

Suppose we want to divide a treatment into N smaller treatments. If d^k is the cumulative dose after k treatments, the problem is to decide how much dose to deliver in subsequent periods. This leads to a discrete-time dynamic system, and if we let u^k be the dose added in period k and w^k be the random error in delivering u^k, then the system is

$$d_{k+1} = d_k + u_k(1 + w_k).$$

The random error is real because the planned dose often deviates from the delivered dose since patient alignment varies from day-to-day. The decision variables u_k must be nonnegative since it is impossible to remove dose after it is delivered. The optimization model used in [16] is

$$\min\{E(\|w^T(d_N - D)\|_1) : \\ d_{k+1} = d_k + u_k(1 + w_k), u_k \geq 0, w_k \in W\}, \tag{5.1}$$

where D is the total dose to deliver, W is the range of the random variable w, and E is the expected value. This model can be approached from many perspectives, and the authors of [16] consider stochastic linear programming, dynamic programming, and neuro-dynamic programming. They suggest that a neuro-dynamic approach is appropriate and experiment with a 1-dimensional problem. Even at this low dimensionality the problem is challenging. They conclude that undertaking such calculations to guide clinical practice is not realistic, but they do use their 1-dimensional model to suggest 'rules-of-thumb.'

Model (5.1) requires a fixed number of divisions, and hence, this problem only address the uniformity of current delivery practices. An interesting question that is not addressed is to decide the number of treatments. If we can solve this problem independent of deciding how the dose is to be delivered, then we can calculate N before solving model (5.1). However, we suggest that it is best to simultaneously make these decisions.

Another delivery question that is currently receiving attention is that of leaf sequencing [3, 13, 26, 27, 49, 50]. This is an important problem, as complicated treatments are possible if we can more efficiently deliver dose. An average treatment lasts from 15 to 30 minutes, and if the leaves of the collimator are adjusted so that the desired dose profile is achieved quickly, then more beams are possible. This translates directly to better patient care because treatment quality improves as the number of beams increases (the same is true for the Gamma Knife as demonstrated in Section 4.4.2). We review the model in [3], which is representative, and encourage interested readers to see the other works and their bibliographies.

Suppose we have solved (in 3-dimensions) Opt($\mathbb{B}, \mathbb{D}, \mathbb{P}$) and that an optimal treatment shows that a patient should be irradiated with the following exposure pattern,

$$\mathcal{I} = \begin{bmatrix} 0 & 0 & 2 & 2 & 2 & 0 \\ 0 & 1 & 1 & 3 & 1 & 0 \\ 0 & 0 & 2 & 2 & 1 & 0 \\ 1 & 2 & 2 & 2 & 1 & 0 \\ 0 & 1 & 2 & 3 & 2 & 1 \\ 0 & 1 & 2 & 2 & 2 & 2 \end{bmatrix}. \tag{5.2}$$

The dose profile $\mathcal{I}$ contains our desired exposure times. Each element of $\mathcal{I}$ represents a rectangular region of the Beam's Eye View — i.e., the view of the patient as one was looks through the gantry. The collimator for this example has 12 leaves (modern collimators have many more), one on the right and left of each row. These leaves can move across the row to shield the patient.

The optimization model in [3] minimizes exposure time by controlling the leaf positions. The treatment process is assumed to follow the pattern: the leaves are positioned, the patient is exposed, the leaves are re-positioned, the patient is exposed, etc..., with the process terminating when the dose profile is attained. For each row i and time step t we let

$$l_{ijt} = \begin{cases} 1, & \text{if the left leaf in row } i \text{ is positioned in column } j \text{ at } t; \\ 0, & \text{otherwise.} \end{cases}$$

$$r_{ijt} = \begin{cases} 1, & \text{if the right leaf in row } i \text{ is positioned in column } j \text{ at } t; \\ 0, & \text{otherwise.} \end{cases}$$

The nonlinear, binary model studied is

$$\min \Bigg\{ \sum_t \alpha_t : \sum_j l_{ijt} = 1, \forall i, t;\ \sum_j r_{ijt} = 1, \forall i, t,$$
$$y_{ijt} = \sum_{k=0}^{j-1} l_{ikt} - \sum_{k=1}^{j} r_{ikt}, \forall t;\ \sum_{k=0}^{j} l_{ikt} \geq \sum_{k=1}^{j} r_{ikt}, \forall t,$$
$$\sum_t \alpha_t y_{ijt} = \mathcal{I}_{ij}, \forall i, j;\ l_{ijt}, r_{ijt}, y_{ijt} \in \{0, 1\}, \forall i, j, t;$$
$$\alpha_t \geq 0, \forall t \Bigg\}. \tag{5.3}$$

Model (5.3) is interpreted as finding a *shape matrix* at each time t. A shape matrix is a binary matrix such that the 1s in every row are contiguous (a row may be void of 1s). Each 1 indicates an unblocked region of the beam, and each shape matrix represents a positioning of the leaves. The y variables in model (5.3) form an optimal collection of shape matrices. For example,

$$\begin{aligned} \mathcal{I} = \begin{bmatrix} 2 & 3 \\ 4 & 2 \end{bmatrix} &= 2 \begin{bmatrix} 1 & 0 \\ 0 & 1 \end{bmatrix} + 4 \begin{bmatrix} 0 & 0 \\ 1 & 0 \end{bmatrix} + 3 \begin{bmatrix} 0 & 1 \\ 0 & 0 \end{bmatrix} \\ &= 2 \begin{bmatrix} 1 & 1 \\ 1 & 1 \end{bmatrix} + 1 \begin{bmatrix} 0 & 1 \\ 1 & 0 \end{bmatrix} + 1 \begin{bmatrix} 0 & 0 \\ 1 & 0 \end{bmatrix}. \end{aligned}$$

The first shape matrix in the first decomposition has $y_{111} = y_{221} = 1$ and $y_{121} = y_{211} = 0$. The total exposure time for the first decomposition is $2+4+3 = 9$ and for the second decomposition the exposure time is $2+1+1 = 4$. So, the second leaf sequence is preferred.

The authors of [3] show that model (5.3) can be re-stated as a network flow problem, and they further develop a polynomial time algorithm to solve the problem. This provides the following theorem (see [13] for related results).

THEOREM 4.3 (BOLAND, HAMACHER, AND LENZEN [3]) *Model (5.3) is solvable in polynomial time.*

We close this section by suggesting a delivery problem that is not addressed in the literature. As Figure 4.19 shows, Gamma Knife treatments improve as the number of shots increases. We anticipate that new technology will permit automated patient movement, which will allow the delivery of treatments with numerous shots. How to move a patient so that shots are delivered as efficiently as possible is related to the traveling salesperson problem, and investigations into this relationship are promising. In the distant future, we anticipate that patients will movement continuously within the treatment machine. This means shots will move continuously through the patient, and finding an optimal path is a control theory problem.

4.6 Conclusion

The goal of this tutorial was to familiarize interested researchers with the exciting work in radiation oncology, and the authors hope that readers have found inspiration and direction from this tutorial. We welcome inquiry and will be happy to answer questions. We conclude with a call to the OR community to vigorously investigate how optimization can aid medical procedures. The management side of health care has long benefited from optimization techniques, but the clinical counterpart has enjoyed much less attention. The focus of this work has been radiation oncology, but there are many procedures where standard optimization routines and sound modeling can make substantial improvements in patient care. This research is mathematically aesthetic, challenging, and intrinsically worthwhile because it aids mankind.

Acknowledgment

The authors thank Roberto Hasfura for his careful editing and support.

References

[1] G. K. Bahr, J. G. Kereiakes, H. Horwitz, R. Finney, J. Galvin, and K. Goode. The method of linear programming applied to radiation treatment planning. *Radiology*, 91:686–693, 1968.

[2] F. Bartolozzi et al. Operational research techniques in medical treatment and diagnosis. a review. *European Journal of Operations Research*, 121(3):435–466, 2000.

[3] N. Boland, H. Hamacher, and F. Lenzen. Minimizing beam-on time in cancer radiation treatment using multileaf collimators. Technical Report Report Wirtschaftsmathematik, University Kaiserslautern, Mathematics, 2002.

[4] T. Bortfeld and W. Schlegel. Optimization of beam orientations in radiation therapy: Some theoretical considerations. *Physics in Medicine and Biology*, 38:291–304, 1993.

[5] Y. Censor, M. Altschuler, and W. Powlis. A computational solution of the inverse problem in radiation-therapy treatment planning. *Applied Mathematics and Computation*, 25:57–87, 188.

[6] D. Cheek, A. Holder, M. Fuss, and B. Salter. The relationship between the number of shots and the quality of gamma knife radiosurgeries. Technical Report 84, Trinity University Mathematics, San Antonio, TX, 2004.

[7] J. Chinneck and H. Greenberg. Intelligent mathematical programming software: Past, present, and future. *Canadian Operations Research Society Bulletin*, 33(2):14–28, 1999.

[8] J. Chinneck. An effective polynomial-time heuristic for the minimum-cardinality iis set-covering problem. *Annals of Mathematics and Artificial Intelligence*, 17:127–144, 1995.

[9] J. Chinneck. Finding a useful subset of constraints for analysis in an infeasible linear program. *INFORMS Journal on Computing*, 9(2):164–174, 1997.

[10] Consortium for Mathematics and Its Applications (COMAP), `www.comap.com`. *Gamma Knife Treatment Planning, Problem B.*

[11] A. Drud. CONOPT: A GRG code for large sparse dynamic nonlinear optimization problems. *Mathematical Programming*, 31:153–191, 1985.

[12] M. Ehrgott and J. Johnston. Optimisation of beam direction in intensity modulated radiation therapy planning. *OR Spectrum*, 25(2):251–264, 2003.

[13] K. Engel. A new algorithm for optimal multileaf collimator field segmentation. Technical report, Operations Research & Radiation Oncology Web Site, w.trinity.edu/aholder/HealthApp/oncology/, 2003.

[14] M. Ferris, J. Lim, and D. Shepard. An optimization approach for the radiosurgery treatment planning. *SIAM Journal on Optimization*, 13(3):921–937, 2003.

[15] M. Ferris, J. Lim, and D. Shepard. Radiosurgery optimization via nonlinear programming. *Annals of Operations Research*, 119:247–260, 2003.

[16] M. Ferris and M. Voelker. Neuro-dynamic programming for radiation treatment planning. Technical Report Numerical Analysis Group Research Report NA-02/06, Oxford University Computing Laboratory, 2002.

[17] S. Gaede, E. Wong, and H. Rasmussen. An algorithm for systematic selection of beam directions for imrt. *Medical Physics*, 31(2):376–388, 2004.

[18] A. Gersho and M. Gray. *Vector Quantization and Signal Processing*. Kluwer Academic Publishers, Boston, MA, 1992.

[19] M. Goitein and A Niemierko. Biologically based models for scoring treatment plans. Scandanavian Symposium on Future Directions of Computer-Aided Radiotherapy, 1988.

[20] H. Greenberg. *A Computer-Assisted Analysis System for Mathematical Programming Models and Solutions: A User's Guide for ANALYZE.* Kluwer Academic Publishers, Boston, MA, 1993.

[21] H. Greenberg. Consistency, redundancy and implied equalities in linear systems. *Annals of Mathematics and Artificial Intelligence*, 17:37–83, 1996.

[22] L. Hodes. Semiautomatic optimization of external beam radiation treatment planning. *Radiology*, 110:191–196, 1974.

[23] A. Holder. Partitioning Multiple Objective Solutions with Applications in Radiotherapy Design. Technical Report 54, Trinity University Mathematics, 2001.

[24] A. Holder. Radiotherapy treatment design and linear programming. Technical Report 70, Trinity University Mathematics, San Antonio, TX, 2002. to appear in the Handbook of Operations Research/Management Science Applications in Health Care.

[25] A. Holder. Designing radiotherapy plans with elastic constraints and interior point methods. *Health Care and Management Science*, 6(1):5–16, 2003.

[26] T. Kalinowski. An algorithm for optimal collimator field segmentation with interleaf collision constraint 2. Technical report, Operations Research & Radiation Oncology Web Site, w.trinity.edu/aholder/HealthApp/oncology/, 2003.

[27] T. Kalinowski. An algorithm for optimal multileaf collimator field segmentation with interleaf collision constraint. Technical report, Operations Research & Radiation Oncology Web Site, w.trinity.edu/aholder/HealthApp/oncology/, 2003.

[28] G. Kutcher and C. Burman. Calculation of complication probability factors for non-uniform normal tissue irradiation. *International Journal of Radiation Oncology Biology and Physics*, 16:1623–30, 1989.

[29] M. Langer, R. Brown, M. Urie, J. Leong, M. Stracher, and J. Shapiro. Large scale optimization of beam weights under dose-volume restrictions. *International Journal of Radiation Oncology, Biology, Physics*, 18:887–893, 1990.

[30] E. Lee, T. Fox, and I. Crocker. Optimization of radiosurgery treatment planning via mixed integer programming. *Medical Physics*, 27(5):995–1004, 2000.

[31] E. Lee, T. Fox, and I. Crocker. Optimization of radiosurgery treatment planning via mixed integer programming. *Medical Physics*, 27(5):995–1004, 2000.

[32] E. Lee, T Fox, and I Crocker. *Integer Programming Applied to Intensity-Modulated Radiation Treatment Planning*. To appear in Annals of Operations Research, Optimization in Medicine.

[33] J. Legras, B. Legras, and J. Lambert. Software for linear and non-linear optimization in external radiotherapy. *Computer Programs in Biomedicine*, 15:233–242, 1982.

[34] G. Leichtman, A. Aita, and H. Goldman. Automated gamma knife dose planning using polygon clipping and adaptive simulated annealing. *Medical Physics*, 27(1):154–162, 2000.

[35] L. Leksell. The stereotactic method and radiosurgery of the brain. *Acta Chirurgica Scandinavica*, 102:316 – 319, 1951.

[36] W. Lodwick, S. McCourt, F. Newman, and S. Humphries. Optimization methods for radiation therapy plans. In C. Borgers and F. Natterer, editors, *IMA Series in Applied Mathematics - Computational, Radiology and Imaging: Therapy and Diagnosis*. Springer-Verlag, 1998.

[37] N. Lomax and S. Scheib. Quantifying the degree of conformality in radiosurgery treatment planning. *International Journal of Radiation Oncology, Biology, and Physics*, 55(5):1409–1419, 2003.

[38] L. Luo, H. Shu, W. Yu, Y. Yan, X. Bao, and Y. Fu. Optimizing computerized treatment planning for the gamma knife by source culling. *International Journal of Radiation Oncology Biology and Physics*, 45(5):1339–1346, 1999.

[39] J. Lyman and A. Wolbarst. Optimization of radiation therapy iii: A method of assessing complication probabilities from dose-volume histograms. *International Journal of Radiation Oncology Biology and Phsyics*, 13:103'109, 1987.

[40] J. Lyman and A. Wolbarst. Optimization of radiation therapy iv: A dose-volume histogram reduction algorithm. *International Journal of Radiation Oncology Biology and Phsyics*, 17:433–436, 1989.

[41] J. Lyman. Complication probability as assessed from dose-volume histrograms. *Radiation Research*, 104:S 13–19, 1985.

[42] S. McDonald and P. Rubin. Optimization of external beam radiation therapy. *International Journal of Radiation Oncology, Biology, Physics*, 2:307–317, 1977.

[43] S. Morrill, R. Lane, G. Jacobson, and I. Rosen. Treatment planning optimization using constrained simulated annealing. *Physics in Medicine & Biology*, 36(10):1341–1361, 1991.

[44] S. Morrill, I. Rosen, R. Lane, and J. Belli. The influence of dose constraint point placement on optimized radiation therapy treatment planning. *International Journal of Radiation Oncology, Biology, Physics*, 19:129–141, 1990.

[45] J. Leong M. Langer. Optimization of beam weights under dose-volume restrictions. *International Journal of Radiation Oncology, Biology, Physics*, 13:1255–1260, 1987.

[46] A. Niemierko and M. Goitein. Calculation of normal tissue complication probability and dose-volume histogram reduction schemes for tissues with critical element architecture. *Radiation Oncology*, 20:20, 1991.

[47] I. Paddick. A simple scoring ratio to index the conformality of radiosurgical treatment plans. *Journal of Neurosurgery*, 93(3):219–222, 2000.

[48] W. Powlis, M. Altschuler, Y. Censor, and E. Buhle. Semi-automatic radiotherapy treatment planning with a mathematical model to satisfy treatment goals. *International Journal of Radiation Oncology, Biology, Physics*, 16:271–276, 1989.

[49] F. Preciado-Walters, M. Langer, R. Rardin, and V. Thai. Column generation for imrt cancer therapy optimization with implementable segments. Technical report, Purdue University, 2004.

[50] F. Preciado-Walters, M. Langer, R. Rardin, and V. Thai. A coupled column generation, mixed-integer approach to optimal planning of intensity modulated radiation therapy for cancer. Technical report, Purdue University, 2004. to appear in Mathematical Programming.

[51] A. Pugachev, A. Boyer, and L. Xing. Beam orientation optimization in intensity-modulated radiation treatment planning. *Medical Physics*, 27(6):1238–1245, 2000.

[52] A. Pugachev and L. Xing. Computer-assisted selection of coplaner beam orientations in intensity-modulated radiation therapy. *Physics in Medicine and Biology*, 46:2467–2476, 2001.

[53] C. Raphael. Mathematical modeling of objectives in radiation therapy treatment planning. *Physics in Medicine & Biology*, 37(6):1293–1311, 1992.

[54] H. Romeijn, R. Ahuja, J.F. Dempsey, and A. Kumar. A column generation approach to radiation therapy treatment planning using aperture modulation. Technical Report Research Report 2003-13, Department of Industrial and Systems Engineering, University of Florida, 2003.

[55] I. Rosen, R. Lane, S. Morrill, and J. Belli. Treatment plan optimization using linear programming. *Medical Physics*, 18(2):141–152, 1991.

[56] D. Shepard, M. Ferris, G. Olivera, and T. Mackie. Optimizing the delivery of radiation therapy to cancer patients. *SIAM Review*, 41(4):721–744, 1999.

[57] D. Shepard, M. Ferris, R. Ove, and L. Ma. Inverse treatment planning for gamma knife radiosurgery. *Medical Physics*, 27(12):2748–2756, 2000.

[58] H. Shu, Y. Yan, X. Bao, Y. Fu, and L. Luo. Treatment planning optimization by quasi-newton and simulated annealing methods for gamma unit treatment system. *Physics in Medicine and Biology*, 43(10):2795–2805, 1998.

[59] H. Shu, Y. Yan, L. Luo, and X. Bao. Three-dimensional optimization of treatment planning for gamma unit treatment system. *Medical Physics*, 25(12):2352–2357, 1998.

[60] G. Starkshcall. A constrained least-squares optimization method for external beam radiation therapy treatment planning. *Medical Physics*, 11(5):659–665, 1984.

[61] J. Stein et al. Number and orientations of beams in intensity-modulated radiation treatments. *Medical Physics*, 24(2):149–160, 1997.

[62] S. Söderström and A. Brahme. Selection of suitable beam orientations in radiation therapy using entropy and fourier transform measures. *Physics in Medicine and Biology*, 37(4):911–924, 1992.

[63] H. Withers, J. Taylor, and B. Maciejewski. Treatment volume and tissue tolerance. *International Journal of Radiation Oncology, Biology, Physics*, 14:751–759, 1987.

[64] A. Wolbarst. Optimization of radiation therapy II: The critical-voxel model. *International Journal of Radiation Oncology, Biology, Physics*, 10:741–745, 1984.

[65] Q. Wu and J. Bourland. Morphology-guided radiosurgery treatment planning and optimization for multiple isocenters. *Medical Physics*, 26(10):2151–2160, 1999.

[66] P. Zhang, D. Dean, A. Metzger, and C. Sibata. Optimization o gamma knife treatment planning via guided evolutionary simulated annealing. *Medical Physics*, 28(8):1746–1752, 2001.

Chapter 5

PARALLEL ALGORITHM DESIGN FOR BRANCH AND BOUND

David A. Bader
Department of Electrical & Computer Engineering, University of New Mexico
dbader@ece.unm.edu

William E. Hart
Discrete Mathematics and Algorithms Department, Sandia National Laboratories
wehart@sandia.gov

Cynthia A. Phillips
Discrete Mathematics and Algorithms Department, Sandia National Laboratories
caphill@sandia.gov

Abstract Large and/or computationally expensive optimization problems sometimes require parallel or high-performance computing systems to achieve reasonable running times. This chapter gives an introduction to parallel computing for those familiar with serial optimization. We present techniques to assist the porting of serial optimization codes to parallel systems and discuss more fundamentally parallel approaches to optimization. We survey the state-of-the-art in distributed- and shared-memory architectures and give an overview of the programming models appropriate for efficient algorithms on these platforms. As concrete examples, we discuss the design of parallel branch-and-bound algorithms for mixed-integer programming on a distributed-memory system, quadratic assignment problem on a grid architecture, and maximum parsimony in evolutionary trees on a shared-memory system.

Keywords: parallel algorithms; optimization; branch and bound; distributed memory; shared memory; grid computing

Introduction

Although parallel computing is often considered synonymous with supercomputing, the increased availability of multi-processor workstations and Beowulf-style clusters has made parallel computing resources available to most academic departments and modest-scale companies. Consequently, researchers have applied parallel computers to problems such as weather and climate modeling, bioinformatics analysis, logistics and transportation, and engineering design. Furthermore, commercial applications are driving development of effective parallel software for large-scale applications such as data mining and computational medicine.

In the simplest sense, parallel computing involves the simultaneous use of multiple compute resources to solve a computational problem. However, the choice of target compute platform often significantly influences the structure and performance of a parallel computation. There are two main properties that classify parallel compute platforms: physical proximity of compute resources and distribution of memory. In *tightly-coupled* parallel computers, the processors are physically co-located and typically have a fast communication network. In *loosely coupled* compute platforms, compute resources are distributed, and consequently inter-process communication is often slow. In *shared memory* systems, all of the RAM is physically shared. In *distributed memory* systems each processor/node controls its own RAM. The owner is the only processor that can access that RAM.

The term *system* usually refers to tightly-coupled architectures. *Shared-memory* systems typically consist of a single computer with multiple processors using many simultaneous asynchronous threads of execution. Most massively-parallel machines are *distributed memory* systems. Parallel software for these machines typically requires explicit problem decomposition across the processors, and the fast communication network enables synchronous inter-processor communication. *Grid* compute platforms exemplify the extreme of loosely-coupled distributed-memory compute platforms. Grid compute platforms may integrate compute resources across extended physical distances, and thus asynchronous, parallel decomposition is best suited for these platforms. Loosely-coupled shared-memory platforms have not proven successful because of the inherent communication delays in these architectures.

This chapter illustrates how to develop scientific parallel software for each of these three types of canonical parallel compute platforms. The *programming model* for developing parallel software is somewhat different for each of these compute platforms. Skillicorn and Talia [103] define a programming model as: “an abstract machine providing certain operations to the programming level above and requiring implementations on all of the architectures below.” Parallel code will perform best if the programming model and the hardware match.

However, in general parallel systems can emulate the others (provide the other required abstractions) with some loss of performance.

As a concrete example, we consider the design of parallel branch and bound. We discuss how the compute platform influences the design of parallel branch and bound by affecting factors like task decomposition and inter-processor coordination. We discuss the application of parallel branch and bound to three real-world problems that illustrate the impact of the parallelization: solving mixed-integer programs, solving quadratic assignment problems, and reconstructing evolutionary trees.

Finally, we provide some broad guidance on how to design and debug parallel scientific software. We survey some parallel algorithmic primitives that form the basic steps of many parallel codes. As with serial software, a working knowledge of these types of primitives is essential for effective parallel software development. Because parallel software is notoriously difficult to debug, we also discuss practical strategies for debugging parallel codes.

5.1 Parallel Computing Systems

Over the past two decades, high-performance computing systems have evolved from special-purpose prototypes into commercially-available commodity systems for general-purpose computing. We loosely categorize parallel architectures as distributed memory, shared memory, or grid; realizing that modern systems may comprise features from several of these classes (for example, a cluster of symmetric multiprocessors, or a computational grid of multithreaded and distributed memory resources). In this section, we briefly describe some theoretical parallel models, the types of high-performance computing architectures available today, and the programming models that facilitate efficient implementations for each platform. For a more details, we refer the reader to a number of excellent resources on parallel computing [32, 33, 49, 58], parallel programming [13, 24, 93, 104, 115], and parallel algorithms [57, 77, 97].

5.1.1 Theoretical Models of Parallel Computers

Theoretical analysis of parallel algorithms requires an abstract machine model. The two primary models roughly abstract shared-memory and distributed-memory systems. For serial computations the RAM model, perhaps augmented with a memory hierarchy, is universally accepted. We are aware of no single model that is both realistic enough that theoretical comparisons carry over to practice and simple enough to allow clean and powerful analyses. However, some techniques from theoretically good algorithms are useful in practice. If the reader wishes to survey the parallel algorithms literature before beginning an implementation, he will need some familiarity with parallel machine models to understand asymptotic running times.

The *Parallel Random Access Machine* (PRAM) [42] has a set of identical processors and a shared memory. At each synchronized step, the processors perform local computation and simultaneously access the shared memory in a legal pattern. In the EREW (exclusive-read exclusive-write) PRAM the access is legal if all processors access unique memory locations. The CRCW (concurrent-read concurrent write) PRAM allows arbitrary access and the CREW PRAM allows only simultaneous reads. The PRAM roughly abstracts shared-memory systems. In Section 5.1.4 we show ways in which it is still unrealistic, frequently fatally so.

The asynchronous LogP model is a reasonable abstraction of distributed-memory systems. It explicitly models communication bandwidth (how much data a processor can exchange with the network), message latency (time to send a message point to point) and how well a system overlaps communication and computation. It can be difficult to compare different algorithms using this model because running times can be complicated functions of the various parameters. The Bulk Synchronous Parallel (BSP) model[113] is somewhat between LogP and PRAM. Processors have local memory and communicate with messages, but have frequent explicit synchronizations.

5.1.2 Distributed Memory Architectures

Overview. In a distributed-memory system each node is a workstation or PC-level processor, possibly even an SMP (see Section 5.1.4). Each node runs its own operating system, controls its own local memory, and is connected to the other processors via a communication network. In this section, we consider distributed-memory *systems* where the processors work together as a single tightly-coupled machine. Such a machine usually has a scheduler that allocates a subset of the nodes on a machine to each user request. However, most of the discussion and application examples for distributed-memory systems also apply to independent workstations on a local-area network, provided the user has access privileges to all the workstations in the computation.

The number of processors and interconnect network topology varies widely among systems. The ASCI Red Storm supercomputer, that is being built by Cray for Sandia National Laboratories, will have 10,368 AMD Opteron processors connected via a three-dimensional mesh with some toroidal wraps. The Earth Simulator at the Earth Simulator Center in Japan has 640 nodes, each an 8-processor NEC vector machine, connected via a crossbar. The networks in both these supercomputers use custom technology. Commercial (monolithic) systems include IBM SP and Blades, Apple G5, Cray systems, and Intel Xeon clusters. However, any department/company can build a Beowulf cluster by buying as many processors as they can afford and linking them using commercial network technology such as Ethernet, Myrinet, Quadrics, or InfiniBand.

These can be quite powerful. For example, the University of New Mexico runs IBM's first Linux supercluster, a 512-processor cluster with Myrinet (LosLobos), allocated to National Science Foundation users, and the Heidelberg Linux Cluster System (HELICS) has 512 AMD Athlon PC processors connected as a Clos network with commercial Myrinet technology. Even at the low end with only a few processors connected by Ethernet, one can benefit from this form of parallelism.

Because distributed-memory systems can have far more processors and total memory than the shared-memory systems discussed in Section 5.1.4, they are well suited for applications with extremely large problem instances.

Programming Models for Distributed-Memory. One programs a distributed-memory machine using a standard high-level language such as C++ or Fortran with explicit message passing. There are two standard Application Programming Interfaces (APIs) for message passing: Message-Passing Interface (MPI [104]) and Parallel Virtual Machine (PVM [92]). MPI is the standard for tightly-coupled large-scale parallel machines because it is more efficient. PVM's extra capabilities to handle heterogeneous and faulty processors are more important for the grid than in this setting.

For either API, the user can assume some message-passing primitives. Each processor has a unique rank or ID, perhaps within a subset of the system, which serves as an address for messages. A processor can send a message directly to another specifically-named processor in a *point-to-point* message. A *broadcast* sends a message from a single source processor to all other processors. In an *all-to-all* message, each of the P processors sends k bytes to all other processors. After the message, each processor has kP bytes with the information from processor i in the ith block of size k. A *reduction* operation (see Section 5.3.1) takes a value from each processor, computes a function on all the values, and gives the result to all processors. For example, a sum reduction gives every processor the sum of all the input values across all processors.

The performance of a distributed-memory application depends critically upon the message complexity: number of messages, message size, and *contention* (number of messages simultaneously competing for access to an individual processor or channel). The bandwidth and speed of the interconnection network determines the total amount of message traffic an application can tolerate. Thus a single program may perform differently on different systems. When a parallel algorithm such as branch and bound has many small independent jobs, the user can tailor the *granularity* of the computation (the amount of work grouped into a single unit) to set the communication to a level the system can support. Some unavoidable communication is necessary for the correctness of a computation. This the user cannot control as easily. Thus a large complex code may require clever problem-dependent message management.

The amount of contention for resources depends in part upon *data layout*, that is, which processor owns what data. The programmer also explicitly manages data distribution. If shared data is static, meaning it does not change during a computation, and the data set is not too large relative to the size of a single processor's memory, it should be *replicated* on all processors. Data replication also works well for data that is widely shared but rarely changed, especially if the computation can tolerate an out-of-date value. Any processor making a change broadcasts the new value. In general, data can *migrate* to another processor as a computation evolves, either for load balancing or because the need to access that data changes. Automated graph-partition-based tools such as Chaco [55] and (Par)METIS [60] assist in initial data assignment if the communication pattern is predictable.

5.1.3 Grid Computing

Overview. *Grid computing* (or "metacomputing") generally describes parallel computations on a geographically-distributed, heterogeneous platform [45, 44]. Grid computing technologies enable coordinated resource sharing and problem solving in dynamic, multi-institutional virtual organizations. Specifically, these tools enable direct access to computers, software, data, and other resources (as opposed to sharing via file exchange). A virtual organization is defined by a set of sharing rules between individuals and/or organizations. These sharing rules define how resource providers coordinate with consumers, what is shared, who can share, and the conditions under which sharing occurs.

Grid computing platforms for scientific applications can use shared workstations, nodes of PC clusters, and supercomputers. Although grid computing methodologies can effectively connect supercomputers, a clear motivation for grid computing is the wide availability of idle compute cycles in personal computers. Large institutions can have hundreds or thousands of computers that are often idle. Consequently, grid computing can enable large-scale scientific computation using existing computational resources without significant additional investment.

Projects like Condor [70], Legion [50] and Globus [43] provide the underlying infrastructure for resource management to support grid computing. These toolkits provide components that define a protocol for interacting with remote resources and an application program interface to invoke that protocol. Higher-level libraries, services, tools and applications can use these components to implement more complex global functionality. For example, various Globus Toolkit components are reviewed by Foster et al. [45]. Many grid services build on the Globus Connectivity and Resource protocols, such as (1) directory services that discover the existence and/or properties of resources, (2) scheduling and brokering services that request the scheduling and allocation of resources, and (3) monitoring and diagnostic services that provide information about the availability and utilization of resources.

Large-scale Grid deployments include: "Data Grid" projects like EU DataGrid (`www.eu-datagrid.org`), the DOE Science Grid (`http://www.doesciencegrid.org`), NASA's Information Power Grid (`www.ipg.nasa.gov`), the Distributed ASCI Supercomputer (DAS-2) system (`www.cs.vu.nl/das2/`), the DISCOM Grid that links DOE laboratories (`www.cs.sandia.gov/discom/`), and the TeraGrid under construction to link major U.S. academic sites (`www.teragrid.org`). These systems integrate resources from multiple institutions using open, general-purpose protocols like the Globus Toolkit.

Programming Model. Grid computing platforms differ from conventional multiprocessor supercomputers and from Linux clusters in several important respects.

- Interprocessor Connectivity: Communication latency (time between message initiation and message delivery) is generally much higher on grid platforms than on tightly-coupled supercomputers and even on Linux clusters. Grid communication latencies are often highly variable and unpredictable.
- Processor Reliability and Availability: Computational resources may disappear without notice in some grid computing frameworks. For example, jobs running on PCs may terminate or halt indefinitely while the PC is used interactively. Similarly, new compute resources may become available during a parallel computation.
- Processor Heterogeneity: Computational resources may vary in their operational characteristics, such as available memory, swap space, processor speed, and operating system (type and version).

It is challenging to divide and robustly coordinate parallel computation across heterogeneous and/or unreliable resources. When resources reside in different administrative domains, they may be subject to different access control policies, and be connected by networks with widely varying performance characteristics.

For simple applications augmented distributed-memory programming models may suffice. For example, MPICH-G2 generalizes MPI to provide a low-level grid programming model that manages heterogeneity directly [59]. Since MPI has proven so effective for tightly-coupled distributed computing environments, MPICH-G2 provides a straightforward way to adapt existing parallel software to grid computing environments. MPICH-G2 is integrated with the Globus Toolkit, which manages the allocation and coordination of grid resources.

Although MPICH-G2 can adapt MPI codes for grid computing environments, the programmer must generally modify the original code for robustness. For example, the new code must monitor node status and reassign work given to killed nodes. Synchronous codes must be made asynchronous for both correctness and performance. When the number of processors vary, it's difficult to tell when all processors have participated in a synchronous step. Since some processor is likely to participate slowly or not at all, each synchronization can be intolerably slow.

A variety of frameworks can facilitate development of grid computing applications. NetSolve [23] provides an API to access and schedule Grid resources in a seamless way, particularly for embarrassingly parallel applications. The Everyware toolkit [117] draws computational resources transparently from the Grid, though it is not abstracted as a programming tool. The MW and CARMI/Wodi tools provide interfaces for programming master-worker applications [48, 90]. These frameworks provide mechanisms for allocating, scheduling computational resources, and for monitoring the progress of remote jobs.

5.1.4 Shared Memory Systems

Overview. Shared-memory systems, often called symmetric multiprocessors (SMPs), contain from two to hundreds of microprocessors tightly coupled to a shared memory subsystem, running under a single system image of one operating system. For instance, the IBM "Regatta" p690, Sun Fire E15K, and SGI Origin, all scale from dozens to hundreds of processors in shared-memory images with near-uniform memory access. In addition to a growing number of processors in a single shared memory system, we anticipate the next generation of microprocessors will be "SMPs-on-a-chip". For example, uniprocessors such as the IBM Power4 using simultaneous multithreading (SMT), Sun UltraSparc IV, and Intel Pentium 4 using Hyper-Threading each act like a dual-processor SMP. Future processor generations are likely to have four to eight cores on a single silicon chip. Over the next five to ten years, SMPs will likely become the standard workstation for engineering and scientific applications, while clusters of very large SMPs (with hundreds of multi-core processors) will likely provide the backbone of high-end computing systems.

Since an SMP is a true (hardware-based) shared-memory machine, it allows the programmer to share data structures and information at a fine grain at memory speeds. An SMP processor can access a shared memory location up to two orders of magnitude faster than a processor can access (via a message) a remote location in a distributed memory system. Because processors all access the same data structures (same physical memory), there is no need to explicitly manage data distribution. Computations can naturally synchronize on data structure states, so shared-memory implementations need fewer explicit synchronizations in some contexts. These issues are especially important for irregular applications with unpredictable execution traces and data localities, often characteristics of combinatorial optimization problems, security applications, and emerging computational problems in biology and genomics.

While an SMP is a shared-memory architecture, it is by no means the Parallel Random Access Memory (PRAM) model (see [57, 97]) used in theoretical work. Several important differences are: (i) the number of processors in real SMP systems remains quite low compared to the polynomial number of processors assumed by theoretic models; (ii) real SMP systems have no lockstep synchronization (PRAMs assume perfect synchronization at the fetch-execute-store level); (iii) SMP memory bandwidth is limited; (iv) real SMPs have caches that require a high degree of spatial and temporal locality for good performance; and (v) these SMP caches must be kept *coherent.* That is, when a processor reads a value of a memory location from its local cache, this value must correspond to the value last written to that location by any processor in the system. For example, a 4-way or 8-way SMP cannot support concurrent read to the same location by a thousand threads without significant slowdown.

The memory hierarchy of a large SMP is typically quite deep, since the main memory can be so large. Thus cache-friendly algorithms and implementations are even more important on large-scale SMPs than on workstations. Very small-scale SMPs maintain cache coherence by a *snoopy protocol* with substantial hardware assistance. In this protocol, whenever a processor writes a memory location, it broadcasts the write on a shared bus. All other processors monitor these broadcasts. If a processor sees an update to a memory location in its cache, it either updates or invalidates that entry. In the latter case, it's not just a single value that's invalidated, but an entire cache line. This can lead to considerable slowdowns and memory congestion for codes with little spatial locality. In the *directory-based* protocol, the operating system records the caches containing each cache line. When a location in the line changes, the operating system invalidates that line for all non-writing processors. This requires less bandwidth than snooping and scales better for larger SMP systems, but it can be slower. *False sharing* occurs when two unrelated data items a and b are grouped in a single cache line. Thus a processor that has the line in its cache to access item a may have the entire line invalidated for an update to item b, even though the processor never accesses b. Eliminating false sharing to improve performance must currently be handled by the compiler or the programmer.

Finally, synchronization is perhaps the biggest obstacle to the correct and efficient implementation of parallel algorithms on shared-memory machines. A theoretic algorithm may assume lockstep execution across all processors down to the level of the fetch-execute-store cycle; for instance, if processor i is to shift the contents of location i of an array into location $i + 1$, each processor reads its array value in lockstep, then stores it in the new location in lockstep. In a real machine, some processors could easily start reading after their neighbor has already completed its task, resulting in errors. The programming solution is to introduce barriers at any point in the code where lack of synchronization could cause indeterminacy in the final answer. However, such a solution is expensive when implemented in software and, if needed on a finely-divided time scale will utterly dominate the running time.

Programming Models. Various programming models and abstractions are available for taking advantage of shared-memory architectures. At the lowest level, one can use libraries, such as POSIX threads (for example, see [62, 89, 108]) to explicitly handle threading of an application. Other standards such as OpenMP (see [24, 85, 93]) require compiler and runtime support. The programmer provides hints to the compiler on how to parallelize sections of a correct sequential implementation. Finally, new high-level languages such as Unified Parallel C (UPC) [21], a parallel extension of C, Co-Array Fortran (CAF) (see `www.co-array.org`) with global arrays, and Titanium [122], are also emerging as new, efficient programming models for multiprocessors.

POSIX threads (often referred to as "pthreads") are native threads of processing that run within a single process/application and can share access to resources and memory at a fine-scale. The programmer explicitly creates and manages threads, with each thread inheriting its parent's access to resources. The programmer can synchronize threads and protect critical sections, such as shared memory locations in data structures and access to I/O resources, via mutual exclusion (or "mutex") locks. These support three operations: lock, unlock, and try, a non-blocking version of lock where a thread either succeeds at acquiring the lock, or resumes execution without the lock. Condition variables suspend a thread until an event occurs that wakes up the thread. These in conjunction with mutex locks can create higher-level synchronization events such as shared-memory barriers. In a threaded code, the programmer can then rely on coherency protocols to update shared memory locations.

OpenMP is a higher-level abstraction for programming shared memory that makes use of compiler directives, runtime systems, and environment variables. The programmer often begins with a working sequential code in C, C++, or Fortran, and inserts directives into the code to guide the compiler and runtime support for parallelization. One can specify, for example, that a loop has no dependencies among its iterations and can be parallelized in a straightforward manner. One can also specify that a loop is a reduction or scan operation that can then be automatically parallelized (see Section 5.3). OpenMP also allows the programmer to mark critical sections and insert synchronization barriers. The programmer can specify how the work varies from iteration to iteration (for example if the work is constant, random, or dependent upon the loop iteration). These hints can improve the scheduling of loop iterations.

UPC [21] is an extension of C that provides a shared address space and a common syntax and semantics for explicitly parallel programming in C. UPC strikes a balance between ease-of-use and performance. The programming model for UPC assumes a number of threads, each with private or shared pointers that can point to local or global memory locations. UPC provides explicit synchronization including barriers. Unlike POSIX threads, UPC provides a library of collective communication routines commonly needed in scientific and technical codes. UPC is emerging as an alternative for parallel programming that builds upon prior languages such as AC and Split-C.

Shared-memory has enabled the high-performance implementation of parallel algorithms for several combinatorial problems that up to now have not had implementations that performed well on parallel systems for arbitrary inputs. We have released several such high-performance shared-memory codes for important problems such as list ranking and sorting [53, 54], ear decomposition [6], spanning tree [4], minimum spanning tree [5], and Euler tour [29]. These parallel codes are freely-available under the GNU General Public License (GPL) from Bader's web site. They use a shared-memory framework for POSIX threads [7].

5.2 Application Examples

In this section, we describe basic branch and bound (B&B) and discuss issues of special consideration in parallelizing B&B applications. We then give three parallel B&B example applications: B&B for mixed-integer programming on a distributed-memory architecture, B&B for the quadratic assignment problem on a grid architecture, and B&B for phylogeny reconstruction on a shared-memory architecture.

5.2.1 Branch and Bound

Branch and bound is an intelligent search heuristic for finding a global optimum to problems of the form $\min_{x \in X} f(x)$. Here X is the *feasible region* and $f(x)$ is the *objective function*. Basic B&B searches the feasible region by iteratively subdividing the feasible region and recursively searching each piece for an optimal feasible solution. B&B is often more efficient than straight enumeration because it can eliminate regions that provably do not contain an optimal solution.

To use B&B for a given problem, one must specify problem-specific implementations of the following procedures. The *bound* procedure gives a lower bound for a problem instance over any feasible region. That is, for instance $\mathcal{I}$ with feasible region $X_{\mathcal{I}}$, the bound procedure returns $b(\mathcal{I})$, such that for all $x \in X_{\mathcal{I}}$ we have $b(\mathcal{I}) \leq f(x)$. The *branch* or *split* procedure breaks the feasible region X into k subregions $X_1, X_2, \ldots, X_k$. In most efficient B&B

implementations, subregions are disjoint ($X_i \cap X_j = \emptyset$ for $i \neq j$), though this need not be the case. The only requirement for correctness is that there exists an $x \in X$ with minimum value of $f(x)$ such that $x \in \bigcup_{i=1}^{k} X_i$ (if we require all optima, then this must be the case for all optimal x). Finally, a *candidate* procedure takes an instance of the problem (a description of X) and returns a feasible solution $x \in X$ if possible. This procedure can fail to return a feasible solution even if X contains feasible solutions. However, if X consists of a single point, then the candidate solution procedure must correctly determine the feasibility of this point. In general, one may have many candidate solution methods. At any point in a B&B search, the best feasible (candidate) solution found so far is called the *incumbent*, denoted x_I.

We now describe how to find a globally optimal solution using B&B given instantiations of these procedures. We grow a search tree with the initial problem r as the root. If the candidate procedure (called on the root) returns a feasible solution, it becomes the first incumbent. Otherwise, we start with no incumbent and define an incumbent value $f(x_I) = +\infty$. We bound the root, which yields a lower bound $b(r)$. If $b(r) = f(x_I)$, then the incumbent is an optimal solution and we are done. Otherwise, we split the root into k subproblems and make each subproblem a child of the root. We process a child subproblem similarly. We bound the subproblem c to obtain a bound $b(c)$. If $b(c) > f(x_I)$, then no feasible solution in subproblem c can be better than the incumbent. Therefore, we can *fathom* subproblem c, meaning we eliminate it from further consideration. If candidate(c) returns a solution x such that $f(x) = b(c)$, then x is an optimal solution for this subproblem. Subproblem c becomes a leaf of the tree (no need to further subdivide it) and solution x replaces the incumbent if $f(x) < f(x_I)$. Otherwise, we split subproblem c and continue. Any subproblem that is not a leaf (still awaiting processing) is called an *open* or *active* subproblem. At any point in the computation, let A be the set of active subproblems. Then $L = \min_{a \in A} b(a)$ is a global lower bound on the original problem. B&B terminates when there are no active subproblems or when the relative or absolute gap between L and $f(x_I)$ is sufficiently small.

A B&B computation is logically divided into two phases: (1) find an optimal solution x^* and (2) prove that x^* is optimal. At any given point in the computation, all active subproblems a such that $b(a) < f(x^*)$ must be processed to prove the optimality of x^*; all other subproblems can be pruned once we have found x^*. Thus for any given bounding and splitting strategy there is a minimum-size tree: that obtained by seeding the computation with the optimal solution or finding the optimal at the root. A candidate procedure that finds near-optimal solutions early in the B&B computation can reduce tree size tree by allowing early pruning.

When the incumbent procedure is weak near the root, the following adaptive branching strategy can quickly identify a feasible solution to enable at least some pruning. Initially apply depth-first search to find a leaf as soon as possible. Given an incumbent, switch to best-first search, which selects the node n with the minimum value of $b(n)$. This hybrid search strategy is particularly applicable to combinatorial problems, since depth-first search will eventually subdivide the region until a feasible solution is found (e.g. after all possible choices have been made for one node). Depth-first search can open up many more subproblems than best-first search with a good early incumbent, but the hybrid strategy can be superior when it is difficult to find an incumbent.

In parallel B&B, one can parallelize the computation of independent subproblems (or subtrees) or parallelize the evaluation of individual subproblems. The latter is better when the tree is small at the start or end of the computation, unless the other processors can be kept busy with other independent work, such as generating incumbents (see Section 5.2.2).

It is important to keep the total tree size close to the size of that generated by a good serial solver. Good branch choices are critical early in the computation and good *load balancing* is critical later. That is, each processor must stay sufficiently busy with high-quality work. Otherwise in pathological cases the parallel computation performs so much more work than its serial counterpart that there is actually a slowdown anomaly [35, 64–67]. That is, adding processors increases wall-clock time to finish a computation.

Parallel platforms of any kind have far more memory than a single computer. A serial B&B computation may be forced to throw away seemingly unfavorable active nodes, thus risking losing the optimal solution. A parallel system is much more likely to have sufficient memory to finish the search.

There are many frameworks for parallel branch and bound including PUBB [102], Bob [12], PPBB-lib [112], PICO [36], Zram [73], and ALPS/BiCePS [96]. The user defines the above problem-specific procedures and the framework provides a parallel implementation. Bob++ [16] and Mallba [71] are even higher-level frameworks for solving combinatorial optimization problems.

5.2.2 Mixed-Integer Programming

A mixed-integer program (MIP) in standard form is:

$$\text{(MIP)} \quad \text{minimize} \quad c^T x \qquad \text{where} \quad \begin{cases} Ax = b \\ \ell \leq x \leq u \\ x_j \in \mathcal{Z} & \forall j \in D \subseteq \{1, \ldots, n\} \end{cases}$$

where x and c are n-vectors, A is an $m \times n$ matrix, b is an m-vector, and $\mathcal{Z}$ is the set of integers. Though in principle all input data are reals, for practical solution on a computer they are rational. Frequently the entries of A, c, and b are integers. We can convert an inequality constraint in either direction to an equality by adding a variable to take up slack between the value of ax and its bound b.

The only nonlinearity in MIP is the integrality constraints. Frequently binary variables represent decisions that must be yes/no (i.e. there can be no partial decision for partial cost and partial benefit). In principle MIPs can express any NP-complete optimization problem. In practice they are used for resource allocation problems such as transportation logistics, facility location, and manufacturing scheduling, or for the study of natural systems such as protein folding.

In practice, MIPs are commonly solved with B&B and its variants. In this section, we consider the application of B&B to solve MIPs on a distributed-memory system. Specifically, we summarize the implementation of Eckstein, Hart, and Phillips, within the Parallel Integer and Combinatorial Optimizer (PICO) system [36].

Branch-and-Bound Strategy. If the integrality constraints are relaxed (removed), a MIP problem becomes a linear program (LP). Solving this LP is the classic bounding procedure for B&B. LPs are theoretically solvable in polynomial time [61] and are usually solved efficiently in practice with commercial tools such as CPLEX [31], XPRESS [119], or OSL [86], or free tools such as COIN-LP [26].

Serial B&B for MIP begins with the MIP as the root problem and bounds the root by computing the LP relaxation. If all integer variables (x_j with $j \in D$) have integer values (within tolerance), then this is a feasible integer solution whose value matches a lower bound, and hence it is an optimal solution. If the LP relaxation is not a feasible solution, then there is some $j \in D$ such that the optimal solution to the LP relaxation x^* has value $x_j^* \notin \mathcal{Z}$. We then create two new sub-MIPs as children: one with the restriction $x_j \leq \lfloor x_j^* \rfloor$ and one with the restriction $x_j \geq \lceil x_j^* \rceil$. For binary variables, one child has $x_j = 0$ and the other has $x_j = 1$. The feasible regions of the two children are disjoint and any solution with $\lfloor x_j^* \rfloor < x_j < \lceil x_j^* \rceil$, including x^*, is no longer feasible in either child. Thus the LP relaxation of a child provides a lower bound on the optimal solution within this subregion, and it will be different from the LP relaxation of the parent.

There are a number of common ways to improve the performance of this standard B&B MIP computation. Most MIP systems apply general and/or problem-specific cutting planes before branching to improve the lower bound on a subproblem while delaying branching as long as possible. Given x^*, an optimal non-integral solution to the LP relation of a (sub)problem, a cutting plane is a constraint $ax = b$ such that $ax' = b$ for all possible (optimal) integer solutions x' but $ax^* \neq b$. Adding this constraint to the system makes the current LP optimal infeasible.

Careful branch selection in MIP solvers can significantly impact search performance. In strong branching, one tries a branch and partially evaluates the subtree to determine a merit rating for the branch. Most or all the work done for strong branching is thrown away, but it can sometimes reduce tree size sufficiently to merit the effort. A less computationally demanding strategy is to maintain gradients for each branch choice. For the simple branching described above, the gradient for a single branch is the change in LP objective value divided by the change in the variable (the latter is always less than 1). When a variable is a branching candidate for the first time, one can initialize its gradient by pretending to branch in each direction. This is much better than, for example, setting an uninitialized gradient to the average of the gradients computed so far [69]. However, each gradient initialization requires two LP bounding operations. Finally, one can compute problem-specific constraints to partition the current feasible region, though this may be expensive.

One can also improve performance of B&B for MIP by finding feasible integer solutions using methods other than finding leaf nodes when an LP relaxation is integer feasible. Since MIPs frequently have combinatorial structure, one can use general heuristics such as evolutionary algorithms or tabu search or problem-specific methods that exploit structure. In particular, there are many approximation algorithms for combinatorial problems that find an LP-relaxation for a MIP and "round" this nontrivially to obtain a feasible integer solution whose objective value is provably close to the LP relaxation bound (see [11, 19, 88] for a tiny sample). The provable bound is limited by the integrality gap of the problem (the ratio between the best integer solution and the best LP solution; this is a measure of the strength of the formulation [22]).

There is a variety of parallel MIP solvers. PARINO [69], SYMPHONY [94], and COIN/BCP [63, 72] are designed for small-scale, distributed-memory systems such as clusters. BLIS [96], under development, is designed as a more scalable version of SYMPHONY and BCP. It will be part of COIN once it is available. See the discussion of all three in [95]. FATCOP [25] is designed for grid systems. PICO's parallel B&B search strategy is particularly well-suited for solving MIPs on tightly-coupled massively-parallel distributed-memory architectures, such as those available at the National Laboratories for solution of national-scale problems. For these architectures, one has exclusive use of perhaps thousands of processors for the entire computation. Thus one major concern is effectively using all these processors during the initial *ramp up* phase, when the search tree is small. After ramp up, PICO enters a *parallel-subproblem* phase managing an asynchronous parallel search using load balancing to ensure that all worker processes are solving interesting subproblems. These two aspects of parallel B&B in PICO are discussed in the next two sections. The combination of strategies can use massive parallelism effectively. In preliminary experiments on some problems PICO had near perfect speed up through 512 processors [36].

Managing Ramp Up. PICO uses an explicit ramp up phase in which all the processors work on a single subproblem, parallelizing the individual subproblem evaluation steps. In particular, PICO supports parallel (1) initialization of gradient estimation (used for branching prioritization), (2) problem-specific preprocessing, (3) root bounding, and (4) incumbent and cutting-plane generation. Also some processors can search for incumbents and cutting planes independently from those growing the tree. For example, since parallel LP solvers do not currently scale well to thousands of processors, excess processors can search for incumbents during root bounding.

PICO parallelizes gradient initialization during ramp up. In general if there are f potential branch variables with uninitialized gradients, each of the P processors initializes the gradients for $\lceil f/P \rceil$ or $\lfloor f/P \rfloor$ variables and sends these values to all processors using all-to-all communication. Though PICO doesn't currently support strong branching, in principle the work for strong branching could be divided among the processors in the same way with the same exchange of branch quality information. One can also parallelize complex custom branching computations.

To parallelize problem-specific preprocessing, processors can cooperate on individual preprocessing steps or they can compute independent separate steps. Good preprocessing is often critical for computing an exact solution to hard combinatorial problems (e.g. those expressed by MIPs). Real-world instances frequently have special structure that is captured by a small upper bound on a parameter k (e.g. the degree of a graph, maximum contention for a resource, etc.). To solve fixed-parameter-tractable problems, one first *kernelizes* the problem, transforming it in polynomial time into a new instance with size bounded by a function of k. These preprocessing steps are frequently local and therefore good candidates for parallelization. See Fellows [39] for an excellent summary of the theory and practice of kernelization.

One can also parallelize the LP bounding procedure. This is particularly desirable at the root because bounding the root problem can be more than an order of magnitude more expensive than bounding subproblems. A subproblem bounding calculation can start from the parent's basis and, since there is typically only one new constraint, the LP is usually quickly re-solved with a few pivots of dual simplex. The root problem frequently starts with nothing, not even a feasible point. The pPCx code [27] is a parallel interior-point LP solver. The core computational problem is the solution of a linear system of the form $AD^2A^Tx = b$ where A is the original constraint matrix and D is a diagonal matrix that changes each iteration. Parallel direct Cholesky solvers are robust, but currently do not provide reasonable speed up beyond a few dozen processors.

Sandia National Laboratories is leading a research effort to find more scalable interior-point solvers using iterative linear systems solvers, but this is still an open research problem. We are not aware of any (massively) parallel dual simplex solvers so even during ramp up, subproblem re-solves are serial for the moment.

PICO must use free LP solvers for serial (and eventually parallel) bounding. Faster commercial LP codes do not have licensing schemes for massively-parallel (MP) machines, and individual processor licenses would be prohibitively expensive. PICO has a coping strategy to avoid slow root solves in MP computations. The user can solve the root LP offline (e.g. using a fast commercial solver) and then feed this LP solution to a subsequent parallel computation.

At any tree node, one can parallelize the generation of cutting planes and the search for feasible integer solutions, by cooperating on the generation of one plane/solution or by generating different planes/solutions in parallel. Heuristic methods such as evolutionary algorithms can effectively use many parallel processors independently from the ramp-up tree growth computation. LP-based methods need the LP solutions and are usually best integrated with the tree evaluation. In some cases, such as alpha-point heuristics for scheduling problems [88], an approximation algorithm has a natural parameter (alpha) whose range of values can be partitioned among the processors. PICO's general cut-pivot-dive heuristic can explore multiple strategy choices in parallel [80].

Managing Independent Subproblems. PICO's ramp-up phase usually terminates when there are enough subproblems to keep most processors busy. After the ramp up phase, PICO switches to a phase where processors work on separate subproblems. During this phase, a number of hubs coordinate the search. Each hub controls a set of worker processors, few enough that the workers are not slowed by contention for attention from the hub.

When ramp up ends, each hub takes control of an equal share of the active subproblems. Each worker has a local pool of subproblems to reduce dependence on its hub. Though PICO has considerable flexibility in subproblem selection criteria, once there is an incumbent both hubs and workers generally use a best-first strategy.

PICO has three load balancing mechanisms. First, if a hub runs out of useful work not delegated to a worker, it can rebalance among its workers by pulling some work back from a worker and giving it to others. Second, when a worker decides not to keep a subproblem locally, it returns it to its hub or probabilistically scatters it to a random hub. The probability of scattering depends upon the load controlled by its hub relative to average system load. The third mechanism is global load balancing done with a rendezvous method as described in Section 5.3.

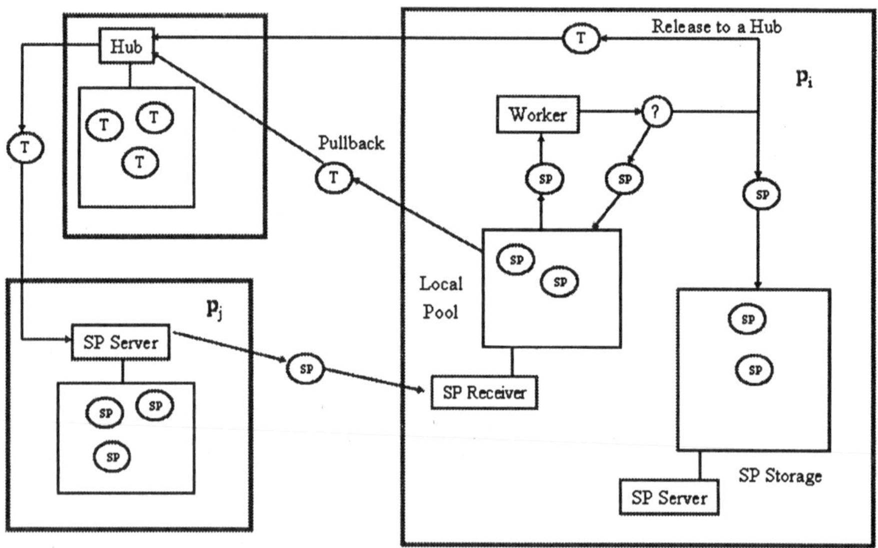

Figure 5.1. PICO's subproblem management. Hubs handle only tokens. When a worker generates a subproblem, it either keeps the subproblem in a local pool or gives control to a hub (not necessarily its own), storing the data to send directly to the worker to which the problem is later assigned. The worker on processor p_j is only partially illustrated. [Figure by Jonathan Eckstein, PICO core designer]

Because hub communication can become a bottleneck, hubs do not handle all the data associated with the subproblems they control. Instead they keep a small *token* containing only the information needed to control the problem. In particular, the token contains the subproblem bound, and the ID of the worker that created the subproblem. When a worker creates a subproblem and relinquishes control to a hub, it stores all the subproblem data (bounds, basis, etc), and sends this small token to the hub (represented by T in Fig. 5.1). When a hub wants to dispatch a subproblem to processor p_i, it sends a message to the processor p_j that created the subproblem telling processor p_j to send the subproblem data to processor p_i (represented by *SP* in Fig. 5.1). Thus the communication pattern in a MIP computation has many small messages going to and from the hubs and, because of the load balancing, long messages going point to point in a reasonably random pattern.

Whenever a processor finds a new incumbent, it broadcasts the new objective value to all processors using a binary tree rooted at that processor. Processors give priority to handling these messages because new incumbents can prune active subproblems. Processors only forward the best incumbent they have seen so far, so if there are multiple incumbent message waves propagating through the system, all dominated ones die as they meet better solutions.

5.2.3 Quadratic Assignment Problem

In this section, we provide another example of B&B applied to the quadratic assignment problem (QAP). Specifically, we summarize the grid-based QAP solver developed by Anstreicher et al. [2]. This solver uses a master-worker paradigm to parallelize B&B. The master-worker paradigm has been widely used to parallelize B&B algorithms [46], and it is very well-suited for grid-based applications. We discuss this programming model, and we provide an overview of the MW framework which was used by Anstreicher et al. [2].

The Master-Worker Paradigm. The master-worker paradigm is a canonical programming paradigm for parallel computing that is particularly well-suited for grid-computing applications. In a master-worker application, all algorithm control is done by a *master* processor. *Worker* processors concurrently execute independent tasks. A wide variety of sequential approaches to large-scale problems map naturally to the master-worker paradigm [48] including tree search algorithms (e.g. for integer programming), stochastic programming, population-based search methods like genetic algorithms, parameter analysis for engineering design, and Monte Carlo simulations.

The master-worker paradigm can parallelize many programs with centralized control using grid computing platforms because it works effectively in dynamic and heterogeneous computing environments. If additional processors become available during the course of a computation, they can be integrated as workers that are given independently executable computational tasks. If a worker fails while executing a task, the master can simply reschedule that portion of the computation. In this manner, the master-worker paradigm is a flexible and reliable tool for grid computing applications. Furthermore, this type of centralized control eases the burden of adapting to a heterogeneous computational environment, since only the master process needs to be concerned with how tasks are assigned to resources.

The basic master-worker paradigm has several limitations. It is not robust if the master fails. However, the programmer can overcome this limitation by using checkpoints to save the state of the master (which implicitly knows the global state of the entire calculation). Another limitation is that the total throughput may become limited because (a) workers can become idle while waiting for work from the master, and (b) precedence constraints between tasks can limit total parallelism of the calculation.

The master-worker paradigm is inherently unscalable because workers can become idle when the master cannot quickly respond to all requests. If the workers finish tasks quickly, there may simply be too many requests for the master to service immediately. The master's response rate is reduced by auxiliary computation (e.g. to prioritize available tasks). In many cases, such bottlenecks can be minimized by adapting the granularity of the tasks. For example, in a tree search the computation required by a task depends on the size/depth of the tasks's subtree. Thus, the master can reduce the rate at which processors make work requests by assigning larger portions of the tree to each processor [48].

A precedence constraint between a pair of tasks (a, b) indicates task a must complete before task b starts. This serialization limits the number of independent tasks at any point in the computation; in the worst case, precedence constraints impose a total linear order on the tasks. On a grid system, with unpredictable processing and communication times, predecessor jobs can delay their successors indefinitely. Some applications permit relaxed precedence. For example, Goux et al. [48] discuss a cutting plane algorithm for stochastic programming for which tasks are weakly synchronized. In this application, the next iteration of the cutting plane algorithm can start after a fraction of the previous iteration's tasks have completed, thereby avoiding some synchronization delays.

Increasing the efficiency of a master-worker algorithm by increasing the grain size or by reducing synchronization can sometimes worsen the basic algorithm. For example, increasing the granularity in a tree search may lead to the exploration of parts of the tree that would have otherwise been ignored. Thus the application developer must balance parallel efficiency (keeping workers busy) with total computational efficiency.

The MW Framework. The MW framework facilitates implementation of parallel master-worker applications on computational grids [48]. The application programming interface of MW is a set of C++ abstract classes. The programmer provides concrete implementation of this abstract functionality for a particular application. These classes define the basic elements of the master controller, how workers are launched and how tasks are executed. The MW framework provides general mechanisms for distributing algorithm control information to the worker processors. Further, MW collects job statistics, performs dynamic load balancing and detects termination conditions. Thus MW handles many difficult metacomputing issues for an application developer, thereby allowing rapid development of sophisticated applications.

MW is a C++ library, and three MW classes must be extended to define a new master-worker application [48]. The master process controls task distribution via the MWDriver class. The MWDriver base class handles workers joining and leaving the computation, assigns tasks to appropriate workers, and rematches running tasks when workers are lost. The programmer must specify how to process commandline information, determine a set of initial jobs, process a completed task, and what information workers need at start up. The MWWorker class controls the worker. The programmer must specify how to process start up data and how to execute a task. The MWTask class describes task data and results. The derived task class must implement functions for sending and receiving this data.

To implement MW on a particular computational grid, a programmer must also extend the MWRMComm class to derive a grid communication object. The initial implementation of MW uses Condor [28] as its resource management system. Condor manages distributively owned collections ("pools") of processors of different types, including workstations, nodes from PC clusters, and nodes from conventional multiprocessor platforms. When a user submits a job, the Condor system discovers a suitable processor for the job in the pool, transfers the executable, and starts the job on that processor. Condor may checkpoint the state of a job periodically, and it migrates a job to a different processor in the pool if the current host becomes unavailable for any reason. Currently, communication between master and workers uses a Condor-enabled version of PVM [92] or Condor's remote system call functionality. MW has also been extended at Sandia National Laboratories to use an MPI-based communication object.

Solving QAP with Branch-and-Bound. The quadratic assignment problem (QAP) is a standard problem in location theory. The QAP in *Koopmans-Beckmann* form is

$$\min_{\pi} \sum_{i=1}^{n} \sum_{j=1}^{n} a_{ij} b_{\pi(i),\pi(j)} + \sum_{i=1}^{n} c_{i,\pi(i)},$$

where n is the number of facilities and locations, a_{ij} is the flow between facilities i and j, b_{kl} is the distance between locations k and l, c_{ik} is the fixed cost of assigning facility i to location k, and $\pi(i) = k$ if facility i is assigned to location k. The QAP is NP-hard. Most exact QAP methods are variants of B&B. Anstreicher et al. [1, 2, 17] developed and applied a new convex quadratic programming bound that provides stronger bounds than previous methods. Because QAP is so difficult, previous exact solutions used parallel high-performance computers. Anstreicher et al. [2] review these results and note that grid computing may be more cost-effective for these problems.

Anstreicher et. al. [2] developed an MW-based branch-and-bound QAP solver (MWQAP). They ran it on a Condor pool communicating with remote system calls. Since Condor provides a particularly dynamic grid computing environment, with processors leaving and entering the pool regularly, MW was critical to ensure tolerance of worker processor failures. To make computations fully reliable, MWQAP uses MW's checkpointing feature to save the state of the master process. This is particularly important for QAP because computations currently require many days on many machines.

The heterogeneous and dynamic nature of a Condor-based computational grid makes application performance difficult to assess. Standard performance measures such as wall clock time and cumulative CPU time do not separate application code performance from computing platform performance. MW supports the calculation of application-specific benchmark tasks to determine the power of each worker so the evaluator can normalize CPU times. For the QAP solver, the benchmark task is evaluating a small, specific portion of the B&B tree.

As we noted earlier, the master-worker paradigm usually requires application-specific tuning for scalability. MWQAP uses coarse granularity. Each worker receives an active node and computes within that subtree independently. If after a fixed number of seconds the worker has not completely evaluated the subtree, it passes the remaining active nodes back to the master. To avoid sending back "easy" subproblems, workers order unsolved nodes based on the relative gap and spend extra time solving deep nodes that have small relative gaps. "Easy" subproblems can lead to bottlenecks at the master because they cannot keep their new workers busy.

The master generally assigns the next worker the deepest active node on its list. However, when the set of active nodes is small, the master dispatches difficult nodes and gives the workers shorter time slices so the workers can return challenging open nodes. This ensures that the master's pool of unsolved subproblems is sufficiently large to keep all available workers busy.

Anstreicher et. al. note that in general B&B, this independent subtree execution could explore many more nodes than its sequential counterpart. However, their QAP calculations are seeded with good known solutions, so many of these nodes are pruned during each worker's search of a subtree. Thus this strategy limits master-worker communication without significantly impairing the overall search performance.

MWQAP has solved instances of the QAP that have remained unsolved for decades, including the nug30 problem defined by Nugent, Vollmand, and Ruml [83]. The Nugent problems are the most-solved set of QAPs, and the solution of these problems has marked advances in processor capability and QAP solution methods. Solution of nug30 required an average of 650 workers for one week when seeded with a previously known solution [2]. The computation halted and restarted five times using MW's checkpointing feature. On average MWQAP solved approximately one million linear assignment problems per second.

5.2.4 Phylogenetic Tree Reconstruction

In this section, we provide an example of B&B applied to reconstructing an evolutionary history (phylogenetic tree). Specifically, we focus on the shared-memory parallelization of the maximum parsimony (MP) problem using B&B based on work by Bader and Yan[10, 78, 120, 121].

Biological Significance and Background. All biological disciplines agree that species share a common history. The genealogical history of life is called phylogeny or an evolutionary tree. Reconstructing phylogenies is a fundamental problem in biological, medical, and pharmaceutical research and one of the key tools in understanding evolution. Problems related to phylogeny reconstruction are widely studied. Most have been proven or are believed to be NP-hard problems that can take years to solve on realistic datasets [20, 87]. Many biologists throughout the world compute phylogenies involving weeks or years of computation without necessarily finding global optima. Certainly more such computational analyses will be needed for larger datasets. The enormous computational demands in terms of time and storage for solving phylogenetic problems can only be met through high-performance computing (in this example, large-scale B&B techniques).

A phylogeny (phylogenetic tree) is usually a rooted or unrooted bifurcating tree with leaves labeled with species, or more precisely with taxonomic units (called *taxa*) that distinguish species [110]. Locating the root of the evolutionary tree is scientifically difficult so a reconstruction method only recovers the topology of the unrooted tree. Reconstruction of a phylogenetic tree is a statistical inference of a true phylogenetic tree, which is unknown. There are many methods to reconstruct phylogenetic trees from molecular data [81]. Common methods are classified into two major groups: criteria-based and direct methods. Criteria-based approaches assign a score to each phylogenetic tree according to some criteria (e.g., parsimony, likelihood). Sometimes computing the score requires auxiliary computation (e. g. computing hypothetical ancestors for a leaf-labeled tree topology). These methods then search the space of trees (by enumeration or adaptation) using the evaluation method to select the best one. Direct methods build the search for the tree into the algorithm, thus returning a unique final topology automatically.

We represent species with binary sequences corresponding to morphological (e. g. observable) data. Each bit corresponds to a feature, called a *character*. If a species has a given feature, the corresponding bit is one; otherwise, it is zero. Species can also be described by molecular sequence (nucleotide, DNA, amino acid, protein). Regardless of the type of sequence data, one can use the same parsimony phylogeny reconstruction methods. The evolution of sequences is studied under a simplifying assumption that each site evolves independently.

The Maximum Parsimony (MP) objective selects the tree with the smallest total evolutionary change. The *edit distance* between two species as the minimum number of evolutionary events through which one species evolves into the other. Given a tree in which each node is labeled by a species, the *cost* of this tree (tree length) is the sum of the costs of its edges. The cost of an edge is the edit distance between the species at the edge endpoints. The *length* of a tree T with all leaves labeled by taxa is the minimum cost over all possible labelings of the internal nodes.

Distance-based direct methods ([37, 38, 68]) require a distance matrix D where element d_{ij} is an estimated evolutionary distance between species i and species j. The distance-based Neighbor-Joining (NJ) method quickly computes an approximation to the shortest tree. This can generate a good early incumbent for B&B. The neighbor-joining (NJ) algorithm by Saitou and Nei [99], adjusted by Studier and Keppler [106], runs in $O(n^3)$ time, where n is the number of species (leaves). Experimental work shows that the trees it constructs are reasonably close to "true" evolution of synthetic examples, as long as the rate of evolution is neither too low nor too high. The NJ algorithm begins with each species in its own subtree. Using the distance matrix, NJ repeatedly picks two subtrees and merges them. Implicitly the two trees become children of a new node that contains an artificial taxon that mimics the distances to the subtrees. The algorithm uses this new taxon as a representative for the new tree. Thus in each iteration, the number of subtrees decrements by one till there are only two left. This creates a binary topology. A distance matrix is *additive* if there exists a tree for which the inter-species tree distances match the matrix distances exactly. NJ can recover the tree for additive matrices, but in practice distance matrices are rarely additive. Experimental results show that on reasonable-length sequences parsimony-based methods are almost always more accurate (on synthetic data with known evolution) than neighbor-joining and some other competitors, even under adverse conditions [98]. In practice MP works well, and its results are often hard to beat.

In this section we focus on reconstructing phylogeny using maximum parsimony (minimum evolution). A brute-force approach for maximum parsimony examines all possible tree topologies to return one that shows the smallest amount of total evolutionary change. The number of unrooted binary trees on n leaves (representing the species or taxa) is $(2n-5)!! = (2n-5)\cdot(2n-7)\cdots 3$. For instance, this means that there are about 13 billion different trees for an input of $n = 13$ species. Hence it is very time-consuming to examine all trees to obtain the optimal tree. Most researchers focus on heuristic algorithms that examine a much smaller set of most promising topologies and choose the best one examined. One advantage of B&B is that it provides instance-specific lower bounds, showing how close a solution is to optimal [56].

The phylogeny reconstruction problem with maximum parsimony (MP) is defined as follows. The input is a set of c characters and a set of taxa represented as length-c sequences of values (one for each character). For example, the input could come from an aligned set of DNA sequences (corresponding elements matched in order, with gaps). The output is an unrooted binary tree with the given taxa at leaves and assignments to the length-c internal sequences such the resulting tree has minimum total cost (evolutionary change). The characters need not be binary, but each usually has a bounded number of states. Parsimony criteria (restrictions on the changes between adjacent nodes) are often classified into Fitch, Wagner, Dollo, and Generalized (Sankoff) Parsimony [110]. In this example, we use the simplest criteria, Fitch parsimony [41], which imposes no constraints on permissible character state changes. The optimization techniques we discuss are similar across all of these types of parsimony.

Given a topology with leaf labels, we can compute the optimal internal labels for that topology in linear time per character. Consider a single character. In a leaf-to-root sweep, we compute for each internal node v a set of labels optimal for the subtree rooted at v (called the Farris Interval). Specifically, this is the intersection of its children's sets (connect children though v) or, if this intersection is empty, the union of its children's sets (agree with one child). At the root, we choose an optimal label and pass it down. Children agree with their parent if possible. Because we assume each site evolves independently, we can set all characters simultaneously. Thus for m character and n sequences, this takes $O(nm)$ time. Since most computers can perform efficient bitwise logical operations, we use the binary encoding of a state in order to implement intersection and union efficiently using bitwise AND and bitwise OR. Even so, this operation dominates the parsimony B&B computation.

The following sections outline the parallel B&B strategy for MP that is used in the GRAPPA (Genome Rearrangement Analysis through Parsimony and other Phylogenetic Algorithms) toolkit [78]. Note that the maximum parsimony problem is actually a minimization problem.

Strategy. We now define the *branch*, *bound*, and *candidate* functions for phylogeny reconstruction B&B. Each node in the B&B tree is associated with either a partial tree or a complete tree. A tree containing all n taxa is a *complete tree*. A tree on the first k ($k < n$) taxa is a *partial tree*. A complete tree is a candidate solution. Tree T is *consistent* with tree T' iff T can be reduced into T'; i.e., T' can be obtained from T by removing all the taxa in T that are not in T'. The subproblem for a node with partial tree T is to find the most parsimonious complete tree consistent with T.

We partition the active nodes into *levels* such that level k, for $3 \leq k \leq n$, contains all active nodes whose partial trees contain the first k taxa from the input. The root node contains the first three taxa (hence, indexed by level 3) since there is only one possible unrooted tree topology with three leaves. The branch function finds the immediate successors of a node associated with a partial tree T_k at level k by inserting the $(k+1)$st taxon at any of the $(2k-3)$ possible places. A new node (with this taxon attached by an edge) can join in the middle of any of the $(2k-3)$ edges in the unrooted tree. For example, in Figure 5.2, the root on three taxa is labeled (A), its three children at level four are labeled (B), (C), and (D), and a few trees at level five (labeled (1) through (5)) are shown. We use depth-first search (DFS) as our primary B&B search strategy, and a heuristic best-first search (BeFS) to break ties between nodes at the same depth. The search space explored by this approach depends on the addition order of taxa, which also influences the efficiency of the B&B algorithm. This issue is important, but not further addressed in this chapter.

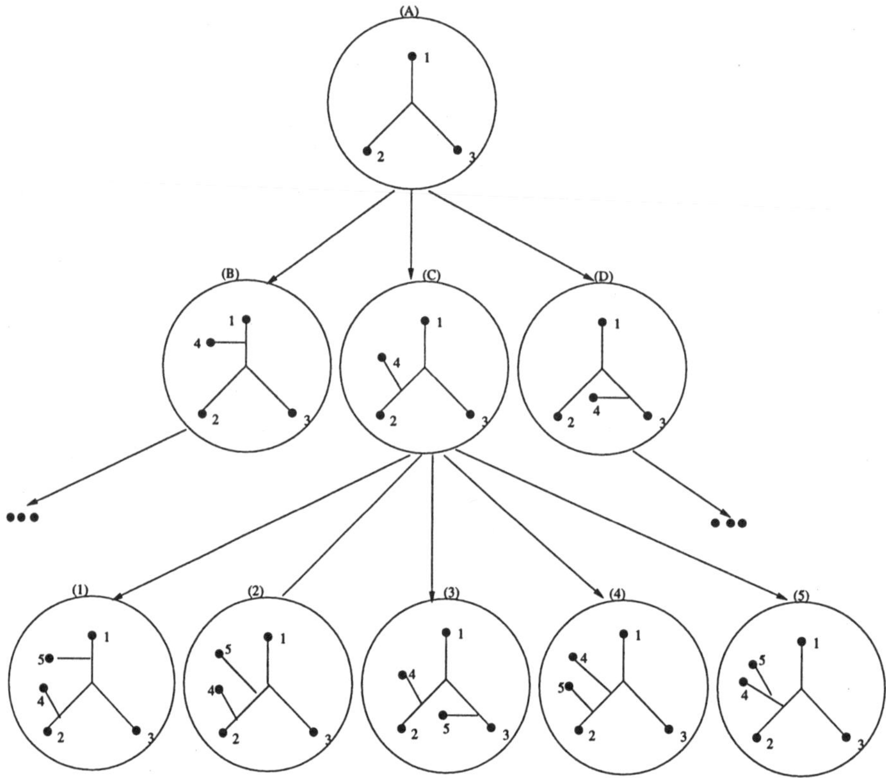

Figure 5.2. Maximum Parsimony B&B search space.

Next we discuss the bound function for maximum parsimony. A node v associated with tree T_k represents the subproblem to find the most parsimonious tree in the search space that is consistent with T_k. Assume T_k is a tree with leaves labeled by $S_1, \ldots, S_k$. Our goal is to find a tight lower bound of the subproblem. However, one must balance the quality of the lower bound against the time required to compute it in order to gain the best performance of the overall B&B algorithm.

Hendy and Penny [56] describe two practical B&B algorithms for phylogeny reconstruction from sequence data that use the cost of the associated partial tree as the lower bound of this subproblem. This traditional approach is straightforward, and obviously, it satisfies the necessary properties of the bound function. However, it is not tight and does not prune the search space efficiently. Purdom et al. [91] use single-character discrepancies of the partial tree as the bound function. For each character one computes a difference set, the set of character states that do not occur among the taxa in the partial tree and hence only occur among the remaining taxa. The single-character discrepancy is the sum over all characters of the number of the elements in these difference sets. The lower bound is therefore the sum of the single-character discrepancy plus the cost of the partial tree. This method usually produces much better bounds than Hendy and Penny's method, and experiments show that it usually fathoms more of the search space [91]. Another advantage of Purdom's approach is that given an addition order of taxa, there is only one single-character discrepancy calculation per level. The time needed to compute the bound function is negligible.

Next we discuss the candidate function and incumbent x_I. In phylogeny reconstruction, it is expensive to compute a meaningful feasible solution for each partial tree, so instead we compute an upper bound on the input using a direct method such as neighbor-joining [99, 106] before starting the B&B search. We call this value the global upper bound, $f(x_I)$, the incumbent's objective function. In our implementation, the first incumbent is the best returned by any of several heuristic methods.

The greedy algorithm [34], an alternative incumbent heuristic, proceeds as follows. Begin with a three-taxa core tree and iteratively add one taxon at a time. For an iteration with a k-leaf tree, try each of the $n-k$ remaining taxon in each of the $2k-3$ possible places. Select the lowest-cost $(k+1)$-leaf tree so formed.

Any program, regardless of the algorithms, requires implementation on a suitable data structure. As mentioned previously, we use DFS as the primary search strategy and BeFS as the secondary search strategy. For phylogeny reconstruction with n taxa, the depth of the subproblems ranges from 3 to n. So we use an array to keep the open subproblems sorted by DFS depth. The array element at location i contains a priority queue (PQ) of the subproblems with depth i, and each item of the PQ contains an external pointer to stored subproblem information.

The priority queues (PQs) support best-first-search tie breaking and allow efficient deletion of all dominated subproblems whenever we find a new incumbent. There are many ways to organize a PQ (see [12] for an overview). In the phylogeny reconstruction problem, most of the time is spent evaluating the tree length of a partial tree. The choice of PQ data structures does not make a significant difference. So for simplicity, we use a D-heap for our priority queues. A heap is a tree where each node has higher priority than any of its children. In a D-heap, the tree is embedded in an array. The first location holds the root of the tree, and locations $2i$ and $2i+1$ are the children of location i.

Parallel framework. Our parallel maximum parsimony B&B algorithm uses shared-memory. The processors can concurrently evaluate open nodes, frequently with linear speedup. As described in Section 5.2.4.0, for each level of the search tree (illustrated in Figure 5.2), we use a priority queue represented by binary heaps to maintain the active nodes in a heuristic order. The processors concurrently access these heaps. To ensure each subproblem is processed by exactly one processor and to ensure that the heaps are always in a consistent state, at most one processor can access any part of a heap at once. Each heap H_i (at level i) is protected by a lock $Lock_i$. Each processor locks the entire heap H_i whenever it makes an operation on H_i.

In the sequential B&B algorithm, we use DFS strictly so H_i is used only if the heaps at higher level (higher on the tree, lower level number) are all empty. In the parallel version, to allow multiple processors shared access to the search space, a processor uses H_i if all the heaps at higher levels are empty or locked by other processors.

The shared-memory B&B framework has a simple termination process. A processor can terminate its execution when it detects that all the heaps are unlocked and empty: there are no more active nodes except for those being decomposed by other processors. This is correct, but it could be inefficient, since still-active processors could produce more parallel work for the prematurely-halted processors. If the machine supports it, instead of terminating, a processor can declare itself idle (e. g. by setting a unique bit) and go to sleep. An active processor can then wake it up if there's sufficient new work in the system. The last active processor terminates all sleeping processors and then terminates itself.

Impact of Parallelization. There are a variety of software packages to reconstruct sequence-based phylogeny. The most popular phylogeny software suites that contain parsimony methods are PAUP* by Swofford [109], PHYLIP by Felsenstein [40], and TNT and NONA by Goloboff [47, 82]. We have developed a freely-available shared-memory code for computing MP, that is part of our software suite, GRAPPA (Genome Rearrangement Analysis through Parsimony and other Phylogenetic Algorithms) [78]. GRAPPA was designed to re-implement, extend, and especially speed up the breakpoint analysis (BPAnalysis) method of Sankoff and Blanchette [100]. Breakpoint analysis is another form of parsimony-based phylogeny where species are represented by ordered sets of genes and distance is measured relative to differences in orderings. It is also solved by branch and bound. Our MP software does not constrain the character states of the input. It can use real molecular data and characters reduced from gene-order data such as Maximum Parsimony on Binary Encodings (MPBE) [30].

The University of New Mexico operates *Los Lobos*, the NSF / Alliance 512-processor Linux supercluster. This platform is a cluster of 256 IBM Netfinity 4500R nodes, each with dual 733 MHz Intel Xeon Pentium processors and 1 GB RAM, interconnected by Myrinet switches. We ran *GRAPPA* on *Los Lobos* and obtained a 512-fold speed-up (linear speedup with respect to the number of processors): a complete breakpoint analysis (with the more demanding inversion distance used in lieu of breakpoint distance) for the 13 genomes in the Campanulaceae data set ran in less than 1.5 hours in an October 2000 run, for a *million-fold* speedup over the original implementation [8, 10]. Our latest version features significantly improved bounds and new distance correction methods and, on the same dataset, exhibits a speedup factor of *over one billion*. In each of these cases a factor of 512 speed up came from parallelization. The remaining speed up came from algorithmic improvements and improved implementation.

5.3 Parallel Algorithmic Primitives

This section describes parallel algorithmic primitives that are representative of the techniques commonly used to coordinate parallel processes. Our three illustrations of parallel B&B use some of these. For a more detailed discussion on algorithm engineering for parallel computation, see the survey by Bader, Moret, and Sanders [9].

5.3.1 Tree-based Reductions

Tree-based reductions are an efficient way to compute the result of applying an associative operator to k values. For example, this is an efficient way to compute the sum or the max of k values. This uses a *balanced binary tree* of processors with k leaves. That is, the heights of the two subtrees rooted at each internal node are approximately equal. Initially each leaf processor holds one of the k input values. Each leaf passes its value to its parent. Each internal node waits to receive the value from each child, applies the operator (e.g. sum or max) and sends the result to its parent. Eventually the root computes the final value. It then sends the results to its children, which in turn propagate the final value, essentially implementing a broadcast from the root. The internal nodes can also have values of their own they add in as the computation proceeds up the tree. In this case, k is the number of nodes rather than the number of leaves. This communication pattern can implement a *synchronization* for example when all processors must wait to write values until all processors have finished reading. Each processor signals its parent when it has finished reading. When the root receives signals from both its children, all processors in the system have finished. When the signal arrives from the root, each processor can continue safely.

5.3.2 Parallel Prefix

The *prefix-sum* operation (also known as the scan operation [14, 15, 57]) takes an array A of length n, and a binary, associative operator $*$, and computes the prefix-sum values $b_i = a_0 * a_1 * \ldots * a_i$, for all $0 \leq i < n$. In parallel, processor i stores value a_i at the beginning and value b_i at the end. There are fast implementations based on balanced binary trees. Suppose an arbitrary subset of processors have a given property (e.g. receiving a 1 from a random bit generator). Each processor knows only whether it has the property or not. We would like to *rank* the processors with the property, giving each a unique number in order starting from 1. This is a prefix operation with $a_i = 1$ if the processor has the property and $a_i = 0$ otherwise. The operator is sum. The ith processor with the property has $b_i = i$. See section 5.3.4 for an example application of ranking. Prefix sums also build other primitives such as array compaction, sorting, broadcasting, and segmented prefix-sums. The latter is a prefix sum where the input array is partitioned into consecutive segments of arbitrary size and the values b_i are computed only within the segments.

5.3.3 Pointer-jumping

Consider a data structure using pointers where each element has at most one outgoing pointer, for example an in-tree or a linked list. In pointer jumping, each element of the data structure begins with its pointer (or a copy of it) and iteratively replaces it with the pointer in the element it's pointing to until it reaches a node with no outgoing pointer. In a linked list, for example, the first element initially points to the second. After one iteration, it points to the fourth, then the eighth, etc. In $\log n$ iterations it points to the nth and final element. This technique is also called path doubling or shortcutting. It can convert trees into strict binary trees (0 or 2 children) by collapsing chains of degree-two nodes. Each node in a rooted, directed forest can find its root quickly. This is a crucial step in handling equivalence classes-such as detecting whether or not two nodes belong to the same component. When the input is a linked list, this algorithm solves the parallel prefix problem.

5.3.4 Rendezvous

In distributed-memory systems, the rendezvous algorithm allows processors to "find" other processors that have similar or complementary properties. The properties evolve during the computation and cannot be predicted a priori. If the properties have unique ranks between 1 and P (the number of processors), then all processors with property type i can "meet" at processor i. Each processor with property i sends a message to processor i with its identifier. Processor i collects all the identifiers sent to it and sends the list to each processor on the list. This effectively "introduces" the processors. As a particular example, consider load balancing in PICO. Each hub computes its local load, a function of the number and quality of subproblems. The hubs all learn the total load via a tree-based reduction (summing the loads). Each hub determines if it is a *donor*, a processor with load sufficiently above average or a *receiver*, a processor with load sufficiently below average. Using a parallel prefix operation on the tree, each donor is assigned a unique rank and each receiver is assigned a unique rank. Then the ith donor and ith receiver rendezvous at processor i. Once they know each other, donor i sends work to receiver i in a point-to-point message.

5.3.5 Advanced Parallel Techniques

An entire family of techniques of major importance in parallel algorithms is loosely termed *divide-and-conquer*—such techniques decompose the instance into smaller pieces, solve these pieces independently (typically through recursion), and then merge the resulting solutions into a solution to the original instance. Such techniques are used in sorting, in almost any tree-based problem, in a number of computational geometry problems (finding the closest pair, computing the convex hull, etc.), and are also at the heart of fast transform methods such as the fast Fourier transform (FFT). A variation on this theme is a *partitioning strategy*, in which one seeks to decompose the problem into independent subproblems—and thus avoid any significant work when recombining solutions; quicksort is a celebrated example, but numerous problems in computational geometry and discrete optimization can be solved efficiently with this strategy (particularly problems involving the detection of a particular configuration in 3- or higher-dimensional space).

Another general technique for designing parallel algorithms is called *pipelining*. A pipeline acts like an assembly line for a computation. Data enters the first stage and proceeds though each stage in order. All stages can compute values for different data simultaneously. Data pumps through the pipeline synchronously, entering the next stage as the previous data exits. Suppose there are k stages, each requiring t time. Then processing each piece of data requires kt time, but a new value arrives every t time units.

Symmetry breaking provides independent work between processors in self-similar problems and allows processors to agree (e. g. on pairings). For example, suppose a set of processors forms a cycle (each processor has two neighbors). Each processor must communicate once with each of its neighbors at the same time that neighbor is communicating with it. In a segment of the ring where processor ranks are monotonic, it's not clear how to do this pairing. One can use previous methods for coloring or maximum independent set or find problem-specific methods.

Tree contraction repeatedly collapses two neighboring nodes in a graph into a supernode until there is only one node left. This algorithm (and its inverse, re-expanding the graph) is part of parallel expression evalution algorithms and other parallel graph algorithms. General graph contraction requires symmetry breaking. Any matching in the graph (a set of edges where no two edges share a vertex) can contract in parallel.

5.3.6 Asynchronous Termination

Proper termination can be tricky in asynchronous distributed-memory computations. Each process p_i has a state s_i that reflects whether the process is active or idle. Active processes can activate idle processes (e.g. by sending the idle process a message). The goal is to terminate when all processes are idle and there are no messages or other pending activations in the system. A process may start a *control wave* which visits all processes and determines whether all processors are idle. A control wave can be implemented by circulating a token sequentially through all processors, or with a tree-based reduction. The control wave also collects information about pending activations. For example, if only messages activate processors, the control wave can collect the number of message sends and receipts at each processor and verify that the totals match globally. However, having all processes idle and no pending activations as inspected by the control wave does not imply all these conditions held *simultaneously*. Processes can be reactivated "behind the back" of the wave, which makes correct termination difficult. For example, *aliasing* may occur when a message receipt substitutes for another. If a sender participates in the wave before it sends a message and the receiver participates after receipt, then only the receipt is recorded. Other messages (e.g. sent but truly undelivered) contribute only to the send count.

Mattern [76] discusses methods to ensure proper termination. PICO uses a variant of the four-counter method [75], which is well-suited to asynchronous contexts in which acknowledgments of "activation" are indirect. PICO uses multiple passes of idleness checks and message balance checks to confirm there is no remaining work in the system. The shared-memory B&B application uses condition variables in its termination procedure. Threads finding no active nodes, and thus wishing to terminate, go to sleep, but can either be awoken with additional work (new active nodes) or terminated by the last working thread.

5.4 Debugging Parallel Software

Debugging parallel software is notoriously difficult. Parallel software coordinates threads of execution across multiple physical processors. Thus parallel software often exhibits programming errors related to timing and synchronization that are not seen in serial codes. Perhaps the most common symptom of a software error is that a code "hangs" and fails to terminate. This occurs when a process is waiting for some event. For example, a process may be waiting for a message that is never sent. This could happen if one process encounters an error condition that other processes do not encounter, thereby leading that process to interrupt its typical flow of communication with the other processes.

Unfortunately it is difficult to robustly reproduce these failures. A parallel bug may not be exposed in repeated executions of a code because of inherent nondeterminism (e.g. nondeterministic message delivery order). Some *race conditions* (order-dependent errors) are exposed only with a rare order of events. Similarly, parallel bugs sometimes disappear when code is inserted to track the state of each process. If such code disrupts the relative rate of computation, the synchronization condition that led to the failure may be difficult to reproduce.

One can develop and debug parallel MPI-based code on a single workstation by running multiple MPI processes on the same machine. These processes share the CPU as independent (communicating) threads. As mentioned later, this can help with debugging. However, because only one thread controls the CPU at once, a code that's fully debugged in this setting can still have bugs (race conditions, synchronization errors) associated with true concurrency.

The remainder of this section considers common approaches to parallel debugging. Specifically, we consider debugging modern programming languages (e.g. Fortran, C and C++) which have compilers and interactive debuggers.

5.4.1 Print Statements

Using print statements to trace interesting events is perhaps the simplest strategy for debugging software. However, there are several caveats for using them for parallel debugging. First, this technique can significantly impact the relative computation rates of processes in a parallel computation. Printing and especially file I/O are often very slow when compared to other computation tasks. Adding printing changes the behavior of asynchronous parallel programs so the precise error condition the debugger is tracking can disappear.

A second caveat for print-based debugging is that the order in which information is presented to a screen may not reflect the true sequence of events. For example, printing I/O for one process may be delayed while a buffer fills, allowing other processes' I/O to be displayed out of order. Explicitly flushing buffers (e.g. using `flush()`), can help, but even then communication delays can affect output ordering.

Finally, it is difficult (and at best inconvenient) to simultaneously display I/O from multiple processes especially for an execution using hundreds to thousands of processors. Operating systems typically interleave the output of multiple processes, which can quickly lead to unintelligible output. One solution to this problem is to stream each process' I/O to a different file. In C, this can be done using `fprintf` statements with a different file descriptor for each process. More sophisticated solutions can be developed in C++ by exploiting the extensibility of stream operators. For example, we have developed a `CommonIO` class in the UTILIB C++ utility library [51] that provides new streams `ucout` and `ucerr`. These streams replace `cout` and `cerr` to control I/O in a flexible manner. The `CommonIO` class ensures that I/O streamed to `ucout` and `ucerr` is printed as a single block, and thus it is unlikely to be fragmented on a screen. Additionally, each line of the output can be tagged with a processor id and line number:

```
[2]-00002 Printing line 2 from process 2
[3]-00007 Printing line 7 from process 3
[3]-00008 Printing line 8 from process 3
[1]-00003 Printing line 3 from process 1
[1]-00004 Printing line 4 from process 1
[0]-00003 Printing line 3 from process 0
```

This facility makes it easy to extract and order output for each process.

5.4.2 Performance Analysis Tools

A natural generalization of print statements is the use of logging, trace analysis, and profiling tools that record process information throughout a parallel run. For example, the MPE library complements the MPI library, adding mechanisms for generating log files. MPE provides simple hooks for opening, closing, and writing to log files. Viewing these logs graphically via the Java-based Jumpshot program provides insight into the dynamic state of a parallel code [79, 123]. The commercial VAMPIR [114] tool provides more general facilities for trace analysis even for multi-threaded applications.

Profiling tools gather statistics about the execution of software. Standard profiling tools like `prof` and `gprof` run on workstations and clusters to identify what parts of the code are responsible for the bulk of the computation. Many commercial workstation vendors provide compilers that support configurable hardware counters which can gather fine-grain statistics about each physical processor's computation. For example, statistics of instruction and data cache misses at various levels of the memory hierarchy and of page faults measure cache performance. These profiling tools are particularly valuable for developing shared-memory parallel software since they help identify synchronization bottlenecks that are otherwise difficult to detect.

5.4.3 Managing and Debugging Errors

A pervasive challenge for software development is the effective management of runtime error conditions (e.g. when an attempt to open a file fails). In parallel codes these include inter-process I/O failures. For example, message passing libraries like MPI include error handler mechanisms. Calls to the MPI library may return errors that indicate simple failure or fatal failure. The process should terminate for fatal failures.

In shared-memory and distributed-computing systems, inter-process communication failures often indicate more critical problems with the operating system or interconnection network. Consequently, it is reasonable to treat these as fatal errors. MPI-based software packages developed for tightly-coupled supercomputers support synchronous communication primitives. They typically assume the network is reliable so that messages sent by one process are received by another. Grid-computing systems usually require asynchronous communication. Inter-process communication failures may reflect problems with the grid computing infrastructure. Tools like Network Weather System (NWS) [84, 116, 118] help identify network-related problems.

One effective strategy for debugging unexpected failures is to generate an `abort()` when an error is detected. This generates a core file with debugging information. When the code runs on a single workstation (preferably with multiple processors), this debugging strategy generates a local core file that contains the call-stack at the point of failure. This debugging strategy can also work for exceptions, which step the computation out of the current context. For example, the UTILIB library includes an `EXCEPTION_MNGR` macro that throws exceptions in a controlled manner. In particular, one configuration of `EXCEPTION_MNGR` calls `abort()` for unexpected exceptions.

5.4.4 Interactive Debuggers

Most commercial workstation vendors provide compilers that naturally support interactive debugging of threaded software. Consequently, mature interactive debuggers are available for shared memory parallel software. There are specialized interactive debuggers for distributed or grid applications running on workstations or clusters. Commercial parallel debugging software tools like Etnus TotalView enable a user to interact with processes distributed across different physical processors [111]. Sun Microsystems offers advanced development, debugging, and profiling tools within their commercial Sun Studio compiler suite and with their native MPI implementation in Sun HPC ClusterTools [107].

On workstations, standard interactive debuggers can attach to a running process given the process identifier. Thus one can debug parallel distributed-memory software by attaching a debugger to each active process. For example, each parallel process prints its process id using the `getpid()` system call and then (a) a master process waits on user I/O (e.g. waiting for a newline) and (b) other processes perform a blocking synchronization; the master performs blocking synchronization after performing I/O. After launching the parallel code, a user can attach debuggers to the given process ids, and then continue the computation. At that point, each debugged process is controlled by an interactive debugger. Setting up debuggers in this manner can be time consuming for more than a few parallel processes, but tools like LAM/MPI support the automatic launching of interactive gdb debuggers [18, 105].

5.5 Final Thoughts

Developing parallel scientific software is typically a challenging endeavor. The choice of target compute platform can significantly influence the manner in which parallelism is used, the algorithmic primitives used for parallelization, and the techniques used to debug the parallel software. Therefore, developing parallel software often requires a greater commitment of time and energy than developing serial counterparts. One must carefully consider the specific benefits of parallelization before developing parallel software. There are a number of issues that might influence the choice of target architecture for parallel software development:

How large are problem instances? How much parallelism is needed? Grid compute platforms deployed within large institutions will likely have a peak compute capacity greater than all but the largest super-computers. However, this scale of parallelization may not be needed in practice, or the granularity of an application might require tighter coupling of resources.

What type of parallelization is needed? If an application has a natural parallelization strategy, this influences the target compute platform. Distributed-memory architectures currently have the greatest capability to model large-scale physical systems (e.g. shock physics). HPC research within DOE has concentrated on tightly-coupled parallel codes on distributed-memory platforms. However, shared-memory platforms are emerging as the system of choice for moderate-sized problems that demand unpredictable access to shared, irregular data structures (e.g. databases, and combinatorial problems).

Who will use the tool, and what type of user support will the software need? The majority of commercial vendors support only parallel software for shared-memory systems. The compute resources of such a system are well defined, and users need minimum training or support to use such parallel tools. For example, a threaded software tool is often transparent.

What are the security requirements of user applications? Security concerns are integral to virtually all business, medical, and government applications, where data has commercial, privacy, or national security sensitivities. The user has control/ownership of the resources and security policy in a distributed-memory system, unlike in a grid system. Even encryption may not provide enough security to run such applications on a grid.

In addition to the issues of decomposition, work granularity and load balancing discussed earlier, the following are algorithmic considerations when developing parallel software:

Managing I/O: I/O is often a bottleneck in parallel computations. Naïve parallelizations of serial software frequently contain hidden I/O (e.g. every processor writes to a log file). Although parallel I/O services are not always available in HPC resources, standards like MPI-IO were developed to meet this need.

Parallel random number generators: When using pseudo-random number generators in parallel, one must ensure the random streams across different processors are statistically independent and not correlated. This is particularly important for parallel computations, such as parallel Monte Carlo simulations, whose performance or correctness depends upon independent random values. Parallel Monte Carlo simulations are estimated to consume over half of all supercomputer cycles. Other applications that use randomized modeling or sampling techniques include sorting and selection [101, 52, 3]. Numerical libraries such as SPRNG [74] help to ensure this independence.

Acknowledgments

This work was performed in part at Sandia National Laboratories. Sandia is a multipurpose laboratory operated by Sandia Corporation, a Lockheed-Martin Company, for the United States Department of Energy under contract DE-AC04-94AL85000. Bader's research is supported in part by NSF Grants CAREER ACI-00-93039, ITR ACI-00-81404, DEB-99-10123, ITR EIA-01-21377, Biocomplexity DEB-01-20709, and ITR EF/BIO 03-31654. We acknowledge the current and former University of New Mexico students who have contributed to the research described in this chapter, including Guojing Cong, Ajith Illendula (Intel), Sukanya Sreshta (OpNet), Nina R. Weisse-Bernstein, Vinila Yarlagadda (Intel), and Mi Yan. We also thank Jonathan Eckstein for his collaboration in the development of PICO.

References

[1] K. M. Anstreicher and N. W. Brixius. A new bound for the quadratic assignment problem based on convex quadratic programming. *Mathematical Programming*, 89:341–357, 2001.

[2] K.M. Anstreicher, N.W. Brixius, J.-P. Goux, and J. Linderoth. Solving large quadratic assignment problems on computational grids. *Mathematical Programming, Series B*, 91:563–588, 2002.

[3] D. A. Bader. An improved randomized selection algorithm with an experimental study. In *Proceedings of the 2nd Workshop on Algorithm Engineering and Experi-*

ments (ALENEX00), pages 115–129, San Francisco, CA, January 2000. www.cs.unm.edu/Conferences/ALENEX00/.

[4] D. A. Bader and G. Cong. A fast, parallel spanning tree algorithm for symmetric multiprocessors (SMPs). In *Proceedings of the International Parallel and Distributed Processing Symposium (IPDPS 2004)*, Santa Fe, NM, April 2004.

[5] D. A. Bader and G. Cong. Fast shared-memory algorithms for computing the minimum spanning forest of sparse graphs. In *Proceedings of the International Parallel and Distributed Processing Symposium (IPDPS 2004)*, Santa Fe, NM, April 2004.

[6] D. A. Bader, A. K. Illendula, B. M. E. Moret, and N. Weisse-Bernstein. Using PRAM algorithms on a uniform-memory-access shared-memory architecture. In G.S. Brodal, D. Frigioni, and A. Marchetti-Spaccamela, editors, *Proceedings of the 5th International Workshop on Algorithm Engineering (WAE 2001)*, volume 2141 of *Lecture Notes in Computer Science*, pages 129–144, Århus, Denmark, 2001. Springer-Verlag.

[7] D. A. Bader and J. JáJá. SIMPLE: A methodology for programming high performance algorithms on clusters of symmetric multiprocessors (SMPs). *Journal of Parallel and Distributed Computing*, 58(1):92–108, 1999.

[8] D. A. Bader and B. M. E. Moret. GRAPPA runs in record time. *HPCwire*, 9(47), November 23, 2000.

[9] D. A. Bader, B. M. E. Moret, and P. Sanders. Algorithm engineering for parallel computation. In R. Fleischer, E. Meineche-Schmidt, and B. M. E. Moret, editors, *Experimental Algorithmics*, volume 2547 of *Lecture Notes in Computer Science*, pages 1–23. Springer-Verlag, 2002.

[10] D. A. Bader, B. M. E. Moret, and L. Vawter. Industrial applications of high-performance computing for phylogeny reconstruction. In H.J. Siegel, editor, *Proceedings of SPIE Commercial Applications for High-Performance Computing*, volume 4528, pages 159–168, Denver, CO, 2001. SPIE.

[11] A. Bar-Noy, S. Guha, J. Naor, and B. Schieber. Approximating the throughput of multiple machines in real-time scheduling. *SIAM Journal on Computing*, 31(2):331–352, 2001.

[12] M. Benchouche, V.-D Cung, S. Dowaji, B. Le Cun, T. Mautor, and C. Roucairol. Building a parallel branch and bound library. In A. Ferreira and P. Pardalos, editors, *Solving Combinatorial Optimization Problems in Parallel:Methods and Techniques*, volume 1054 of *Lecture Notes in Computer Science*, pages 201–231. Springer-Verlag, 1996.

[13] R. H. Bisseling. *Parallel Scientific Computation: A Structured Approach using BSP and MPI.* Oxford University Press, 2004.

[14] G. E. Blelloch. Scans as primitive parallel operations. *IEEE Transactions on Computers*, C-38(11):1526–1538, 1989.

[15] G. E. Blelloch. Prefix sums and their applications. In J. H. Reif, editor, *Synthesis of Parallel Algorithms*, pages 35–60. Morgan Kaufman, San Mateo, CA, 1993.

[16] BOB++. http://www.prism.uvsq.fr/~blec/Research/BOBO/.

[17] N. W. Brixius and K. M. Anstreicher. Solving quadratic assignment problems using convex quadratic programming relaxations. *Optimization Methods and Software*, 16:49–68, 2001.

[18] G. Burns, R. Daoud, and J. Vaigl. LAM: An open cluster environment for MPI. In *Proceedings of Supercomputing Symposium*, pages 379–386, 1994.

[19] G. Calinescu, H. Karloff, and Y. Rabani. An improved approximation algorithm for multiway cut. In *Proceedings of the 30th Annual ACM Symposium on Theory of Computing*, pages 48–52, 1998.

[20] A. Caprara. Formulations and hardness of multiple sorting by reversals. In *3rd Annual International Conference on Computational Molecular Biology (RECOMB99)*, pages 84–93. ACM, April 1999.

[21] W. W. Carlson, J. M. Draper, D. E. Culler, K. Yelick, E. Brooks, and K. Warren. Introduction to UPC and language specification. Technical Report CCS-TR-99-157, IDA Center for Computing Sciences, Bowie, MD, May 1999.

[22] R. D. Carr and G. Konjevod. Polyhedral combinatorics. In H. J. Greenberg, editor, *Tutorials on Emerging Methodologies and Applications in Operations Research*. Kluwer Academic Press, 2004.

[23] H. Casanova and J. Dongarra. NetSolve: Network enabled solvers. *IEEE Computational Science and Engineering*, 5(3):57–67, 1998.

[24] R. Chandra, L. Dagum, D. Kohr, D. Maydan, J. McDonald, and R. Menon. *Parallel programming in OpenMP*. Academic Press, 2001.

[25] Q. Chen and M. C. Ferris. FATCOP: A fault tolerant Condor-PVM mixed integer programming solver. *SIAM Journal on Optimization*, 11(4):1019–1036, 2001.

[26] Computational Infrastructure for Operations Research, home page, 2004. `http://www-124.ibm.com/developerworks/opensource/coin/`.

[27] T. Coleman, J Czyzyk, C. Sun, M. Wager, and S. Wright. pPCx: Parallel software for linear programming. In *Proceedings of the Eighth SIAM Conference on Parallel Processing for Scientific Computing*, 1997.

[28] The Condor Project Homepage, 2004. `http://www.cs.wisc.edu/condor/`.

[29] G. Cong and D. A. Bader. The Euler tour technique and parallel rooted spanning tree. Technical report, Electrical and Computer Engineering Department, The University of New Mexico, Albuquerque, NM, February 2004.

[30] M. E. Cosner, R. K. Jansen, B. M. E. Moret, L.A. Raubeson, L.-S. Wang, T. Warnow, and S. Wyman. An empirical comparison of phylogenetic methods on chloroplast gene order data in Campanulaceae. In D. Sankoff and J. Nadeau, editors, *Comparative Genomics: Empirical and Analytical Approaches to Gene Order Dynamics, Map Alignment, and the Evolution of Gene Families*, pages 99–121. Kluwer Academic Publishers, Dordrecht, Netherlands, 2000.

[31] ILOG, CPLEX home page, 2004. `http://www.ilog.com/products/cplex/`.

[32] D. Culler and J. P. Singh. *Parallel Computer Architecture: A Hardware/Software Approach*. Morgan Kaufmann, 1998.

[33] J. Dongarra, I. Foster, G. Fox, W. Gropp, K. Kennedy, L. Torczon, and Andy White, editors. *The Sourcebook of Parallel Computing*. Morgan Kaufmann, 2002.

[34] R. V. Eck and M. O. Dayhoff. *Atlas of Protein Sequence and Structure*. National Biomedical Research Foundation, Silver Spring, MD, 1966.

[35] J. Eckstein. Parallel branch-and-bound algorithms for general mixed integer programming on the CM-5. *SIAM Journal on Optimization*, 4(4):794–814, 1994.

[36] J. Eckstein, W. E. Hart, and C. A. Phillips. PICO: An object-oriented framework for parallel branch-and-bound. In *Inherently Parallel Algorithms in Feasibility and Optimization and Their Applications*, Elsevier Scientific Series on Studies in Computational Mathematics, pages 219–265, 2001.

[37] D. P. Faith. Distance method and the approximation of most-parsimonious trees. *Systematic Zoology*, 34:312–325, 1985.

[38] J. S. Farris. Estimating phylogenetic trees from distance matrices. *The American Naturalist*, 106:645–668, 1972.

[39] M. Fellows. Parameterized complexity. In R. Fleischer, E. Meineche-Schmidt, and B. M. E. Moret, editors, *Experimental Algorithmics*, volume 2547 of *Lecture Notes in Computer Science*, pages 51–74. Springer-Verlag, 2002.

[40] J. Felsenstein. PHYLIP – phylogeny inference package (version 3.2). *Cladistics*, 5:164–166, 1989.

[41] W. M. Fitch. Toward defining the course of evolution: Minimal change for a specific tree topology. *Systematic Zoology*, 20:406–416, 1971.

[42] S. Fortune and J. Willie. Parallelism in random access machines. In *Proceedings of the 10th Annual ACM Symposium on Theory of Computing*, pages 114–118, 1978.

[43] I. Foster and C. Kesselman. Globus: A metacomputing infrastructure toolkit. *International Journal of SuperComputer Applications*, 11(2):115–128, 1997.

[44] I. Foster and C. Kesselman. *The Grid: Blueprint for a New Computing Infrastructure*. Morgan Kaufmann, 1999.

[45] I. Foster, C. Kesselman, and S. Tuecke. The anatomy of the grid: Enabling scalable virtual organization. *International Journal of High Performance Computing Applications*, 15(3):200–222, 2001.

[46] B. Gendron and T. G. Crainic. Parallel branch and bound algorithms: Survey and synthesis. *Operations Research*, 42:1042–1066, 1994.

[47] P. A. Goloboff. Analyzing large data sets in reasonable times: Solutions for composite optima. *Cladistics*, 15:415–428, 1999.

[48] J.-P. Goux, S. Kulkarni, J. T. Linderoth, and M. E. Yoder. Master-Worker: An enabling framework for applications on the computational grid. *Cluster Computing*, 4:63–70, 2001.

[49] A. Grama, A. Gupta, G. Karypis, and V. Kumar. *An Introduction to Parallel Computing, Design and Analysis of Algorithms*. Addison-Wesley, second edition, 2003.

[50] A. Grimshaw, A. Ferrari, F. Knabe, and M. Humphrey. Legion: An operating system for wide-area computing. Technical report, 1999. Available at `http://legion.virginia.edu/papers/CS-99-12.ps.Z`.

[51] W. E. Hart. UTILIB user manual version 1.0. Technical Report SAND2001-3788, Sandia National Laboratories, 2001. Available for download at `http://software.sandia.gov/Acro/UTILIB/`.

[52] D. R. Helman, D. A. Bader, and J. JáJá. A randomized parallel sorting algorithm with an experimental study. *Journal of Parallel and Distributed Computing*, 52(1):1–23, 1998.

[53] D. R. Helman and J. JáJá. Sorting on clusters of SMP's. In *Proceedings of the 12th International Parallel Processing Symposium*, pages 1–7, Orlando, FL, March/April 1998.

[54] D. R. Helman and J. JáJá. Designing practical efficient algorithms for symmetric multiprocessors. In *Algorithm Engineering and Experimentation*, volume 1619 of *Lecture Notes in Computer Science*, pages 37–56, Baltimore, MD, January 1999. Springer-Verlag.

[55] B. Hendrickson and R. Leland. The Chaco user's guide: Version 2.0. Technical Report SAND94-2692, Sandia National Laboratories, 1994.

[56] M. D. Hendy and D. Penny. Branch and bound algorithms to determine minimal evolutionary trees. *Mathematical Biosciences*, 59:277–290, 1982.

[57] J. JáJá. *An Introduction to Parallel Algorithms*. Addison-Wesley Publishing Company, New York, 1992.

[58] H. F. Jordan and G. Alaghband. *Fundamentals of Parallel Processing*. Prentice Hall, 2003.

[59] N. Karonis, B. Toonen, and I. Foster. MPICH-G2: A Grid-enabled implementation of the Message Passing Interface. *Journal of Parallel and Distributed Computing*, 63(5):551–563, 2003.

[60] G. Karypis and V. Kumar. *MeTiS: A Software Package for Partitioning Unstructured Graphs, Partitioning Meshes, and Computing Fill-Reducing Orderings of Sparse Matrices*. Department of Computer Science, University of Minnesota, version 4.0 edition, September 1998.

[61] L. Khachian. A polynomial time algorithm for linear programming. *Soviet Mathematics, Doklady*, 20:191–194, 1979.

[62] S. Kleiman, D. Shah, and B. Smaalders. *Programming with Threads*. Prentice Hall, Englewood Cliffs, NJ, 1996.

[63] L. Ladányi. BCP (Branch, Cut, and Price). Available from `http://www-124.ibm.com/developerworks/opensource/coin/`.

[64] T.-H. Lai and S. Sahni. Anomalies in parallel branch-and-bound algorithms. *Communications of the ACM*, 27:594–602, 1984.

[65] T.-H. Lai and A. Sprague. Performance of parallel branch-and-bound algorithms. *IEEE Transactions on Computing*, C-34:962–964, 1985.

[66] G.-J. Li and B. Wah. Coping with anomalies in parallel branch-and-bound algorithms. *IEEE Transactions on Computing*, C-35:568–573, 1986.

[67] G.-J. Li and B. Wah. Computational efficiency of parallel combinatorial OR-tree searches. *IEEE Transactions on Software Engineering*, 18:13–31, 1990.

[68] W.-H. Li. Simple method for constructing phylogenetic trees from distance matrices. *Proceedings of the National Academy of Sciences USA*, 78:1085–1089, 1981.

[69] J. T. Linderoth. *Topics in Parallel Integer Optimization*. PhD thesis, Georgia Institute of Technology, Department of Industrial and Systems Engineering, 1998.

[70] M. Livny, J. Basney, R. Raman, and T. Tannenbaum. Mechanisms for high throughput computing. *SPEEDUP*, 11, 1997. Available from `http://www.cs.wisc.edu/condor/doc/htc_mech.ps`.

[71] Mallba library v2.0. `http://neo.lcc.uma.es/mallba/easy-mallba/`.

[72] F. Margot. BAC: A BCP based branch-and-cut example. Technical Report RC22799, IBM, 2003.

[73] A. Marzetta. *ZRAM*. PhD thesis, ETH Zurich, Institute of Theoretical Computer Science, 1998.

[74] M. Mascagni and A. Srinivasan. Algorithm 806: SPRNG: A scalable library for pseudorandom number generation. *ACM Transactions on Mathematical Software*, 26:436–461, 2000. Available for download from `http://sprng.cs.fsu.edu/`.

[75] F. Mattern. Algorithms for distributed termination detection. *Distributed Computing*, 2:161–175, 1987.

[76] F. Mattern. Distributed termination detection with sticky state indicators. Technical report, 200/90, Department of Computer Science, University of Kaiserslautern, Germany, 1990.

[77] R. Miller and L. Boxer. *Algorithms Sequential and Parallel: A Unified Approach*. Prentice Hall, 2000.

[78] B. M. E. Moret, S. Wyman, D. A. Bader, T. Warnow, and M. Yan. A new implementation and detailed study of breakpoint analysis. In *Proceedings of the 6th Pacific Symposium on Biocomputing*, pages 583–594, Hawaii, 2001.

[79] Performance visualization for parallel programs, 2004. `http://www-unix.mcs.anl.gov/perfvis/publications/index.htm`.

[80] M. Nediak and J. Eckstein. Pivot, cut, and dive: A heurstic for mixed 0-1 integer programming. Technical Report RRR 53-2001, RUTCOR, October 2001.

[81] M. Nei and S. Kumar. *Molecular Evolution and Phylogenetics*. Oxford University Press, Oxford, UK, 2000.

[82] K.C. Nixon. The parsimony ratchet, a new method for rapid parsimony analysis. *Cladistics*, 15:407–414, 1999.

[83] C. E. Nugent, T. E. Vollman, and J. Ruml. An experimental comparison of techniques for the assignment of facilities to locations. *Operations Research*, pages 150–173, 1968.

[84] Network Weather Service WWW Page, 2004. `http://nws.cs.ucsb.edu/`.

[85] OpenMP Architecture Review Board. OpenMP: A proposed industry standard API for shared memory programming. `www.openmp.org`, October 1997.

[86] IBM Optimization Solutions and Library, home page, 2004. `http://www-306.ibm.com/software/data/bi/osl/`.

[87] I. Pe'er and R. Shamir. The median problems for breakpoints are NP-complete. Technical Report 71, Electronic Colloquium on Computational Complexity, November 1998.

[88] C. Phillips, C. Stein, and J. Wein. Minimizing average completion time in the presence of release dates. *Mathematical Programming B*, 82:199–223, 1998.

[89] POSIX. *Information technology—Portable Operating System Interface (POSIX)—Part 1: System Application Program Interface (API)*. Portable Applications Standards Committee of the IEEE, 1996-07-12 edition, 1996. ISO/IEC 9945-1, ANSI/IEEE Std. 1003.1.

[90] J. Pruyne and M. Livny. Interfacing Condor and PVM to harness the cycles of workstation clusters. *Journal on Future Generation of Computer Systems*, 12(53-65), 1996.

[91] P. W. Purdom, Jr., P. G. Bradford, K. Tamura, and S. Kumar. Single column discrepancy and dynamic max-mini optimization for quickly finding the most parsimonious evolutionary trees. *Bioinfomatics*, 2(16):140–151, 2000.

[92] PVM: Parallel Virtual Machine, 2004. `http://www.csm.ornl.gov/pvm/pvm_home.html`.

[93] M. J. Quinn. *Parallel Programming in C with MPI and OpenMP*. McGraw-Hill, 2004.

[94] T. K. Ralphs. Symphony 4.0 users manual, 2004. Available from `www.branchandcut.org`.

[95] T. K. Ralphs, L. Ladányi, and M. J. Saltzman. Parallel branch, cut, and price for large-scale discrete optimization. *Mathematical Programming*, 98(1-3), 2003.

[96] T. K. Ralphs, L. Ladányi, and M. J. Saltzman. A library for implementing scalable parallel search algorithms. *The Journal of SuperComputing*, 28(2):215–234, 2004.

[97] J. H. Reif, editor. *Synthesis of Parallel Algorithms*. Morgan Kaufmann Publishers, 1993.

[98] K. Rice and T. Warnow. Parsimony is hard to beat. In *Computing and Combinatorics*, pages 124–133, August 1997.

[99] N. Saitou and M. Nei. The neighbor-joining method: A new method for reconstruction of phylogenetic trees. *Molecular Biological and Evolution*, 4:406–425, 1987.

[100] D. Sankoff and M. Blanchette. Multiple genome rearrangement and breakpoint phylogeny. *Journal of Computational Biology*, 5:555–570, 1998.

[101] H. Shi and J. Schaeffer. Parallel sorting by regular sampling. *Journal of Parallel and Distributed Computing*, 14:361–372, 1992.

[102] Y. Shinano, M. Higaki, and R. Hirabayashi. Generalized utility for parallel branch and bound algorithms. In *Proceedings of the Seventh Symposium on Parallel and Distributed Processing*, 1995.

[103] D. Skillicorn and D. Talia. Models and languages for parallel computation. *Computing Surveys*, 30(2):123–169, 1998.

[104] M. Snir, S. Otto, S. Huss-Lederman, D. Walker, and J. Dongarra. *MPI: The Complete Reference*. MIT Press, Inc., second edition, 1998.

[105] J. M. Squyres and A. Lumsdaine. A component architecture for LAM/MPI. In *Proceedings of the 10th European PVM/MPI Users' Group Meeting*, number 2840 in Lecture Notes in Computer Science, Venice, Italy, September / October 2003. Springer-Verlag.

[106] J. A. Studier and K. J. Keppler. A note on the neighbor-joining method of Saitou and Nei. *Molecular Biological and Evolution*, 5:729–731, 1988.

[107] Sun Microsystems, Inc., WWW Page, 2004. `www.sun.com`.

[108] Sun Microsystems, Inc. POSIX threads. WWW page, 1995. `www.sun.com/developer-products/sig/threads/posix.html`.

[109] D. L. Swofford and D. P. Begle. *PAUP: Phylogenetic analysis using parsimony*. Sinauer Associates, Sunderland, MA, 1993.

[110] D. L. Swofford, G. J. Olsen, P. J. Waddell, and D. M. Hillis. Phylogenetic inference. In D. M. Hillis, C. Moritz, and B. K. Mable, editors, *Molecular Systematics*, pages 407–514. Sinauer, Sunderland, MA, 1996.

[111] Etnus, L.L.C. TotalView WWW Page, 2004. `http://www.etnus.com/Products/TotalView/index.html`.

[112] S. Tschöke and T. Polzer. Portable parallel branch-and-bound library, PPBB-Lib, user manual, library version 2.0. Technical report, University of Paderborn Department of Computer Science, 1996.

[113] L. G. Valiant. A bridging model for parallel computation. *Commununications of the ACM*, 33(8):103–111, 1990.

[114] Intel GmbH VAMPIR WWW Page, 2004. `http://www.pallas.com/e/products/vampir/`.

[115] B. Wilkinson and M. Allen. *Parallel Programming: Techniques and Applications Using Networked Workstations and Parallel Computers*. Prentice Hall, second edition, 2004.

[116] R. Wolski. Dynamically forecasting network performance using the Network Weather Service. *Journal of Cluster Computing*, 1:119–132, January 1998.

[117] R. Wolski, J. Brevik, C. Krintz, G. Obertelli, N. Spring, and A. Su. Running Everyware on the computational grid. In *SC99 Conference on High Performance Computing*, 1999. Available from `http://www.cs.utk.edu/~rich/papers/ev-sc99.ps.gz`.

[118] R. Wolski, N. Spring, and J. Hayes. The Network Weather Service: A distributed resource performance forecasting service for metacomputing. *Journal of Future Generation Computing Systems*, 15(5–6):757–768, 1999.

[119] Dash Optimization, XPRESS-MP, 2004. `http://www.dashoptimization.com/`.

[120] M. Yan. *High Performance Algorithms for Phylogeny Reconstruction with Maximum Parsimony*. PhD thesis, Electrical and Computer Engineering Department, University of New Mexico, Albuquerque, NM, January 2004.

[121] M. Yan and D. A. Bader. Fast character optimization in parsimony phylogeny reconstruction. Technical report, Electrical and Computer Engineering Department, The University of New Mexico, Albuquerque, NM, August 2003.

[122] K. A. Yelick, L. Semenzato, G. Pike, C. Miyamoto, B. Liblit, A. Krishnamurthy, P. N. Hilfinger, S. L. Graham, D. Gay, P. Colella, and A. Aiken. Titanium: A high-performance Java dialect. *Concurrency: Practice and Experience*, 10(11-13):825–836, 1998.

[123] O. Zaki, E. Lusk, W. Gropp, and D. Swider. Toward scalable performance visualization with Jumpshot. *High Performance Computing Applications*, 13(2):277–288, 1999.

Chapter 6

COMPUTER-AIDED DESIGN FOR ELECTRICAL AND COMPUTER ENGINEERING

John W. Chinneck
Department of Systems and Computer Engineering, Carleton University
chinneck@sce.carleton.ca

Michel S. Nakhla
Department of Electronics, Carleton University
msn@doe.carleton.ca

Q.J. Zhang
Department of Electronics, Carleton University
qjz@doe.carleton.ca

Abstract Computer-Aided Design (CAD) in Electrical and Computer Engineering abounds with modeling, simulation and optimization challenges that are familiar to operations researchers. Many of these problems are on a far larger scale and of much greater complexity than usual (millions of variables and constraints), so CAD researchers have of necessity developed techniques for approximation and decomposition in order to cope. The goal of this article is to bridge the gap between the two communities so that each can learn from the other. We briefly review some of the most common O.R.-related problems in CAD, and sketch some CAD techniques for handling problems of extreme scale. We also mention emerging CAD topics that are in particular need of assistance from operations researchers.

Keywords: computer-aided design, operations research, VLSI, electrical and computer engineering, optimization, simulation

Introduction

Computer-aided design (CAD) in Electrical and Computer Engineering (ECE) is a rich source of the kind of complex, large-scale problems that are best tackled using the approaches, tools, and techniques of operations research (OR). Unfortunately, the ECE-CAD and OR communities are mostly disjoint. The purpose of this tutorial is to introduce ECE-CAD to the OR community. Better interaction between the two communities has two main advantages:

- The OR community becomes aware of a fertile area for application of OR skills and tools, leading to better CAD results,
- CAD problems are often of extreme scale and complexity. The OR community can benefit from approaches taken by CAD researchers to deal with models and solutions at this scale, leading to better OR tools and techniques.

"CAD" in ECE generally refers to the design of very large scale integrated (VLSI) circuits, e.g. computer central processors, mobile telephone chips, controllers, etc., as shown in Figure 6.1 for example. VLSI chips have millions of devices and interconnections packed into extremely small areas. The process of design is complex, including choices of materials, selection and sizing of devices, layout, interconnection routing and so on, all subject to objectives such as minimizing area, power consumption, or cost, or maximizing reliability, performance etc. In addition, the physical devices are very difficult to model accurately, generally having highly nonlinear behavior. Given the scale and complexity of the problem, computer assistance is absolutely necessary. The ECE community has been developing CAD tools for many years, and has learned valuable practical lessons along the way. The OR community would do well to absorb those hard-won lessons, but at the same time has its own lessons to offer towards improving the state of the art in the field.

The CAD process involves modeling, simulation (in the sense of accurately predicting the behavior of complex nonlinear integrated circuits), and optimization. One common theme is the interaction between these three activities. For example, how a system is modeled impacts the choice of optimization tools. It may be preferable to use a simpler model (derived from simulation studies) that has properties that make it amenable to solution via commercial nonlinear programming tools. The solution returned by this simple model may then start anew the cycle of simulation, approximation, and optimization.

Figure 6-1. Example VLSI chip.

Table 6-1 illustrates the extent to which common techniques of operations research are used in ECE. The Institute of Electrical and Electronics Engineers, the world's largest professional organization for electrical and computer engineers, maintains a bibliographic database (IEEE Xplore [48]) that can be searched by keyword. Table 6-1 shows the results of a search for a selection of operations research related keywords in paper abstracts. The search covers IEEE and IEE journals, conference proceedings, and standards documents in recent years (2000 to present). It is interesting to note the very large number of papers whose abstracts mention the very generic terms "simulate (or simulation)" and "optimization (or optimize)". This contrasts quite markedly with the small number of papers whose abstracts mention such bread and butter OR techniques as "linear program(ming)", "branch and bound", "nonlinear program(ming)" or "nonlinear optimization".

In contrast, "neural network" and "genetic algorithm" appear relatively frequently. This partly reflects the scale of the problems dealt with in VLSI CAD. Neural networks are commonly used in model simplification, and genetic algorithms are one of the few effective ways to deal with the large-scale mixed-integer nonlinear programs that constitute the core of CAD. However this also reflects the general underutilization of many of the basic tools of OR.

Table 6-1. IEEE Xplore bibliographic database (2000 to present)

Search term	**papers**
Simulate or simulation	46,725
Optimization or optimize	14,216
Neural network	6,251
Genetic algorithm	2,603
Linear program(ming)	576
Simulated annealing	448
Branch and bound	199
Nonlinear program(ming) or nonlinear optimization	159
Integer program(ming)	152
Mathematical program(ming)	68
Operations research	37
PERT	17

It is also interesting to note the relatively frequent appearance of the simulated annealing heuristic, which is mentioned almost as often as linear programming. Prior to the appearance of genetic algorithms, simulated annealing was the main algorithm that demonstrated practical success in solving large-scale mixed-integer nonlinear CAD design problems. Genetic algorithms have now become more popular, but simulated annealing is still used in some CAD tools. Table 6-2 shows the popularity of the two heuristics in four-year increments.

Table 6-2. IEEE Xplore abstract citations for genetic algorithms and simulated annealing

heuristic	1984-87	1988-91	1992-95	1996-99	2000-03
genetic algorithm	0	111	1103	2458	2603
simulated annealing	8	281	456	519	448

The relative frequency of "genetic algorithm" and "simulated annealing" highlights the tendency in the CAD community to concentrate on a few dominant solution techniques and to be generally unaware of other, perhaps better, techniques that may be available in the OR community. It is our hope that this tutorial will spark the interest of operations researchers in transferring knowledge to the CAD community.

6.1 Computer-Aided Design in Electrical and Computer Engineering

The process of design in electrical and computer engineering is nowadays substantially computer-based. Computer-aided design techniques for microelectronics, VLSI, and microwave circuits and systems play an especially important role in realizing new computer and telecommunication systems. CAD techniques are needed at many levels: at the semiconductor process level, transistor device level, and circuit and systems levels.

There are several categories of design, including architectural, functional, and physical design. *Architectural design* involves early decisions concerning the structural and technology aspects of the overall problem such that costly structural rework in later design stages can be avoided or minimized. *Functional design* aims to achieve required functional relationships using a mixture of electronic hardware and software implementations, involving aspects such as hardware/software co-design and synthesis from functional requirements to gate or transistor level realizations. *Physical design* involves the determination of physical and geometric parameters during physical layout, placement, packaging, interconnection, and so on. The functionality of the overall system often involves digital, analog and mixed digital/analog circuits. A simplified view of the design process is shown in Figure 6.2. Although the design process is often described in a linear top-down fashion for simplicity, in reality there is much iteration back and forth. Both top-down and bottom-up approaches are needed and are combined in an iterative fashion.

A CAD process involves 3 types of computer-based techniques: modeling, simulation and design. *Modeling* establishes a computational representation of the electrical or physical behavior of various components in the overall problem with respect to physical or electrical parameters. The computational part of the model does not have to be the first-principle theory of the original problem: empirical, semi-analytical and equivalent representations are often used. *Simulation* is used to compute the electrical or physical responses of an overall component, circuit or system, typically based on first principle formulas or variations as the underlying framework for computation. *Design* adjusts circuit or component parameters in order to make the circuit or system satisfy certain design objectives. Design is typically an iterative process where simulations are used at each iteration to provide the circuit or system behavior. Each simulation in turn requires the use of various models of components. This process is illustrated in Figure 6.3.

There has been a tremendous amount of research in the past several decades leading to substantial automation of many aspects of circuit design, especially in digital design. However analog-related problems remain the most difficult facing CAD researchers today, such as analog electrical and electromagnetic phenomenon in high-speed digital circuits, high-frequency RF (radio-frequency)/microwave circuits such as those in the front-end of

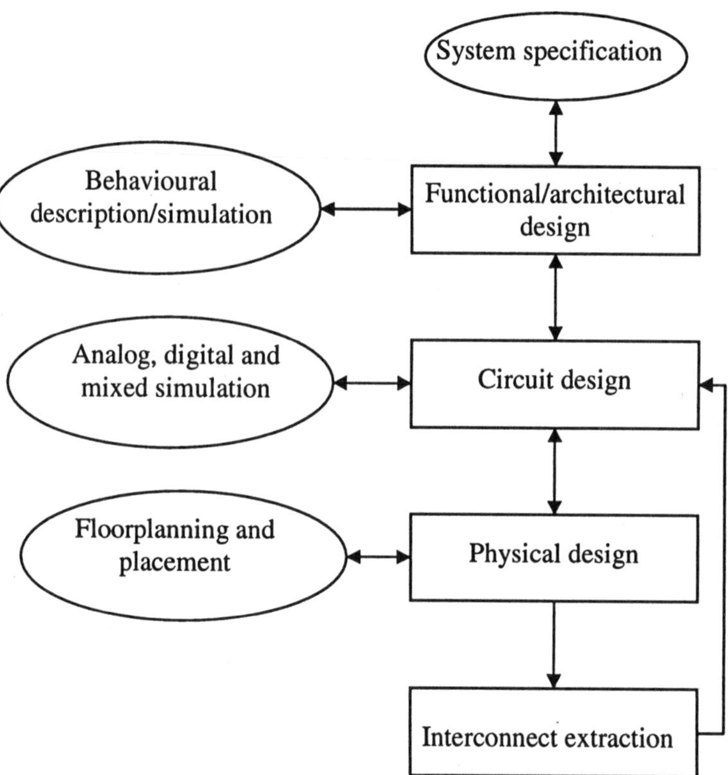

Figure 6-2. Simplified view of the design process.

wireless communication systems. The difficulty of these problems is compounded by the fact that they typically involve both electrical and physical design parameters. The design of high-speed and high-frequency circuits and systems is emerging as one of the most active areas of CAD research today.

CAD for electrical and computer engineering is a very broad area. This chapter focuses on several important themes and directions that are of current relevance and also potentially of interest to operations researchers, such as physical design, circuit optimization, and CAD for high-speed/high-frequency circuits and systems.

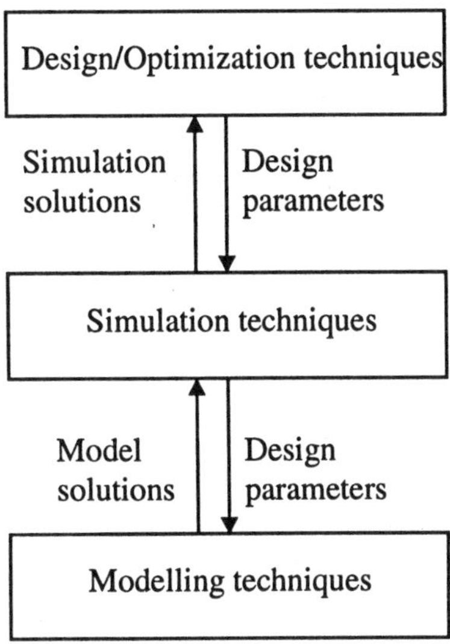

Figure 6-3. Typical 3-level CAD hierarchy with modelling techniques at the lowest level and design/optimization techniques at the highest level. Higher level techniques use lower level techniques during computation. The parameters in the lower level may be dictated by the higher level.

6.2 Physical Design

Physical Design is a recognized, mature area in CAD. It deals with the physical design and layout of VLSI circuits, for example synthesizing the initial abstract design, then placing and sizing the devices and connecting wires. In abstract terms, this is a mixed-integer nonlinear optimization problem of colossal scale involving discrete decisions such as choices of technology as well as complex nonlinear optimizations to size and place the constituent devices. In practice the problem is typically broken into a series of phases, with frequent iteration between phases. A typical breakdown of phases is [44]:

1. *Conceptual design.* The product is conceptualized and project management goals are set. General simulation tools such as Matlab might be used.

2. *System design.* The overall architecture is designed and partitioned. Tasks to be carried out by hardware and software are defined. Implementation issues are decided, such as the selection of the chip materials (silicon, gallium arsenide, etc.), package design, and overall testing strategy. Hardware and software simulation tools may be used. The output hardware design is normally a schematic logic diagram.

3. *Architectural design.* The logic diagram is decomposed into functional blocks (or cells), including decisions on which blocks will be analog and which will be digital. Hardware and software simulation tools are again used.

4. *Cell design.* This is the first level of detailed design, whose output is a detailed circuit schematic. Circuit topology is decided and the circuit parameters are sized. Larger blocks are further decomposed into sub-blocks. Tolerances are taken into account to make sure that the circuit can be manufactured with high yields. Detailed circuit simulation tools such as SPICE [80] are used.

5. *Cell layout.* The schematic is converted into IC layouts that can be manufactured. Layouts are multi-layered and include the locations and sizes of the constituent devices, and the routing of the interconnections between devices. A major consideration is minimizing chip size and power consumption. Optimization tools are much used, as are detailed simulations.

6. *System layout.* Given that the individual cells have been designed, a similar layout and routing problem exists at the system level. Optimization and detailed simulation are again the major tools.

Backtracking between stages and re-design is common since the decisions made at each stage affect the stages both before and after. Note how early decisions (such as the decomposition into cells) impose limits on the later decisions (such as cell design and layout). This arrangement helps to manage the complexity of the process, but at the same time it limits the solution space. Sub-optimal decisions may be made at earlier stages, leading to designs that are not as good as they could be. Even with the reduction in complexity afforded by this approach, computer-aided tools are still necessary at every stage – even the reduced problems are extremely difficult to solve to optimality.

6.2.1 Physical Design Problems

There are numerous specific problems to be solved in physical design. Brief outlines of the most common tasks are given below, following the excellent book by Sherwani [86]. Sherwani also provides a very good tutorial overview of the huge range of special-purpose heuristic optimization algorithms applied to these tasks. Our goal here is give the general idea of the optimization problems posed by physical design, so details of the model formulations and many algorithms used in their solutions are delegated to several excellent tutorial textbooks (Sherwani [86], Sait and Youssef [82], Drechsler [35], etc.).

Partitioning

Partitioning is the process of breaking down the overall circuit into manageable subsystems, each of which is then individually designed. Because of the size of the overall problem, this is often a hierarchical process. The original circuit is broken into modules or blocks, usually with the goal of minimizing the number of interconnections between modules. Partitioning happens at the system level to divide the circuit into printed circuit boards, at the board level to divide the printed circuit boards into chips and at the chip level to divide the chip into manageable components.

At all levels the input is a set of components and their interconnections and the output is a set of subcomponents. The objective function may be to minimize the number of interconnections between subcomponents, or to minimize the maximum number of times that a path through the overall circuit crosses partition boundaries. The latter objective arises because the delay in signal propagation between components is significantly larger than the delay within a component. The constraints may include the maximum area occupied by a subcomponent, the maximum number of terminals, or the maximum number of subcomponents. Of course, the division into subcomponents must also not affect the functioning of the overall circuit.

Partitioning is usually viewed as a graph-partitioning problem, which is known to be NP-complete [43], so a wide variety of heuristics have been developed. The two main classes of heuristics include improvement algorithms that start with a constructed solution that is then improved by swapping pairs of vertices between partitions, and simulated annealing and evolution based algorithms. There is also a wide variety of other approaches. An example of partitioning a logic circuit is shown in Figure 6.4.

Placement, Floorplanning, and Pin Assignment

Once the individual components have been identified by partitioning, they must be assigned a shape and positioned. Components cannot overlap, and sufficient space must be left to place the interconnecting "wires" (this is done during the routing phase). Major considerations include positioning to minimize the total area consumed and deciding the location of the interconnection terminals or pins. Some standard components may have known sizes and shapes, but others need to be designed.

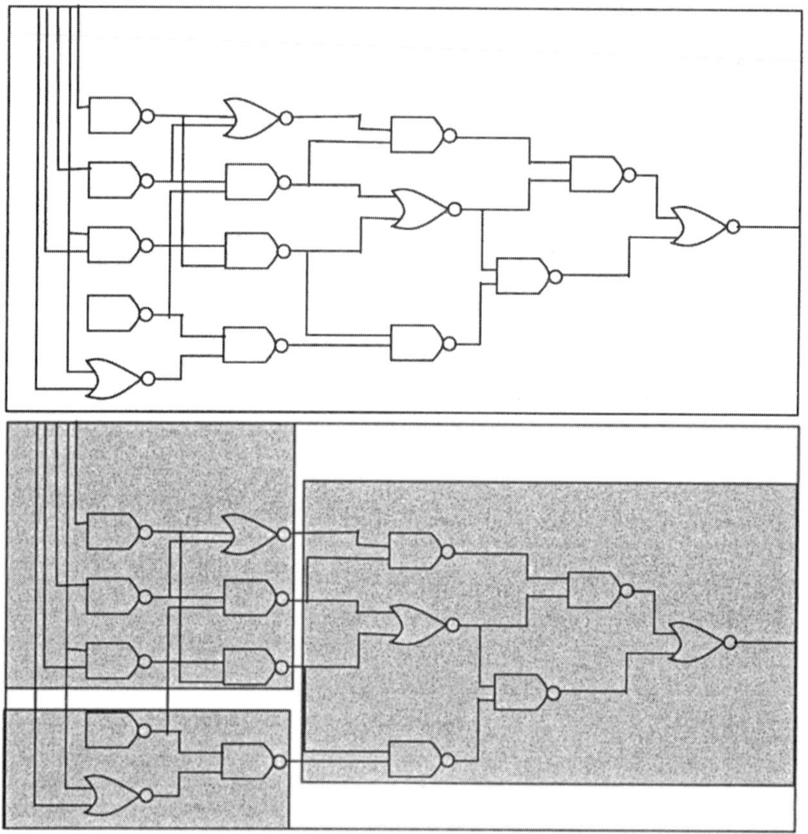

Figure 6-4. Partitioning a logic circuit (above) into three components (below).

If the component shapes are known, this is called a *placement* problem. If the component shapes are not known then this is called a *floorplanning* problem. The positioning of the terminals of a component, called *pin assignment* normally happens after the components have been laid out.

Considerations in solving these problems include:

- *Shape of the components*. Rectangular shapes are often assumed, but L-shapes and more complex shapes can be considered, though the resulting floorplanning or placement problem becomes more difficult.
- *Routing*. While this happens in the next phase, the component placement must leave enough room for the interconnections. At the same time we do not want to leave too much room since an important objective is area minimization.
- *Performance*. There are often timing limitations on circuits, so critical paths must be carefully laid out to minimize the time consumed.
- *Packaging*. The hotter components must be carefully distributed so that heat dissipation is uniform. This requirement may conflict with some of the performance objectives.
- *Pre-placed components*. The placement may have to work around components that have been pre-placed for technical reasons.

A wide range of algorithms have been applied to the placement problem, including simulated annealing, genetic algorithms, and "force-directed" heuristics. The latter uses an analogy to spring forces between components that should be connected, and tries to find a solution that achieves equilibrium among the forces. A variety of partitioning-based algorithms are also used. These subdivide the component iteratively and at each iteration also subdivide the layout area, continuing until each individual device has been assigned a specific position. Other methods include the opposite bottom-up cluster growth approach, a quadratic assignment formulation, and branch and bound.

When performance is the major consideration, different placement algorithms may be used. The physical length of the route of the critical path in the circuit may be minimized, or there may be timing limitations placed on signal propagation through individual components and these components are then laid out in such a way as to meet these restrictions.

Floorplanning is more difficult, since the shapes of the components must be determined (though rectangles are often assumed). Typical input to the problem is the set of components with their associated areas. The output is the shape of each component and its placement. Again, a wide range of heuristic algorithms are used. In constraint-based floorplanning, rectangular components are assumed, and a large set of constraints on the horizontal and vertical placement of the components is generated. An iterative series of steps removes redundant constraints and looks at the horizontal and vertical paths to reshape the components looking for a better floorplan.

Integer programming has also been applied to floorplanning, as has a partitioning-based approach known as "rectangular dualization".

Pin assignment deals with the physical placement of the component interconnection points. Pins are *functionally equivalent* if exchanging their signals does not affect the circuit, and are *equipotential* if both are internally connected. Pin assignment then optimizes the assignment of subcomponents within equipotential or functionally equivalent groups. Correct assignment reduces the amount of congestion and crossover in the subsequent wire routing.

There are numerous geometric heuristics for pin assignment. Dynamic programming has also been applied.

Figure 6.5 gives a simple example of a set of blocks and pins and one possible floorplan.

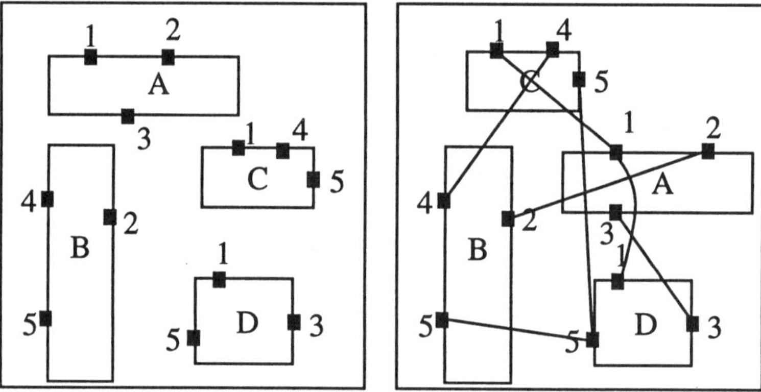

Figure 6-5. Rectangular floorplanning and pin assignment. Left: a set of blocks and pins. Right: a possible arrangement of the blocks and connecting networks (prior to routing).

Global Routing

Laying out the interconnecting "wires" between components is known as routing. Routing takes place in the spaces left between the components and must be done such that the capacity of the spaces is not violated and so that the wires do not short-circuit. The objective is usually to minimize the total wire length, or perhaps to minimize the longest wire length to obtain the best performance. There may be different treatments of routing to clock components, and power and ground circuits.

Routing channels between components vary in their capacity depending on dimensions (height, number of layers) and wire features (width and minimum separation). Global routing assigns a sequence of routing channels to each wire, making sure to respect the channel capacities.

The general routing problem is NP-complete, so it has traditionally been subdivided into an initial global routing phase that is more approximate, and a subsequent detailed routing phase. See Figure 6.6 for an example of approximate global routing.

Global routing is abstracted as various kinds of graph models which are then treated by a wide array of heuristic methods. Some take a sequential approach in routing one interconnection at a time, while others take a concurrent approach. Algorithms make use of various graph algorithms including shortest path routines and routines for finding Steiner trees. Integer programming has also been applied.

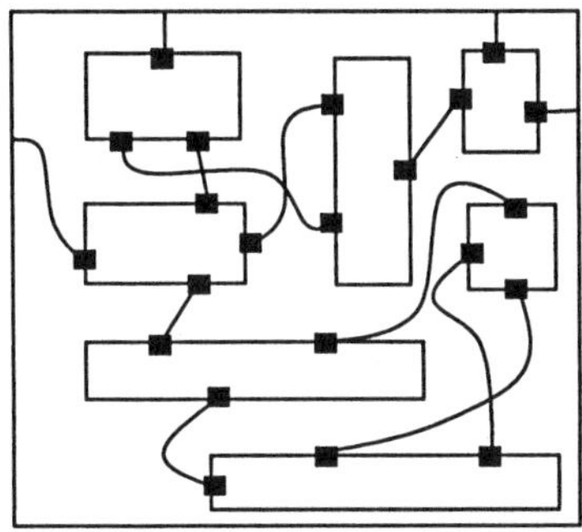

Figure 6-6. Example of global routing.

Detailed Routing

Global routing does not actually lay out the individual wires – it simply defines a set of more restricted detailed routing problems, which must then be solved during detailed routing. Characteristics of a detailed routing problem include:

- *Number of terminals.* Most connections are made between two terminals, but some connections are made between many terminals, e.g. connecting the clock.
- *Number of layers.* Wires can be laid down on several levels. More layers increases the complexity of the solution, as well as the manufacturing cost.
- *Connection width.* This is influenced by the layer the connection is assigned to and how much current it must conduct.
- *Interlayer connections.* The number of interlayer connections (vias) is usually kept to a minimum since these require great precision in laying down the layers and are a source of manufacturing difficulties.

- *Boundary shape.* The boundaries of the channels are normally straight lines. Routing is much more difficult when irregular channel boundaries are involved.
- *Type of connection.* Power, ground, and clock connections are critical. Power and ground connections are normally wider, and clock connections get preferential treatment since they can affect the performance of the entire chip.

Some approaches use a grid-based model to limit the possible wire routes; more complex models may be gridless. Wires may be pre-assigned to a layer, or a layer may be unreserved.

In channel-routing problems, the terminals of two components are located on opposite sides of the space between them. The two component faces are assumed to be linear and parallel, creating the intervening channel. The problem is to determine how to interconnect the terminals as necessary for proper functioning of the circuit, but in such a way that the channel has minimum width. Even with grid-based models, this is quite challenging, involving the selection of grid based routes on several levels and the location of vias. Even single-layer routing is known to be NP-complete in the general case.

Switchboxes are places where channels intersect. Switchbox routing is more complicated because terminals can be located on all four sides of the intersection.

A wide variety of algorithms have been developed for detailed routing. Many involve graph-theoretic approaches. Some are specialized for one, two, three, or multi-layer arrangements. Some try to minimize channel "congestion", others try to minimize the number of turns ("doglegs") in the connections, some use solution backtracking (called "rip-up"). This is a very fundamental step in VLSI chip design, so it has been intensively studied, but no one algorithm dominates.

Via Minimization and Over-the-Cell Routing

Vias are not desirable from the point of view of reliable chip manufacture, electrical performance, and reduction in available wire routing space, so the detailed routing is often revisited by specialized algorithms in order to minimize the number of vias. Also, the placement of components in standard cell designs may be such that reductions in channel size are possible only if layers available *over* the components are used. Hence, channel routing can also be re-visited with the goal of obtaining further channel size reductions. Vias are illustrated in Figure 6.7.

In constrained via minimization, the wire routes are known, but the layers have not been assigned. The objective is then to complete the routing by assigning the wires to layers so that the number of vias is minimized. A variety of algorithms have been developed. Under certain conditions the two-layer problem can be solved optimally. In unconstrained via minimization, neither the wire routes nor the layers are known beforehand.

Since over-the-cell routing is a generalization of channel routing, it is also NP-complete. Graph aspects are heavily involved in most solution approaches, e.g. the minimum density spanning forest problem.

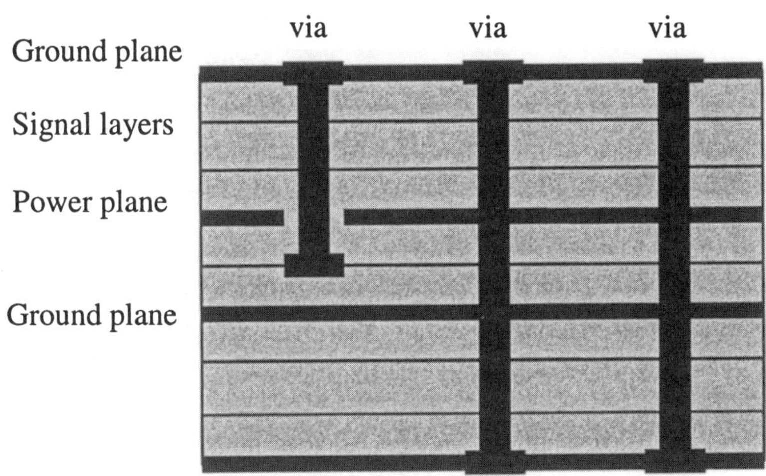

Figure 6-7. Cross-sectional view of via holes in a multi-layer printed circuit. Vias are used to connect signal or power or ground between layers.

Specialized Routing

The chip clock and power connections require special routing algorithms. The maximum length of any connection from the clock to another component affects the maximum frequency for the entire chip. Considerations in clock routing include the electrical characteristics of the metal layers, interference signals from neighbouring wires, and the type of load. Clock signals must arrive nearly simultaneously at all of the connected components. One extremely simple approach is the "H-tree" algorithm, shown in Figure 6.8. More sophisticated algorithms, of which there are several, are more commonly used.

Power and ground routing must also be provided to all components. Considerations include current density and total area used by these wider wires. The width of power and ground wires varies over the circuit according to component power needs and is to be minimized as much as possible.

Both problems are solved in practice using graph-based heuristics.

Compaction

At this stage, the chip design could actually be manufactured. However, because so many of the algorithms used so far are heuristics that do not guarantee optimality, there are still improvements that can be made. The compaction step looks for opportunities to reduce the size of the chip even further by bringing components closer together, reducing their size, or by reshaping them.

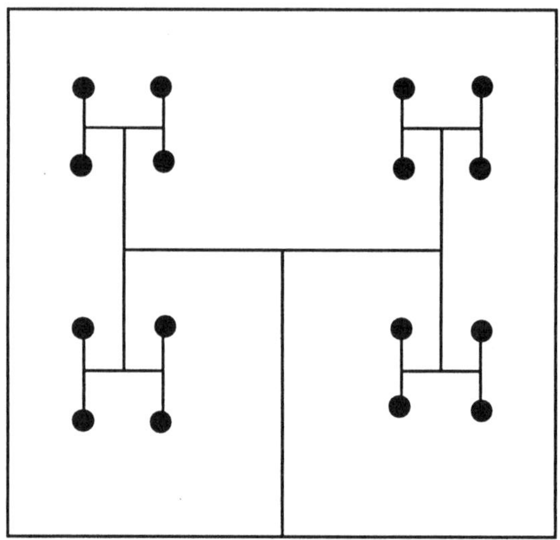

Figure 6-8. Example H-tree for clock routing [86].

Numerous special-purpose algorithms are applied. One approach constructs "constraint graphs" in which each vertex is a component and each arc is either a physical separation constraint or a connectivity constraint. This graph is then used in numerous ways, such as finding critical paths or strongly-connected components which can then be optimized. Another approach involves the use of a virtual grid to look for compaction opportunities.

6.2.2 Top Ten Current Physical Design Challenges

There is still a great deal of opportunity for the development of improved methodologies for physical design, mostly improved optimization heuristics. The Semiconductor Research Corporation has membership from key semiconductor companies. Its Task Force on Physical Design monitors the research and development needs of member companies in the Physical Design area. The Task Force publishes a list of the top ten physical design problems that require better solutions [85], as summarized in the following sections. These are obvious topics for research, all with a definite operations research flavor. Notice also the significant overlap with the "classical" list of physical design problems in the previous section: these problems are not yet solved to a satisfactory degree.

Block Level Placement

Better algorithms for the placement of blocks ("components") are still needed. The Task Force places some emphasis on the interactions between the levels in the partitioning hierarchy so that a better final solution is obtained. Components at the various levels are referred to as giga-blocks, mega-blocks, blocks, and cells. Constraints on area, timing and coupling or shielding and placement of repeaters should be considered. Better estimates of the area and delay characteristics of blocks are needed; this may involve some iteration between levels. Noise and power should also be considered along with the usual delay and area minimization objectives.

To give an idea of the scale of the problem, the suggested benchmark is an existing microprocessor or other large chip that has more than 25 Mega blocks and more than 25 blocks per Mega block where the number of cells within a block should be greater than 10,000. This highlights the need for a hierarchical decomposition to create sub problems of manageable scale.

New Physical Design Tools for Silicon on Insulator Technology

Chip technology is constantly evolving. There are a number of chip materials in addition to basic CMOS, each of which has its own advantages and drawbacks. A recent promising addition is silicon on insulator technology. Because the physical characteristics of this material are different than for plain CMOS, the design approaches and algorithms cannot be directly re-used for the new technology. New design tools are needed. As for any chip design, the objective is to optimize area, timing, coupling, and power. The new design algorithms should be careful to account for floating body effects that occur in silicon on insulator technology such as hysteretic behavior, frequency-dependent delay time, and pulse stretching.

Interconnect Driven Floorplanning

Classical floorplanning is completely separated from interconnection planning, whereas these two problems do impact each other greatly. The Task Force suggests that floorplanning algorithms should also consider interconnection issues in an integrated manner. Area, shape, proximity, alignment, timing, coupling, noise and shielding constraints all need to be considered, as well as the insertion of repeaters. Interconnection issues are a dominant aspect of the overall design, so they must be considered at every stage. The objective function deals with minimizing area or signal propagation times, as usual.

High Speed Clock Interconnect Design with Power and Skew Constraints

As previously mentioned, interconnect routing involving the clock is particularly important – and difficult. The problem is not yet solved to industrial satisfaction. The clock signal must reach each component with minimum jitter (variability in signal propagation time) and skew (difference in arrival time compared to arrival time at other components) and within the power limitations. This is difficult because clock interconnects usually have the greatest fan-out and the signals travel the farthest and operate at the highest speeds. Long clock interconnects are a special problem because long lines have increased resistance.

Floorplanner Shape Generator

Achieving an early floorplan is essential to completing a chip design quickly. New floorplanning tools need to consider more aspects of the problem other than just the estimated area of the resulting block. It should also consider the type of block, the number of cells, the interconnections, the process technology and the possible shapes for the block. The algorithm should output the area and the shape of the blocks, considering designs other than rectangular. As in classical floorplanning the objective is to minimize the total area consumed.

Layout Driven Synthesis

In the classical approach, the logic of the circuit is first decided, and then it is laid out. However interconnect delay has come to dominate physical design in recent years, so it may be possible to get better designs by considering layout during logic synthesis. The goal is a logical circuit that has better delay characteristics when ultimately laid out.

Placement Techniques for Large Blocks

Chips are continually incorporating more and more components. The challenge now is to find placement techniques for around 1,000,000 cells per block. The placement should be completed within a typical 12-16 hours of computation time. An optimization priority of timing, then area, then noise, then power is suggested. The ability to manufacture the resulting design with acceptable yields should also be considered. Note that hierarchical approaches are still considered, but blocks must be as large as possible so the block placement problem is not too difficult. The Task Force considers it imperative that the million components placement problem be solved within next two years to be able to take advantage of the capabilities of the coming next generation technology, and to be able to generate the larger designs that will then be possible.

RLC Routing

The electrical properties resistance (R), inductance (L) and coupling capacitance (C) need to be taken into account during routing. Inductance in particular is increasingly a problem in chips due to wide and long interconnect wires and increased operating speed. Coupling capacitance is also more and more of a problem as wires are packed more tightly. Traditional routing algorithms have not considered these aspects. The width and spacing of the wires must be arranged to avoid deleterious results from these effects. The nonlinear inductance and coupling capacitance must be estimated by the routing algorithms because they are a function of the geometry. The classical routing objectives and constraints continue to apply.

Variable Accuracy Inductance Extraction

The same inductance and coupling capacity effects are felt at the chip level as well. Accurate inductance calculation requires a three-dimensional analysis and is very time consuming. The Task Force recommends the development of inductance modeling techniques that can provide inductance estimates in a range of accuracies depending on the purpose at hand and the amount of computing time available.

Coupling Aware Static Timing Enhancements

Signal travel time estimates are fundamental to a correct chip design. Current timing analysis methods do not consider coupling capacitance, so the estimates can be quite poor. Interaction effects between neighbouring wires can be quite large: switching pattern of neighbouring wires can affect the delay through a wire significantly. Routing and other algorithms need to consider these effects.

6.2.3 Summary

The physical design problem, viewed in its entirety, is an enormous mixed-integer nonlinear optimization problem. Traditional methods use a hierarchical decomposition approach to break the problem in smaller and more manageable sub problems, both by decomposing parts of the logic diagram, and by separating aspects of the problem such as block placement vs. wire routing. A wide variety of partitioning, placement, and routing problems results, and an extremely long list of specialized optimization algorithms is applied.

Less than optimum solutions are a natural outcome of this approach. As a general principle of optimization, a better solution would result if the complete problem could be tackled in its entirety. It is interesting to note that the list of the top ten physical design challenges produced by the Task Force on Physical Design includes several challenges that involve the idea of re-integrating problems that have traditionally been separated (e.g. interconnect-driven floorplanning, layout driven synthesis, and placement techniques for large blocks).

Another common theme in the top ten list is better modeling of physical effects in the optimization algorithms (e.g. RLC routing, variable accuracy inductance extraction, and coupling aware static timing enhancements). These all introduce nonlinear effects which make the associate optimization problems more difficult to solve.

Both themes in the top ten list implicitly move the solution techniques towards a more integrated – and larger – nonlinear mixed-integer formulation. Improvements from the operations research arena can undoubtedly provide assistance in this movement, while at the same time operations researchers can learn important lessons about dealing with optimization problems of extremely large scale and complexity.

6.3 Circuit Optimization and Applications in Design Of High-speed IC Packages

Optimization is an important step in designing electronic circuits and systems. Standard applications of optimization are filter design, linear and nonlinear circuit performance optimization, yield optimization and more [9]. With the rapid increase in operating frequency in today's electronic systems such as computers and digital communication systems, signal integrity in VLSI packages and interconnects has become one of the critical issues in an overall system design. Improperly designed packages and interconnects lead to signal integrity degradations such as signal delay, crosstalk and ground noise, limiting the overall system performance. With subnanosecond rise times, the electrical length of the interconnects becomes a significant fraction of a signal wavelength, necessitating the use of transmission line models. For designing high-speed electronic circuits and systems, it becomes very important to perform signal integrity based optimization [109].

Straight use of conventional circuit optimization approaches is not always suitable due to several difficulties encountered in signal integrity oriented design. In reality the number of interconnects in a printed circuit board is usually large and design specifications at many interconnection ports need to be satisfied. Conventional optimization techniques may balk at such large scale problems. Another difficulty is the highly repetitive simulation of distributed transmission line models or even EM models since interconnect geometry may be adjusted during optimization. The hierarchical nature of signal integrity which exists at chip, multichip module, and printed circuit board levels, and which depends on interconnect parameters at all levels also poses challenges to standard optimization approaches. In addition, as the signal spectrum including harmonic frequencies enters the microwave frequency range, electromagnetic (EM) effects in the passive components must be taken into account. However EM simulation is computationally intensive, making EM optimization an extremely challenging task. Signal integrity and electromagnetic oriented optimizations are presently one of the most active topics in circuit optimization [9, 11, 90]. This section highlights the major issues in this direction.

6.3.1 Optimization Objectives

Consider a VLSI interconnect network excited by a trapezoidal signal. Let $\boldsymbol{\phi}$ be a vector of design variables. Let $v_j(\boldsymbol{\phi}, t)$ be the response signal at circuit node j and time t. Suppose $\mathbf{J}_1$ is an index set containing all nodes of interest at which the desired response is a signal corresponding to the excitation. Let w denote a positive weighting factor. Suppose the signal propagation delay is described by the time at which the transient signal reaches a threshold value v_T. Let $\tau_{j,max}$ be the upper specification for the propagation delay at node j. The error function e for delay minimization can be equivalently described by

$$-w(v_j(\boldsymbol{\phi}, t) - v_T), \text{ for } j \in \mathbf{J}_1$$

In a similar way, a variety of time-domain error functions based on the performances of transient responses such as delay, rise/fall times, crosstalk and reflection need to be formulated [97]. A minmax or generalized least *p*th objective function combining all the error functions becomes the overall objective function for optimization. Another type of design objective is skew reduction, i.e. to minimize the difference in the delay of several signal paths. In this case the delay specifications for those paths should be same and may contain both upper and lower specifications simultaneously.

An alternative method for formulating error functions for optimization is to use frequency domain formulation. For example, a frequency domain concept called group delay can be used to indirectly optimize signal delay. By reducing group delay and the spectrum of crosstalks, and improving the flatness of group delay and gain slope responses, the transient signal delay, crosstalk and signal distortions are indirectly optimized. The actual optimization is performed in the frequency domain.

A proper formulation of the error functions requires efficient representation of how the design specifications are imposed and how the circuit performances are interpreted. Various formulations such as the time- and frequency-domain formulations as well as other formulations such as ground noise minimization, multidisciplinary thermal and electrical design have been developed [109].

6.3.2 Optimization Variables and Constraints

The optimization variables may include parameters in the interconnections and/or their terminations since all such parameters have an impact on interconnect performance. The terminations typically represent basic circuit blocks such as drivers/receivers and matching networks or additional terminating elements. The interconnect parameters can be the physical/geometrical parameters of the interconnects, such as the length of the transmission line, the distance between the coupled conductors, and the width of the conductors. Other parameters such as the thickness of the circuit board and the dielectric parameters which also affect the signal integrity are typically selected from a set of standard and predetermined values. Another type of interconnect variable is the equivalent electrical model parameters, e.g., resistance-inductance-capacitance-conductance (RLCG) parameters which are often used for defining transmission lines during simulation. However, the result of optimized equivalent electrical parameters must be converted into feasible physical/geometrical parameters. This concept for selecting optimization variables can be generalized for general EM based optimization, including either physical/geometrical design variables and/or equivalent circuit parameters [11, 90].

Design variables must be subject to design rules. Simple lower and upper bounds on variables should be applied according to the design profile. Conflicting factors in design are also due to additional constraints. For example, the total length of several interconnects are limited by the physical dimensions of the circuit chip and/or the circuit board. In this case, shortening an interconnect will make another interconnect longer. Also the total separations between several coupled conductors are constrained by the geometrical space available to them. In this case reducing coupling by distancing a pair of conductors will force other conductors to be closer. Other factors such as manufacturing constraints, e.g., component clearance and height restrictions, may also need to be addressed [97].

6.3.3 Circuit Optimization Formulations and Algorithms

Performance Optimization: The basic form of circuit optimization is to enhance the performance of the circuit or to ensure that design specifications are met. Here we formulate error functions according to different design specifications and carry out optimization by minimizing the maximium or least *p*th objective functions derived from these error functions. This type of optimization is fundamental in both electrical and physical design of electronic circuits.

Yield Optimization: With the industrial drive for reduced time to market of new products, the ability of computer-aided design tools to capture various factors affecting yield during the design stage becomes very important. One aspect of such a design methodology is manufacturability driven design which takes into account manufacturing tolerances, material variations and other uncertainties as part of the design optimization formulation. Yield optimization and design with tolerances become necessary.

Multidisciplinary Optimization: With the advancement of technology, there has been a significant increase in the complexity of the designs leading to tighter design margins and specifications. Thermal management and signal integrity are the two prominent issues of concern in the design of high speed systems. Multidisciplinary optimization techniques are being developed for optimizing the physical partitioning and placement of packages on multichip modules and printed circuit boards, taking into account the thermal and electrical performance as well as the manufacturability of the design.

Optimization Methods in Use: The optimization problem is typically a nonlinear continuous optimization problem with or without constraints. For performance based design optimization involving many error functions, we typically have nonlinear minimax, least-*p*th, or generalized least-*p*th optimization problems in order to simultaneously deal with various error functions for delay, crosstalk and reflections at various network output ports. Efficient multistage optimization algorithms such as the 2-stage algorithms [9] combine the robustness of a first-order method of the Gauss-Newton type with the speed of the Quasi-Newton method. The optimizer requires user-supplied first order derivatives of error functions with respect to optimization variables. The optimizer automatically generates approximate second order derivatives by BFGS (Broyden-Fletcher-Goldfarb-Shanno) update to provide fast convergence. Sensitivity analysis techniques for use in gradient based interconnect optimization have been developed, e.g., [62]. Other optimization techniques such as Gauss-Marquardt algorithms, Sequential Quadratic Programming approach, Greedy algorithm and heuristic algorithms have also been used [109].

6.4 Advanced Optimization Techniques

A practical printed circuit board often has a large number of interconnects and specifications/constraints at many interconnection ports required in design. Conventional optimization approaches, while efficient for problems of ordinary size, will balk at such large problems. Another difficulty with interconnect optimization is the computational expense of interconnect simulations involving lossy coupled transmission line models or even EM models. Circuit design involving EM effects requires the solution of the EM equations, i.e., the Maxwell equations. Such EM simulation must solve a 2- or 3-dimensional field problem using numerical techniques such as finite element methods, finite-difference-time-domain methods, methods of moments etc, and is computationally expensive. Repetitive simulations required during optimization make the overall design optimization very time consuming.

6.4.1 Methods based on Speeding up Simulations

The most direct approach to improving optimization efficiency for the challenges discussed above is to speed up the simulations. One example is parallel simulation in which multiple simulations required in different error functions of the optimization objective are computed simultaneously. The computation time for the overall objective function is reduced to that of one simulation instead of multiple simulations originally required in evaluating different error functions.

6.4.2 Methods based on Use of Fast Models of the Original Simulation

Instead of using original and slow simulations directly inside the optimization loop, we can develop faster models to represent the original simulation results and use this fast model inside the optimization. This is an effective direction and is a very active research topic today. Efficient model reduction methods and neural network modeling methods are two recent breakthroughs in this area. This important direction is further elaborated in Section 5.

6.4.3 Decomposition and Multi-level Optimization Technique

When the number of optimization variables and error functions become large, a considerable fraction of CPU time will also be spent in executing the optimization algorithm. In this case speeding up circuit simulation alone is not adequate. Large-scale optimization techniques are needed. An important concept addressing large scale optimization is decomposition where the overall large optimization problem is decomposed into a sequence of small optimization problems each containing a subset of error functions and subset of optimization variables. Multilevel optimization approach exploiting the circuit hierarchy as part of the decomposition can be used [97].

6.4.4 Space Mapping Optimization Methods

Space Mapping is a recent optimization concept, allowing expensive EM optimization to be performed more efficiently with the help of fast and approximate "coarse" or surrogate models [8, 10]. It has been applied with great success to otherwise expensive direct EM optimizations of microwave components and circuits with substantial computation speedup. Recent research is focused on mathematical motivation to place Space Mapping into the context of classical optimization. The aim of Space Mapping is to achieve a satisfactory solution with a minimal number of computationally expensive "fine" model evaluations. Space Mapping procedures iteratively update and optimize surrogates based on fast physically-based "coarse" models.

6.5 Emerging Areas for CAD of High-Speed/high-frequency circuits

The essential and basic building blocks of all electronic systems such as computers, communication networks, and wireless telecommunications are electronics components or subsystems such as integrated circuits (IC), chip packages, multichip modules (MCM), printed circuit boards (PCB), power/ground planes and backplanes (BP). As the signal speed and frequency increase, the dimensions of PCB, MCM and chip packages become a significant fraction of the signal wavelength. The conventional electrical models for the components are not valid anymore. Models with physical/geometrical information, and including electromagnetic (EM) effects, become necessary. Modeling and design methods involving microwave modeling and design concepts become important. Unfortunately, direct EM-based design optimization is extremely expensive using current design methodologies. Performance optimization requires a large number of function evaluations. Each function evaluation in turn requires accurate evaluation of the circuit response at hundreds and sometimes thousands of time points. Hence the success of the optimization process depends on two critical factors: modeling accuracy and simulation efficiency. This section discusses some of the emerging directions in high-frequency electronic CAD. Specifically, in Section 5.1, the major challenges in accurate modeling of high-speed circuits and interconnects are described. Overviews of model-order reduction and artificial neural networks techniques are presented in Sections 5.2-5.5 as viable means to speed up the simulation and optimization tasks.

6.5.1 Simulation and Optimization of High-Speed Interconnects

Taking a broader perspective, a “high-speed interconnect” is one in which the time taken by the propagating signal to travel between its end points cannot be neglected. An obvious factor which influences this definition is the physical extent of the interconnect: the longer the interconnect, the more time the signal takes to travel between its end points. Smoothness of signal propagation suffers once the line becomes long enough for a signal’s rise/fall times to roughly match its propagation time through the line. As a result, the interconnect electrically isolates the driver from the receivers, which no longer function directly as loads to the driver. Instead, within the time taken for the signal’s transition between its high and low voltage levels, the impedance of the interconnect becomes the load for the driver and also the input impedance to the receivers. This leads to various transmission line effects, such as reflections, overshoot, undershoot, and crosstalk. Modeling of these requires the blending of EM and circuit theory [7, 22, 26-27, 39, 45, 52, 66-67, 75, 79].

Alternatively, ‘high-speed’ can be defined in terms of the frequency content of the signal. At low frequencies an ordinary wire (an interconnect in other words) will effectively short two connected circuits. However, this is not the case at higher frequencies. The same wire, which is so effective at lower frequencies for connection purposes, has too much inductive/capacitive effect to function as a short at higher frequencies. Faster clock speeds tend to add more and more high-frequency content.

An important criterion used for classifying interconnects is its *electrical length.* An interconnect is considered to be *"electrically short"* if, at the highest operating frequency of interest, the interconnect length is physically shorter than approximately one-tenth of the wavelength. Otherwise, the interconnect is referred as *"electrically long"*. In most digital applications, the desired highest operating frequency (which corresponds to the minimum wavelength) of interest is governed by the rise/fall time of the propagating signal. Electrically short interconnects can be represented by lumped models whereas electrically long interconnects need distributed or fullwave models.

6.5.1.1 High-Speed Interconnect Models

Depending on the operating frequency, signal rise times, and nature of the structure, the interconnects can be modeled as lumped or distributed circuits.

Lumped Models

At lower frequencies, the interconnect circuits can be represented by a combination of lumped resistive (R) and capacitive (C) elements or a combination of lumped resistive, capacitive, and inductive (L) elements. RC circuit responses are monotonic in nature. However, in order to account for ringing in signal waveforms, RLC circuit models may be required. Lumped interconnect circuits extracted from layouts usually have a large number of nodes, which makes the simulation highly CPU intensive.

Distributed Transmission Line Models

At relatively higher signal speeds, the electrical length of interconnects becomes a significant fraction of the operating wavelength, giving rise to signal distorting effects that do not exist at lower frequencies. Consequently, the conventional lumped impedance interconnect models become inadequate and transmission line models based on quasi-TEM assumptions are needed. The TEM (Transverse Electromagnetic Mode) approximation represents the ideal case, where both electric field (E) and magnetic field (H) fields are perpendicular to the direction of propagation, which is valid under the condition that the line cross-section is much smaller than the wavelength. However, the inhomogeneities in practical wiring configurations give rise to E or H fields in the direction of propagation. If the line cross-section or the extent of these nonuniformities remains a small fraction of the wavelength in the frequency range of interest, the solution to Maxwell's equations are given by the so called quasi-TEM modes and are characterized by distributed resistance, inductance, capacitance and conductance per unit length [75].

Distributed Models with Frequency-Dependent Parameters

At low frequencies, the current in a conductor is distributed uniformly throughout its cross section. However, as the operating frequency increases, the current distribution becomes uneven and it begins to concentrate more and more near the surface or edges of the conductor. This phenomenon can be categorized as follows: skin, edge and proximity effects [31, 56, 75, 101]. The skin effect causes the current to concentrate in a thin layer near the conductor surface and this reduces the effective cross-section available for signal propagation. This leads to an increase in the resistance to signal propagation and other related effects [27]. The edge effect causes the current to concentrate near the sharp edges of the conductor. The proximity effect causes the current to concentrate in the sections of ground plane that are close to the signal conductor. To account for these effects, modelling based on frequency-dependent per-unit-length parameters may be necessary.

6.5.1.2 Interconnect Simulation

Transmission line characteristics are in general described by partial differential equations known as Telegrapher's equations [75]. The major difficulty in simulating high-frequency models such as distributed transmission lines is due to the fact that they are best represented in the frequency-domain. On the other hand, nonlinear devices can only be described in the time-domain. These simultaneous formulations are difficult to handle by a traditional ordinary differential equation solver such as SPICE [80].

The objectives of recent interconnect simulation algorithms are to address both the mixed frequency/time problem as well as to handle large linear circuits without excessive CPU expense. There have been several algorithms proposed for this purpose. In general, the partial differential equations describing the interconnects are translated into a set of ordinary differential equations (known as macromodels) through some kind of discretization.

Uniform Lumped Segmentation

The conventional approach [29, 75] for macromodelling of distributed interconnects is to divide the line into segments of a length chosen to be small fraction of the wavelength. If each of these segments is electrically small at the frequencies of interest, then each segment can be replaced by a lumped model.

One of the major drawbacks of this discretization is that it requires a large number of sections, especially for circuits with high operating speeds and sharper rise times. This leads to large circuit sizes and the simulation becomes CPU inefficient

Method of Characteristics

The method of characteristics (MC) [15, 20] transforms the partial differential equations representing a transmission line into ordinary differential equations containing time-delayed controlled sources.

Transfer Function Approximation

Least-square approximation-based techniques [13] derive a transfer-function representation for the frequency response of transmission line subnetworks. The method fits data from sample frequency points to a complex rational function.

Exponential Matrix-Rational Approximation

This algorithm directly converts partial differential equations into time-domain macromodels based on Padé rational approximations of exponential matrices [32, 34]. In this technique, coefficients describing the macromodel are computed *a priori* and analytically, using a closed-form Padé approximation of exponential matrices. Since closed-form relations are used, this technique doesn't suffer from the usual ill-conditioning experienced with the direct application of Padé approximations, hence it allows a higher order of approximation. It also guarantees the passivity of the resulting macromodel.

6.5.1.3 Full-Wave Interconnect Characterization

At further sub-nanosecond rise times, the line cross-section or the nonuniformities become a significant fraction of the wavelength and field components in the direction of propagation can no longer be neglected. Consequently, full-wave models, which take into account all possible field components and satisfy all boundary conditions, are required to give an accurate estimation of high frequency effects. However, circuit simulation of full-wave models is highly involved. The information that is obtained through a full-wave analysis is in terms of field parameters such as propagation constant, characteristic impedance etc. A circuit simulator requires the information in terms of currents, voltages and circuit impedances. This demands a generalized method to combine modal results into circuit simulators in terms of full-wave stamps. References [50, 51, 64, 95] provide solution techniques and moment generation schemes for such cases.

6.5.1.4 Measured data

In practice, it may not be possible to obtain accurate analytical models for interconnects because of the geometric inhomogeneity and associated discontinuities. To handle such situations modeling techniques based on measured data have been proposed in the literature. In general, the behavior of high-speed interconnects can easily be represented by measured frequency-dependent scattering parameters or time-domain terminal measurements. However, handling measured data in circuit simulation is a tedious and a computationally expensive process. References [1, 19, 21, 24, 25, 77, 84, 94, 110] address such cases.

6.5.1.5 EMI Subnetworks

Electrically long interconnects function as spurious antennas to pick up emissions from other nearby electronic systems. This makes susceptibility to emissions a major concern to designers of high-frequency products. For this reason, the availability of interconnect simulation tools including the effect of incident fields is becoming an important design requirement. See [5, 12, 16, 37, 49, 53, 57, 65, 70-74, 76, 89, 93, 98] for analysis techniques for interconnects subjected to external EM interference and also for radiation analysis of interconnects.

6.5.2 Model-Order Reduction Techniques

Circuits generally tend to have a large number of natural modes spread over a wide frequency range. These natural modes are directly related to the poles of the circuit. Poles control the rate of decay of the impulse response of the circuit. Poles far from the imaginary axis will have a fast decaying transient response and will not significantly affect the circuit response. Even though the majority of these poles would normally have very little effect on simulation results, they make the simulation CPU-expensive by forcing the simulator to take smaller step sizes. Dominant poles are those that are close to the imaginary axis and significantly influence the time as well as the frequency characteristics of the system. Moment-matching techniques (MMTs) capitalize on the fact that, irrespective of the presence of a large number of poles in a system, just the dominant poles are sufficient to accurately characterize a given system.

Several algorithms can be found in the literature for reduction of large interconnect subnetworks [2-4, 6, 14, 17-18, 23, 33, 36, 40-42, 46-47, 54-55, 58-61, 63, 68-69, 78, 81, 83, 87-88, 91-92, 99, 102-103]. They can be broadly classified into two categories: explicit and indirect MMTs.

Explicit MMTs

These techniques employ Padé approximation, based on explicit moment-matching to extract the dominant poles and residues of a given system. However, due to the inherent limitations of Padé approximants, MMTs based on a single expansion often give inaccurate results. The number of accurate poles that can be extracted from any expansion point is generally fewer than 10 poles. In addition, there is no guarantee that the reduced model obtained as above is passive. Passivity implies that a network cannot generate more energy than it absorbs, and no passive termination of the network will cause the system to go unstable. The loss of passivity can be a serious problem because transient simulations of reduced networks may encounter artificial oscillations.

Indirect MMTs

These techniques are generally based on Krylov subspace methods [36, 40-42, 55, 68, 87-88]. The numerical advantage of these algorithms stems from using the moments implicitly instead of computing them explicitly. Guaranteeing the passivity of the reduced-order models is achieved via orthogonal projection algorithms based on a congruence transform [68].

6.5.3 Artificial Neural Networks for EM modeling

In the recent years, an EM-oriented CAD approach based on Artificial Neural Networks (ANN) has gained recognition. Accurate and fast neural models can be developed from measured or simulated EM data. These neural models can be used in place of CPU-intensive detailed EM models to significantly speed up electronic circuit design, while maintaining EM-level accuracies. ANN has been applied to modeling and design of a variety of components and circuits such as bends, embedded passive components, transmission line components, vias, coplanar waveguide (CPW) components, spiral inductors, filters and so on [105-108].

A variety of EM-ANN based modeling methods have been developed in the RF/microwave CAD community. The basic ANN structure called multi-layer perceptron (MLP) neural networks were the popular choice for many frequency domain EM neural models, such as CPW bends and junctions. Using such models, accurate and fast EM based design and optimization techniques for microwave circuits have been demonstrated. A recent trend in the microwave-ANN area is knowledge-based approaches to efficient EM based modeling and design, which advocates the use of existing knowledge such as conventional empirical models together with neural networks [96]. Pioneering works in this direction include knowledge based neural networks, difference networks, prior knowledge input networks and space mapped neural networks. Neural model development using these knowledge approaches requires fewer EM simulations (or training data) and shorter training times, thereby facilitating cost-effective neural based high-frequency circuit design and optimization [106].

The need for large-signal analysis of microwave circuits consisting of both active devices and passive components demands powerful nonlinear CAD methodologies. EM based neural models of passive components and their applications to high-frequency and high-speed nonlinear circuit optimizations in both frequency- and time-domains, are being developed, including the component's geometrical/physical parameters as optimization variables [30]. EM based neural models in both frequency- and time-domain formats are realized, which are compatible with the framework of existing CAD tools.

6.5.4 ANN Models for Nonlinear Devices and Circuits

Nonlinear device modeling is an important area of CAD, and many device models have been developed. Examples of active devices for high-frequency performance are Gallium Arsenide (GaAs) Field Effect Transistors (FET), Heterojunction Bipolar Transistors (HBT), High Electron-Mobility Transistor (HEMT) and so on. Due to rapid technology development in the semiconductor industry, new devices constantly evolve. Models that were developed to fit previous devices may not fit new devices well, so there is an ongoing need for new models. Conventional methods for modeling such devices use equivalent circuits to fit device measurement data. With the demand for shortened design cycles, physics-based device simulation becomes important in providing accurate solutions of device behavior. However, such simulations are CPU intensive since they are based on first principle methods that involve numerically solving differential or integral equations. ANN based approaches have been developed recently for modeling nonlinear devices. Neural network models can represent the currents and capacitive charges of each terminal of the transistor devices as functions of terminal voltages. Such a model is trained using transistor device data (from detailed simulation or measurement). Pure neural network models have been used to represent the direct-current (DC) and small-signal behaviors of devices. The ANN models can also be combined with existing empirical equivalent circuit models of devices to provide automatic modification of the model to match the required device data [28].

Another type of modeling is behavioral modeling of nonlinear circuits or subsystems. With the increasing size and complexity of today's electronic systems, CAD at the system level becomes necessary. Simulation and design for systems with mixed digital/analog circuits require accurate behavioral models for the digital and analog parts. Behavioral modeling for analog circuits is more difficult and remains an active research topic. For accurate representation of the full analog behavior of a nonlinear circuit, a fundamental approach is to directly model the relationship between the dynamic input and output signals of the circuit. Since in the time domain the circuit output signals are not algebraic functions of the inputs, recurrent neural networks (RNN) and dynamic neural networks (DNN) are proposed to address the problem. Recurrent neural networks represent the circuit dynamic behavior in such a way that the output signals of the model at a time instance are functions of the past history of the circuit input and output signals. Dynamic neural networks use differential equations in their neurons such that the output signals of the model are related to the input signals and the time-derivatives of input and output signals. Such RNN and DNN based neural network models have been used to model analog circuits such as amplifiers and mixers and have been used to perform system level simulation [38, 100].

6.5.5 ANN-Based Circuit Optimization

After the neural models are trained and verified to be accurate, they can then be used in circuit design and optimization. If the neural models have been trained from EM/physics data, their use in place of the original EM/physics models improves design speed while maintaining accuracy. To achieve this benefit, we first incorporate the trained neural models into circuit or system simulators. A set of neural models for various passive and active circuit components can be incorporated such that at the circuit or system simulation level, all the neural models as well as conventional models of circuit components are accessible. A neural model can be connected to other neural models or any other models in the simulator to form a high-level circuit. During simulation, the circuit simulator passes input variables such as frequency and the physical parameters of a component to the neural model; the neural model then computes and returns the corresponding outputs of the neural network back to the simulator. For DNN neural models, an equivalent circuit representation can be formulated that will exactly interpret the DNN dynamics using resistor/capacitor/inductor models and controlled sources, incorporating the neural network formulas in the controlled sources.

With the neural models embedded in the circuit simulation environment, the components they represent can be optimized along with the rest of the circuit with the model parameters as optimization variables. The approach treats the neural network inputs $\boldsymbol{x}$ as optimization variables, e.g., physical/geometrical parameters of the device or circuit. The objective for circuit performance optimization is to modify the circuit design parameters such that the circuit responses best meet or exceed design specifications subject to electrical or physical/geometrical constraints on the circuit elements. Advanced optimization objectives such as yield optimization can be similarly formulated using neural network models in the circuit simulation. Such design optimization tasks involve highly repetitive computations. Neural networks can significantly improve the optimization speed because the computationally expensive EM/physics models required inside the repetitive computational loops have been replaced by fast neural models [104,109].

6.5.6 Summary

Optimization is an important step in designing electronic circuits and systems. With the rapid increase in operating frequency in today's electronic systems such as computers and digital communication systems, signal integrity in VLSI packages and interconnects has become one of the critical issues in an overall system design. Straight use of conventional circuit simulation and optimization approaches will not always be suitable due to several difficulties encountered in signal integrity oriented design. In reality the number of interconnects in a printed circuit board is usually large and design specifications at many interconnection ports need to be satisfied. Conventional optimization techniques may balk at such large-scale problems. Another difficulty is due to the highly repetitive simulation of distributed transmission line models or even EM models since interconnect geometry may be adjusted during optimization. Model-order reduction and artificial neural networks are two new emerging promising techniques that offer substantial efficiency gains with very little accuracy degradation.

6.6 Conclusions

The general CAD problem amounts to an enormous mixed-integer nonlinear program, and hence is extremely difficult to solve. The traditional approach in the CAD community has been to subdivide the problem into more manageable pieces and to solve these in sequence. While this makes the scale manageable, it also makes it harder to find a very good solution. As computer power has grown in recent years, the trend has been towards re-integrating some of the previously separated subproblems in the quest for better solutions. Meanwhile the subproblems themselves remain difficult enough to solve and provide much scope for input from operations researchers.

Another approach taken by CAD researchers to deal with very large problems is systematic simplification. Lumped models, moment-matching techniques, and neural networks are all used to produce simplified models that can actually be solved. The solutions produced in this manner may then be used to produce better models, which are then re-solved in an ever-improving cycle. Approaches like these may prove useful in other large-scale optimization applications and should be more generally known among operations researchers.

The operations research and CAD communities have much to teach each other. It is our hope that this tutorial will encourage further interaction between the two groups.

References

[1] R. Achar and M. Nakhla. Efficient transient Simulation of embedded subnetworks characterized by S-parameters in the presence of nonlinear elements. *IEEE Transactions on Microwave Theory and Techniques*, 46(12):2356-2363, 1998.

[2] R. Achar and M. Nakhla. *Signal Propagation on Interconnects.* In H. Grabinski, editor, *Minimum Realization of Reduced-Order Models of High-Speed Interconnect Macromodels*, Boston, 1998. Kluwer Academic Publishers.

[3] R. Achar, M. Nakhla, P. Gunupudi and E. Chiprout. Passive interconnect reduction algorithm for distributed/measured networks. *IEEE Transactions on Circuits and Systems-II*,47(4):287-301, 2000.

[4] R. Achar, M. Nakhla and Q. J. Zhang. Full-wave analysis of high-speed interconnects using complex frequency hopping. *IEEE Transactions on Computer-Aided Design*, 17(10)997-1016, Oct. 98.

[5] N. Ari and W. Blumer. Analytic formulation of the response of a two-wire transmission line excited by a plane wave. *IEEE Transactions on Electromagnetic Compatibility*, 30(4):437-448, 1988.

[6] G.A. Baker Jr. *Essential of Padé Approximants*. Academic, New York, 1975.

[7] H.B. Bakoglu. *Circuits, Interconnections and packaging for VLSI.* Addison-Wesley, Reading MA, 1990.

[8] J.W. Bandler, R.M. Biernacki, S.H. Chen, P.A. Grobelny and R.H. Hemmers. Space mapping technique for electromagnetic optimization. *IEEE Transactions on Microwave Theory and Techniques*, 42(12):2536-2544, 1994.

[9] J.W. Bandler and S.H. Chen. Circuit optimization: the state of the art. *IEEE Transactions on Microwave Theory and Techniques*, vol. 36(2):424-443, 1988.

[10] J.W. Bandler, Q.S. Cheng, S. Dakroury, A.S. Mohamed, M.H. Bakr, K. Madsen and J. Søndergaard. Space mapping: the state of the art. *IEEE Transactions on Microwave Theory and Techniques,* 52(10):337-361, 2004.

[11] J.W. Bandler and Q.J. Zhang. Next generation optimization methodologies for wireless and microwave circuit design. In *IEEE Microwave Theory and Techniques Society International Topical Symposium on Technologies for Wireless Applications Digest,* Vancouver, B.C., pages 5-8, Feb. 1999.

[12] P. Bernardi, R. Cicchetti and C. Pirone. Transient response of a microstrip line circuit excited by an external electromagnetic source. *IEEE Transactions on Electromagnetic Compatibility*, 34(2):100-108, 1992.

[13] W.T. Beyene, and J.E. Schutt-Aine. Efficient transient simulation of high-speed interconnects characterized by sampled data. *Components, Packaging and Manufacturing Technology, Part B: Advanced Packaging*, 21(1):105-113, 1998.

[14] J.E. Bracken, V. Raghavan, and R.A. Rohrer. Interconnect simulation with asymptotic waveform evaluation (AWE). *IEEE Transactions on Circuits and Systems*, 39(11):869-878, 1992.

[15] F.H. Branin, Jr. Transient analysis of lossless transmission lines. *Proceedings of the IEEE,* 55(12):2012-2013, 1967.

[16] G.J. Burke, E. K. Miller, and S. Chakrabarti. Using model based parameter estimation to increase the efficiency of computing electromagnetic transfer functions. *IEEE Transactions on Magnetics*, 25(4):2087-2089, 1989.

[17] A.C. Cangellaris, S. Pasha, J. L. Prince and M. Celik. A new discrete time-domain model for passive model order reduction and macromodelling of high-speed interconnections. *IEEE Transactions on Components, Packaging and Manufacturing Technology, Part B: Advanced Packaging*, 22(2):356-364, 1999.

[18] M. Celik and A. C. Cangellaris. Simulation of dispersive multiconductor transmission lines by Padé approximation via Lanczos process. *IEEE*

Transactions on Microwave Theory and Techniques, 44(12):2525-2535, 1996.

[19] M. Celik, A. C. Cangellaris and A. Deutsch. A new moment generation technique for interconnects characterized by measured or calculated S-parameters. In *IEEE International Microwave Symposium Digest*, pp. 196-201, June 1996.

[20] F.Y. Chang. Transient analysis of lossless coupled transmission lines in a nonhomogeneous medium. *IEEE Transactions on Microwave Theory and Techniques*, 18(9):616-626, Sept. 1970.

[21] P.C. Cherry and M.F. Iskander. FDTD analysis of high frequency electronic interconnection effects. *IEEE Transactions on Microwave Theory and Techniques*, 43(10):2445-2451, 1995.

[22] E. Chiprout and M. Nakhla. *Asymptotic Waveform Evaluation and Moment Matching for Interconnect Analysis.* Kluwer Academic Publishers, Boston, 1993.

[23] E. Chiprout, M. Nakhla. Analysis of interconnect networks using complex frequency hopping. *IEEE Transactions on Computer-Aided Design*, 14(2):186-199, 1995.

[24] B.J. Cooke, J.L. prince and A.C. Cangellaris. S-parameter analysis of multiconductor integrated circuit interconnect systems. *IEEE Transactions on Computer-Aided Design*, 11(3):353-360, Mar. 1992.

[25] S.D. Corey and A.T. Yang. Interconnect characterization using time-domain reflectometry. *IEEE Transactions on Microwave Theory and Techniques*, 43(9):2151-2156, 1995.

[26] W.W.M. Dai (guest editor). Special issue on simulation, modeling, and electrical design of high-speed and high-density interconnects. *IEEE Transactions on Circuits and Systems*, 39(11):857-982, 1992.

[27] A. Deustsch. Electrical characteristics of interconnections for high-performance systems. *Proceedings of the IEEE,* 86(2):315-355, 1998.

[28] V. Devabhaktuni, M.C.E. Yagoub, Y. Fang, J.J. Xu and Q.J. Zhang. Neural networks for microwave modeling: model development issues and nonlinear techniques. *International Journal on Radio Frequency and Microwave Computer-Aided Engineering*, 11(1):4-21, 2001.

[29] T. Dhane and D.D. Zutter. Selection of lumped element models for coupled lossy transmission lines. *IEEE Transactions on Computer-Aided Design*, 11(7):959-967, 1992.

[30] X. Ding, V.K. Devabhaktuni, B. Chattaraj, M.C.E. Yagoub, M. Doe, J.J. Xu and Q.J. Zhang. Neural network approaches to electromagnetic based

modeling of passive components and their applications to high-frequency and high-speed nonlinear circuit optimization. *IEEE Transactions on Microwave Theory and Techniques*, 52(1):436-449, 2004.

[31] A.R. Djordjević and T.K. Sarkar. Closed-form formulas for frequency-dependent resistance and inductance per unit length of microstrip and strip transmission lines. *IEEE Transactions on Microwave Theory and Techniques*, 42(2):241-248, 1994.

[32] A. Dounavis, R. Achar and M. Nakhla. Efficient passive circuit models for distributed networks with frequency-dependent parameters. *IEEE Transactions on Components, Packaging and Manufacturing Technology, Part B: Advanced Packaging*, 23(3):382-392, 2000.

[33] A. Dounavis, E. Gad, R. Achar and M. Nakhla. Passive model-reduction of multiport distributed networks including frequency-dependent parameters. *IEEE Transactions on Microwave Theory and Techniques*, 48(12):2325-2334, 2000.

[34] A. Dounavis, X. Li, M. Nakhla and R. Achar. Passive closed-loop transmission line model for general purpose circuit simulators. *IEEE Transactions on Microwave Theory and Techniques*, 47(12):2450-2459, Dec. 1999.

[35] R. Drechsler. *Evolutionary Algorithms for VLSI CAD*. Kluwer Academic Publishers, Boston, 1998.

[36] I.M. Elfadel and D.D. Ling. A block rational Arnoldi algorithm for multiport passive model-order reduction of multiport RLC networks. *Proceedings of ICCAD-97*, pp. 66-71, Nov. 1997.

[37] I. Erdin, R. Khazaka and M. Nakhla. Simulation of high-speed interconnects in the presence of incident field. *IEEE Transactions on Microwave Theory and Techniques*, 46(12):2251-2257, 1998.

[38] Y. Fang, M.C.E. Yagoub, F. Wang and Q.J. Zhang. A new macromodeling approach for nonlinear microwave circuits based on recurrent neural networks. *IEEE Transactions on Microwave Theory and Techniques*, 48(12):2335-2344, 2000.

[39] J.B. Faria. *Multiconductor Transmission Line Structures*. John Wiley and Sons Inc., New York, 1993.

[40] P. Feldmann and R.W. Freund. Efficient linear circuit analysis by Padé via Lanczos process. *IEEE Transactions on Computer-Aided Design*, 14(5):639-649, 1995.

[41] P. Feldmann and R.W. Freund. Reduced order modeling of large linear subcircuits via a block Lanczos algorithm. Proceedings of the Design Automation Conference, pp. 474-479, June 1995.

[42] R.W. Freund. Reduced-order modelling techniques based on Krylov subspace and their use in circuit simulation. Technical Memorandum, Lucent Technologies, 1998.

[43] M.R. Garey and D.S. Johnson. *Computers and Intractability: A Guide to the Theory of NP-Completeness*. Freeman, San Francisco, 1979.

[44] G.G.E. Gielen and R.A. Rutenbar. Computer-Aided Design of Analog and Mixed-Signal Integrated Circuits. *Proceedings of the IEEE,* 88(12):1825-1852, 2000.

[45] R. Goyal. Managing signal integrity [PCB Design]. *IEEE Spectrum,* 31(3):54-58, 1994.

[46] R. Griffith, E. Chiprout, Q. J. Zhang, and M. Nakhla. A CAD framework for simulation and optimization of high-speed VLSI interconnections. *IEEE Transactions on Circuits and Systems*, 39(11):893-906, 1992.

[47] P. Gunupudi, M. Nakhla and R. Achar. Simulation of high-speed distributed interconnects using Krylov-subspace techniques. *IEEE Transactions on CAD of Integrated Circuits and Systems*, 19(7):799-808, 2000.

[48] IEEE, IEEE Xplore, World Wide Web, http://ieeexplore.ieee.org/Xplore/, 2004.

[49] I. Ierdin, M. Nakhla, and R. Achar. Circuit analysis of electromagnetic radiation and field coupling effects for networks with embedded full-wave modules. *IEEE Transactions on Electromagnetic Compatibility*, 42(4):449-460, 2000.

[50] T. Itoh and R. Mittra. Spectral Domain approach for calculating the dispersion Characteristics of microstrip lines. *IEEE Transactions on Microwave Theory and Techniques,* 21(2):496-499, 1973.

[51] R.H. Jansen. Spectral Domain Approach for microwave integrated circuits. *IEEE Transactions on Microwave Theory and Techniques.* 33(2):1043-1056, 1985.

[52] H.W. Johnson and M. Graham. *High-speed Digital Design*. Prentice-Hall, New Jeersey, 1993.

[53] Y. Kami and R. Sato. Circuit-concept approach to externally excited transmission lines. *IEEE Transactions on Electromagnetic Compatibility,* 27(4):177-183, 1985.

[54] M. Kamon, F. Wang and J. White. Generating nearly optimally compact models from Krylov-subspace based reduced-order models. *IEEE Transactions on Circuits and Systems,* 47(4):239-248, 2000.

[55] K.J. Kerns and A.T. Yang. Preservation of passivity during RLC network reduction via split congruence transformations. *IEEE Transactions on Computer-Aided Design*, 17(17):582-591, 1998.

[56] R. Khazaka, E. Chiprout, M. Nakhla and Q. J. Zhang. Analysis of high-speed interconnects with frequency dependent parameters. *Proceedings of the International Symposium on Electromagnetic Compatibility,* pp.203-208, Zurich, March 1995.

[57] R. Khazaka and M. Nakhla. Analysis of high-speed interconnects in the presence of electromagnetic interference. *IEEE Transactions on Microwave Theory and Techniques,* 46(6):940-947, July 1998.

[58] S.Y. Kim, N. Gopal, and L.T. Pillage. Time-domain macromodels for VLSI interconnect analysis. *IEEE Transactions on Computer-Aided Design,* 13(10):1257-1270, 1994.

[59] M.A. Kolbehdari, M. Srinivasan, M. Nakhla, Q.J. Zhang and R. Achar. Simultaneous time and frequency domain solution of EM problems using finite element and CFH techniques. *IEEE Transactions on Microwave Theory and Techniques*, 44(9):1526-1534, Sept. 1996.

[60] S. Kumashiro, R. A. Rohrer, A. J. Strojwas. Asymptotic waveform evaluation for transient analysis of 3-D interconnect structures. *IEEE Transactions on Computer-Aided Design*, 12(7):988-996, 1993.

[61] S. Lin and E. S. Kuh. Transient simulation of lossy interconnects based on the recursive convolution formulation. *IEEE Transactions on Circuits and Systems,* 39(11):879-892, 1992.

[62] S. Lum, M.S. Nakhla and Q.J. Zhang. Sensitivity analysis of lossy coupled transmission lines with nonlinear terminations. *IEEE Transactions on Microwave Theory and Techniques*, 42(4):607-615, 1994.

[63] J.H. McCabe. A formal extension of the Padé table to include two point Padé quotients. *J. Inst. Math. Applic.*, 15:363-372, 1975.

[64] D. Mirshekar-Syahkal. *Spectral Domain Method for Microwave Integrated Circuits*. Wiley & Sons Inc., New York, 1990.

[65] E.S.M. Mok, G.I. Costache. Skin-effect considerations on transient response of a transmission line excited by an electromagnetic wave. *IEEE Transaction on Electromagnetic Compability,* 34(3):320-329, 1992.

[66] M. Nakhla and R. Achar. Chapter XVII: Interconnect Modelling and Simulation. In W.-K. Chen, editor, *The VLSI Handbook,* pp. 17.1-17.29, Boca Raton, 2000. CRC Press.

[67] M. Nakhla and A. Ushida (Guest Editors). Special issue on modelling and simulation of high-speed interconnects. *IEEE Transactions on Circuits and Systems*, 39(11):857-982, 2000.

[68] A. Odabasioglu, M. Celik and L. T. Pillage. PRIMA: Passive Reduced-Order Interconnect Macromodeling Algorithm. *IEEE Transactions on Computer-Aided Design*, 17(8):645-654, Aug. 1998.

[69] A. Odabasioglu, M. Celik and L. T. Pillage. Practical considerations for passive reduction of RLC circuits. *Proceedings of the Design Automation Conference*, pp. 214-219, June 1999.

[70] F. Olyslager, D .D. Zutter, and K. Blomme. Rigorous analysis of the propagation characteristics of general lossless and lossy multiconductor transmission lines in multi-layered media. *IEEE Transactions on Microwave Theory and Techniques*, 41(1):79-88, 1993.

[71] F. Olyslager, D. D. Zutter, and A. T. de Hoop. New reciprocal circuit model for lossy waveguide structures based on the orthogonality of the eigenmodes. *IEEE Transactions on Microwave Theory and Techniques,* 42(12):2261-2269, 1994.

[72] C.R. Paul. Frequency response of multiconductor transmission lines illuminated by an incident electromagnetic field. *IEEE Transactions on Microwave Theory and Techniques*, 22(4):454-457, 1976.

[73] C.R. Paul. A comparison of the Contributions of Common-Mode and Differential-Mode Currents in Radiated Emissions. *IEEE Transactions on Electromagnetic Compatibility*, 31(2):189-193, 1989.

[74] C.R. Paul. A SPICE model for multiconductor transmission lines excited by an incident electromagnetic field. *IEEE Transactions on Electromagnetic Compatibility*, 36(4):342-354, Nov. 1994.

[75] C.R. Paul. *Analysis of Multiconductor Transmission Lines*. John Wiley and Sons Inc., New York, 1994.

[76] C.R. Paul. Literal solutions for the time-domain response of a two-conductor transmission line excited by an incident electromagnetic field. *IEEE Transactions on Electromagnetic Compatibility*, 37(2):241-251, 1995.

[77] M. Picket-May, A. Taflove and J. Baron. FD-TD modeling of digital signal propagation in 3-D circuits with passive and active loads. *IEEE Transactions on Microwave Theory and Techniques*, 42(8):1514-1523, 1994.

[78] L.T. Pillage and R.A. Rohrer. Asymptotic waveform evaluation for timing analysis. *IEEE Transactions on Computer-Aided Design*, 9(4):352-366, Apr. 1990.

[79] R.K. Poon. *Computer Circuits Electrical Design*, Prentice-Hall, New Jersey, 1995.

[80] T.L. Quarles. *The SPICE3 Implementation Guide.* Technical Report ERL-M89/44, University of California, Berkeley, 1989.

[81] V. Raghavan, J. E. Bracken, and R. A. Rohrer. AWESpice: A general tool for accurate and efficient simulation of interconnect problems. In *Proceedings of the ACM/IEEE Design Automation Conference*, pp. 87-92, June 1992.

[82] S.M. Sait and H. Youssef. *VLSI Physical Design Automation: Theory and Practice.* IEEE Press, New York, 1995.

[83] R. Sanaie, E. Chiprout, M. Nakhla, and Q. J. Zhang. A fast method for frequency and time domain simulation of high-speed VLSI interconnects. *IEEE Transactions on Microwave Theory and Techniques*, 42(12):2562-2571, 1994.

[84] J.E. Schutt-Aine and R. Mittra. Scattering parameter transient analysis of transmission lines loaded with nonlinear terminations. *IEEE Transactions on Microwave Theory and Techniques*, 36(3):529-536, 1988.

[85] Semiconductor Research Corporation. *SRC Physical Design Top Ten Problems.* World Wide Web, http://www.src.org/member/sa/cadts/pd.asp, 2004.

[86] N. Sherwani. *Algorithms for VLSI Physical Design Automation.* Kluwer Academic Publishers, Boston, 1993.

[87] L.M. Silviera, M. Kamen, I. Elfadel and J. White. A coordinate transformed Arnoldi algorithm for generating guaranteed stable reduced-order models for RLC circuits. *Technical Digest of the International Conference on Computer-Aided Design*, pp. 2288-294, Nov. 1996.

[88] L.M. Silveira, M. Kamen and J. White. Efficient reduced-order modelling of frequency-dependent coupling inductances associated with 3-D interconnect structures. *IEEE Transactions on Components, Packaging, and Manufacturing Technology Part B: Advanced Packaging*, 19(2):283-288, 1996.

[89] A.A. Smith. A more convenient form of the equations for the response of a transmission line excited by nonuniform fields. *IEEE Transactions on Electromagnetic Compatibility*, 15(3):151-152, Aug. 1973.

[90] M.B. Steer, J.W. Bandler and C.M. Snowden. Computer aided design of RF and microwave circuits and systems. *IEEE Transactions on Microwave Theory and Techniques*, 50(3):996-1005, 2002.

[91] T. Tang and M. Nakhla. Analysis of high-speed VLSI interconnect using asymptotic waveform evaluation technique. *IEEE Transactions on Computer-Aided Design*, 11(3):341-352, 1992.

[92] T. Tang, M. Nakhla and Richard Griffith. Analysis of lossy multiconductor transmission lines using the asymptotic waveform evaluation technique. *IEEE Transactions on Microwave Theory and Techniques*, 39(12):2107-2116, 1991.

[93] C.D. Taylor, R.S. Satterwhite, and C.W. Harrison. The response of a terminated two-wire transmission line excited by a nonuniform electromagnetic field. *IEEE Transactions on Antennas and Propagation*, 13(11):987-989, 1965.

[94] L.P. Vakanas, A.C. Cangellaris and O.A. Palusinski. Scattering parameter based simulation of transients in lossy, nonlinearly terminated packaging interconnects. *IEEE Transactions on Components, Packaging and Manufacturing Technology, Part B: Advanced Packaging*, 17(4):472-479, 1994.

[95] R. Wang, and O. Wing. A circuit model of a system of VLSI interconnects for time response computation. *IEEE Transactions on Microwave Theory and Techniques*, 39(4):688-693, April 1991.

[96] F. Wang and Q.J. Zhang. Knowledge based neural models for microwave design. *IEEE Transactions on Microwave Theory and Techniques*, 45(12):2333-2343, 1997.

[97] Y.J. Wei, Q.J. Zhang and M.S. Nakhla. Multilevel optimization of high-speed VLSI interconnect networks by decomposition. *IEEE Transactions on Microwave Theory and Techniques*, 42(9):1638-1650, 1994.

[98] I. Wuyts and D. De Zutter. Circuit model for plane-wave incidence on multiconductor transmission lines. *IEEE Transaction on Electromagnetic Compatibility*, 36(3):206-212, 1994.

[99] D. Xie and M. Nakhla. Delay and crosstalk simulation of high speed VLSI interconnects with nonlinear terminations. *IEEE Transactions on Computer-Aided Design*, 12(11):1798-1811, 1993.

[100] J.J. Xu, M.C.E. Yagoub, R. Ding and Q.J. Zhang. Neural-based dynamic modeling of nonlinear microwave circuits. *IEEE Transactions on Microwave Theory and Techniques*, 50(12):2769-2780, 2002.

[101] C. Yen, Z. Fazarinc, and R.L. Wheeler. Time-Domain Skin-Effect Model for Transient Analysis of Lossy Transmission Lines. *Proceedings of the IEEE*, 70(5):750-757, 1982.

[102] Q. Yu and E.S. Kuh. Exact moment-matching model of transmission lines and application to interconnect delay estimation. *IEEE Transactions on VLSI,* 3(2):311-322, 1995.

[103] Q. Yu, J. M. L. Wang and E. S. Kuh. Passive multipoint moment-matching model order reduction algorithm on multiport distributed interconnect networks. *IEEE Transactions on Circuits and Systems - I*, 46(1):140-160, 1999.

[104] A.H. Zaabab, Q.J. Zhang, and M.S. Nakhla. A neural network modeling approach to circuit optimization and statistical design. *IEEE Transactions on Microwave Theory and Techniques*, 43(6):1349-1358, 1995.

[105] Q.J. Zhang and G.L. Creech (Guest Editors). Special Issue on Applications of Artificial Neural Networks for RF and Microwave Design. *International Journal of RF and Microwave Computer-Aided Engineering*, Wiley, New York, 1999.

[106] Q.J. Zhang and K.C. Gupta. *Neural Networks for RF and Microwave Design*, Artech House, Boston, 2000.

[107] Q.J. Zhang, K.C. Gupta and V.K. Devabhaktuni. Artificial neural networks for RF and microwave design: from theory to practice. *IEEE Transactions on Microwave Theory and Techniques*, 51(4):1339-1350, 2003.

[108] Q.J. Zhang and M. Mongiardo (Guest Editors). 2nd Special Issue on Application of ANN for RF and Microwave Design. *International Journal of RF and Microwave Computer-Aided Engineering*, Wiley, New York, 2002.

[109] Q.J. Zhang, F. Wang and M.S. Nakhla. Optimization of high-speed VLSI interconnects: a review. *International Journal on Microwave and Millimeter-wave Computer-Aided Engineering, Special Issue on Optimization Oriented Microwave CAD*, 7(1):83-107, 1997.

[110] G. Zheng, Q. J. Zhang, M. Nakhla and R. Achar. An efficient approach for simulation of measured subnetworks with complex frequency hopping. In *Proceedings of the IEEE/ACM International Conference on Computer Aided Design*, pp. 23-26, Nov. 1996, San Jose, CA.

Chapter 7

NONLINEAR PROGRAMMING AND ENGINEERING APPLICATIONS

Robert J. Vanderbei
Operations Research and Financial Engineering Department, Princeton University
rvdb@princeton.edu

Abstract The last decade has seen dramatic strides in ones ability to solve nonlinear programming problems. In this chapter, we review a few applications of nonlinear programming to interesting, and in some cases important, engineering problems.

Introduction

Modern interior-point methods for nonlinear programming have their roots in linear programming and most of this algorithmic work comes from the operations research community which is largely associated with solving the complex problems that arise in the business world. However, engineers and scientists also need to solve nonlinear optimization problems. While it is true that many engineers are well aware of the activities of SIAM and therefore are at least somewhat familiar with the advances that have taken place in optimization in the past two decades, there are many engineers and scientists who think that the only tool for solving an optimization problem is to code up something for MATLAB to solve using a genetic algorithm. Often the results are extraordinarily slow in coming and sometimes not even the results that one was looking for.

Having been immersed for more than a decade in the development of algorithms, and software, for solving nonlinear optimization problems, I eventually developed a desire to see the other side, the user's side. I became quite interested in applications not just as little toys to illustrate the power of my pet algorithms but rather as important problems that require solution and for which some of my tools might be useful. In this chapter I will chronicle some of these applications and illustrate how modern interior-point methods can be very useful to scientists and engineers.

The next section provides a brief description of an interior-point algorithms for nonlinear programming. Then, in subsequent sections we will discuss the following four application areas:

- Finite Impulse Response (FIR) filter design
- Telescope design—optics
- Telescope design—truss structure
- Stable orbits for the n-body problem

7.1 LOQO: An Interior-Point Code for NLP

Operations Research, our field of study, got its start more than 50 years ago roughly when George Dantzig invented the simplex method for solving linear programming problems. The simplex method together with the development of the computer provided a new extremely powerful tool for solving complex decision making problems. Even today, the simplex method is an indispensable tool to the operations researcher.

Of course, it was fairly soon after the invention that people began to realize that the linear programming problem was too restrictive for most of the real-world problems that needed to be solved. Many problems have the extra constraint that some or all of the variables need to be integer valued. Thus was born the field of integer programming. Even more problems involve nonlinear functions in the objective function and/or in the constraints and so there also arose the subject of nonlinear programming. The simplex method has played a critical role in both of these directions of generalization. For integer programming, the simplex method is used as a core engine in cutting-plane, branch-and-bound, and branch-and-cut algorithms. Some of these algorithms have proved to be very effective at solving some amazingly difficult integer programming problems. For nonlinear programming, the ideas behind the simplex method, namely the idea of active and inactive variables, were extended to this broader class of problems. For many years, the software package called MINOS, which implemented these ideas, was the best and most-used software for solving constrained nonlinear optimization problems. Its descendent, SNOPT, remains a very important tool even today.

In the mid 1980's, N. Karmarkar [9] invented a new algorithm for linear programming. It was totally unlike the simplex method. He proved that this algorithm has polynomial time worst-case complexity—something that has not yet been established for any variant of the simplex method. Furthermore, he claimed that the algorithm would also be very good in its average-case performance and that it would compete with, perhaps even replace, the simplex method as the method of choice for linear programming. True enough, this new class of algorithms, which we now call *interior-point methods*, did prove to be competitive. But, the new algorithm did not uniformly dominate the old simplex method and even today the simplex method, as embodied by the commercial software package called CPLEX, remains the most-used method for solving linear programming problems. Furthermore, interior-point methods have not proved to be effective for solving integer programming problems. The tricks that allow one to use the simplex method to solve integer programming problems depends critically on being able to solve large numbers of similar linear programming problems very quickly. The simplex method has the nice feature that solving a second instance of a problem starting from the solution to a first instance is often orders of magnitude faster than simply solving the second instance from scratch. There is no analogous property for interior-point methods and so today the simplex method remains the best method for solving integer programming problems.

So, do interior-point methods have a natural extension to nonlinear programming and, if so, how do they compare to the natural extension of the simplex method to such problem? Here, the answers are much more satisfactory. The answer is: yes, there is a very natural extension and, yes, the methods perform very well in this context. In fact, interior-point methods are really best understood as methods for constrained convex nonlinear optimization.

I have for many years been one of the principle developers of a particular piece of software, called LOQO, which implements an interior-point algorithm for nonlinear programming. In the remainder of this section, I will give a brief review of the algorithm as implemented in this piece of software. The basic family of problems that we wish to solve are given by

$$\begin{array}{ll} \text{minimize} & f(x) \\ \text{subject to} & b \leq h(x) \leq b + r, \\ & l \leq x \leq u, \end{array}$$

where b, h, and r take values in $\mathbb{R}^m$ and l, x, and u take values in $\mathbb{R}^m$. We assume that the functions $f(x)$ and $h(x)$ must be twice differentiable (at least at points of evaluation) but not necessarily convex or concave.

The standard *interior-point paradigm* can be described as follows:

- Add slacks thereby replacing all inequalities with nonnegativities.

- Replace nonnegativities with logarithmic barrier terms in objective; that is, terms of the form $-\mu \log(s)$ where μ is a positive *barrier parameter* and s is a slack variable.

- Write first-order optimality conditions for the (equality constrained) barrier problem.

- Rewrite the optimality conditions in primal-dual symmetric form (this is the only step that requires linear programming for its intuition).

- Use Newton's method to derive search directions. Here is the resulting linear system of equations:

$$\begin{bmatrix} -H(x,y) - D & A^T(x) \\ A(x) & E \end{bmatrix} \begin{bmatrix} \Delta x \\ \Delta y \end{bmatrix} = \begin{bmatrix} \nabla f(x) - A^T(x)y \\ -h(x) + \mu Y^{-1}e \end{bmatrix}.$$

  Matrices D and E are diagonal matrices involving slack variables,

$$H(x,y) = \nabla^2 f(x) - \sum_{i=1}^{m} y_i \nabla^2 h_i(x) + \lambda I, \text{ and } A(x) = \nabla h(x),$$

  where λ is chosen to ensure appropriate descent properties.

- Compute a step length that ensures positivity of slack variables.

- Shorten steps further to ensure a reduction in either infeasibility or in the barrier function.

- Step to new point and repeat.

Further details about the algorithm can be found in [13, 16].

7.2 Digital Audio Filters

A digital audio signal is a stream of integers, which represents a discretization both in time and in amplitude of an analog signal. For CD-quality sound, there are 44,100 samples per second each sample being a short integer (i.e., an integer between $-32,768$ and $+32,767$). Of course, for stereo there are two such streams of integers. Figure 7.1 shows an example of a very short stretch of music.

A digital audio signal is read off from a CD-ROM device, converted back to analog, amplified, and then sent to a speaker where the induced displacement creates a sound wave that is a replication of the original signal. A speaker that accurately reproduces low frequency signals generally does a bad job at high frequencies and vice versa. Therefore, modern audio equipment generally splits a signal into two, or more often three, frequency ranges and sends them to different speakers. Speakers designed for low frequencies are called *woofers*, those for a middle range of frequencies are called *midranges*, and those for high frequencies are called *tweeters*.

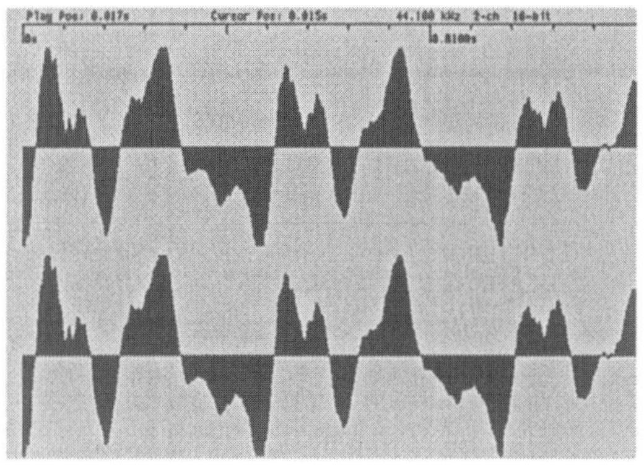

0	-32768	8	-23681	16	12111	24	31919	32	28095
1	-32768	9	-18449	17	17311	25	32767	33	28399
2	-32768	10	-11025	18	21311	26	32767	34	28751
3	-30753	11	-6913	19	23055	27	32767	35	28751
4	-28865	12	-4337	20	23519	28	32767	36	26911
5	-29105	13	-1329	21	25247	29	32031	37	24063
6	-29201	14	1743	22	27535	30	29759	38	21247
7	-26513	15	6223	23	29471	31	28399	39	18415

Figure 7.1. A stereo audio signal digitized.

Traditionally, the three speakers—woofer, midrange, and tweeter—were housed in the same physical box and the amplified signal was split into three parts using analog filtering components built into the speaker. However, there are limitations to this design. First of all, it is easy to determine the direction of a high frequency signal but not a low frequency one. Hence, the placement of the tweeters is important but the woofer can be put anywhere within hearing range. Furthermore, for stereo systems it is not even necessary to have two woofers—the low frequency signal can be combined into one. Also, woofers are physically large whereas tweeters can be made very small. Hence in a home theater system, one puts small tweeters in appropriate locations on either side of a video screen but puts a single woofer somewhere out of the way, such as under a coffee table.

Once the notion of physically separating the components of a speaker system is introduced, it is natural to consider also using separate amplifiers for each component. The amplifiers can then be designed to work optimally in a narrower range of frequencies. For example, it takes a lot of power to amplify a low frequency signal and much less power to do the same for high frequencies. Hence, using a hefty amplifier on the high frequency component of a signal is wasteful.

Finally, given that the signal is to be split before being amplified it is now possible to consider splitting it even before converting it from digital to analog. At the digital level one has much more control over how the split is accomplished than can be achieved with an analog signal. The most common type of digital filter is a finite impulse response filter, which we describe next.

7.2.1 Finite Impulse Response (FIR) Filters

A *finite impulse response (FIR) filter* is given by a finite sequence of real (or complex) numbers $h_{-n}, \ldots, h_{-1}, h_0, h_1, \ldots, h_n$. This sequence transforms an input signal, x_k, $k \in \mathbb{Z}$, into an output signal, y_k, $k \in \mathbb{Z}$, according to the following convolution formula:

$$y_k = \sum_{i=-n}^{n} h_i x_{k-i}, \qquad k \in \mathbb{Z}.$$

Since the sequence of filter coefficients is finite, the sum is finite too. Typically n is a small number (less than 100) and so the output signal at any given point in time depends on the values of the input signal in a very narrow temporal range symmetric around this time. With 44,100 samples per second, $n = 100$ corresponds to a time interval that is only a small fraction of a second long. To implement the filter there must be at least this much delay between the input and output signals. Since this delay is small, it is generally unnoticeable.

Of course the filter coefficients h_i must be determined when the system is designed, which is long before any specific input signal is decided upon. Hence, one treats the input signal as a random process which will only be realized in the future but whose statistical properties can be used to design the filter. To this end, we assume that x_k is a stationary second-order random process. This means that each x_k is random, has mean zero, finite variance, and a covariance structure that is temporally homogeneous. This last property means that the following covariances depend on the difference between the sample times but not on the time itself:

$$s_k = \mathbf{E} x_i \bar{x}_{i+k}.$$

(The bar on the x_{i+k} denotes complex conjugate—most of our processes are real-valued in which case conjugation will plays no role.) The sequence s_k characterizes the input signal and its Fourier transform

$$S(\nu) = \sum_k s_k e^{2\pi jk\nu}$$

($j = \sqrt{-1}$) characterizes it in the frequency domain. The function $S()$ is called the *spectral density*. It is periodic in ν with period 1 and so its domain is usually taken to be $[-1/2, 1/2)$. Values $\nu \in [-1/2, 1/2)$ are called *frequencies*. They

Linear Phase Filters. For simplicity, we assume that the filter coefficients are real and symmetric about zero: $h_{-i} = h_i$. Such a filter is said to be *linear phase*. >From these properties it follows that the function $H()$ defined by

$$H(\nu) = \sum_{k=-n}^{n} h_k e^{2\pi jk\nu}$$

is real-valued and symmetric about zero: $H(-\nu) = H(\nu)$. Indeed,

$$H(\nu) = h_0 + 2\sum_{k=1}^{n} h_k \cos(2\pi k\nu).$$

We then see that

$$\begin{aligned} G(\nu) &= \sum_k g_k e^{2\pi jk\nu} = \sum_{i,k} h_i h_{i+k} e^{2\pi jk\nu} \\ &= \sum_{i,k} h_{-i} e^{-2\pi ji\nu} h_{i+k} e^{2\pi j(i+k)\nu} = H(\nu)^2 \end{aligned}$$

and the transfer equation can be written in terms of $H()$:

$$R(\nu) = H(\nu)^2 S(\nu).$$

Power. For stationary signals, the power is defined as the expected value of the square of the signal at any moment in time. So, the input power is

$$P_{\text{in}} = \mathbf{E}|u_0|^2 = s_0 = \int_{-1/2}^{1/2} S(\nu)d\nu$$

and the output power is

$$P_{\text{out}} = \mathbf{E}|y_0|^2 = r_0 = \int_{-1/2}^{1/2} R(\nu)d\nu = \int_{-1/2}^{1/2} H(\nu)^2 S(\nu)d\nu.$$

A signal that is uniformly distributed over low frequencies, say from $-a$ to a has a spectral density given by

$$S(\nu) = 1_{[-a,a]}(\nu).$$

For such a signal, the input and output powers are given by

$$\begin{aligned} P_{\text{in}} &= 2a \\ P_{\text{out}} &= \int_{-a}^{a} H(\nu)^2 d\nu \qquad (7.2) \\ &= \sum_{k,k'} h_k h_{k'} \int_{-a}^{a} e^{2\pi j(k-k')\nu} d\nu \\ &= \sum_{k,k'} 2a h_k h_{k'} \text{sinc}(2\pi(k-k')a), \qquad (7.3) \end{aligned}$$

can be converted to the usual scale of cycles-per-second (Hz) using the sample rate but for our purposes we will take them as numbers in $[-1/2, 1/2)$.

An Example. Consider the simplest input process—a complex signal which is a pure wave with known frequency ν_0 and an unknown phase shift:

$$x_k = e^{2\pi j(k+\theta)\nu_0}.$$

Here, θ is a random variable uniformly distributed on $[-1/2, 1/2)$. For this process, the autocorrelation function is easy to compute:

$$s_k = \mathbf{E}x_i\bar{x}_{i+k} = \mathbf{E}e^{2\pi j(i+\theta)\nu_0}e^{-2\pi j(i+k+\theta)\nu_0} = e^{-2\pi jk\nu_0}.$$

The spectral density is given by

$$S(\nu) = \begin{cases} \infty & \nu = \nu_0 \\ 0 & \text{else.} \end{cases}$$

The Transfer Function. We are interested in the spectral properties of the output process. Hence, we introduce the autocorrelation function for y_k

$$r_k = \mathbf{E}y_i\bar{y}_{i+k}$$

and its associated spectral density function

$$R(\nu) = \sum_k r_k e^{2\pi jk\nu}.$$

Substituting the definition of the output process y_i into the formula for the autocorrelation function, it is easy to check that

$$r_k = \sum_l g_l s_{k-l},$$

where

$$g_k = \sum_i h_i h_{i+k}.$$

Similarly, it is easy to relate the output spectral density $R(\nu)$ to the input spectral density $S(\nu)$:

$$R(\nu) = G(\nu)S(\nu), \tag{7.1}$$

where

$$G(\nu) = \sum_k g_k e^{2\pi jk\nu}.$$

Equation (7.1) is called the *transfer equation* and $G()$ is the *transfer function*.

where

$$\operatorname{sinc}(x) = \begin{cases} \frac{\sin x}{x} & x \neq 0 \\ 1 & \text{else.} \end{cases}$$

Passbands. In some cases, it is desirable to have the output be as similar as possible to the input. That is, we wish the difference process,

$$z_k = y_k - x_k,$$

to have as little energy as possible. Let q_k denote the autocorrelation function of the difference process:

$$q_k = \mathbf{E} z_i \bar{z}_{i+k}$$

and let $Q()$ denote the corresponding spectral density function. It is then easy to check from the definitions that

$$Q(\nu) = (H(\nu) - 1)^2 S(\nu).$$

The output power for the difference process is then given by

$$P_{\text{diff\_out}} = \int_{-1/2}^{1/2} (H(\nu) - 1)^2 S(\nu) d\nu.$$

As before, if the input spectral density $S()$ is a piecewise constant even function, then this output power can be expressed in terms of the sinc function.

7.2.2 Coordinated Woofer–Midrange–Tweeter Filtering

Having covered the basics of FIR filters, we return now to the problem of designing an audio system based on three filters: woofer, midrange, and tweeter. There are four power measurements that we want to be small: for each filter we want the output to be small if the input is uniformly distributed over a range of frequencies *outside* of the desired frequency range and finally when added together the difference between the summed signal and the original signal should be small over the entire input spectrum. Let

$$\begin{aligned} \mathcal{T} &= (-1/2, -b_t) \cup (b_t, 1/2), \\ \mathcal{M} &= (-b_m, -a_m) \cup (a_m, b_m), \\ \mathcal{W} &= (-a_w, a_w) \end{aligned}$$

denote the design frequency ranges for the tweeter, midrange, and woofer, respectively. Of course, we assume that the three ranges cover the entire available spectrum:

$$\mathcal{T} \cup \mathcal{M} \cup \mathcal{W} = (-1/2, 1/2)$$

(or, in other words, that $a_m < a_w$ and $b_t < b_m$). Each speaker has its own filter which is defined by its filter coefficients

$$h_k^{(j)}, \quad k = -n, -n+1, \ldots, n-1, n, \quad j \in \{t, m, w\}$$

and associated spectral density function $H_j(\nu), j \in \{t, m, w\}$. The three constraints which say that for each filter the output power per unit of input power is smaller than some threshold ρ can now be written as

$$\begin{aligned} \frac{1}{|\mathcal{T}^c|} \int_{\mathcal{T}^c} H_t^2(\nu) d\nu &\leq \rho, \\ \frac{1}{|\mathcal{M}^c|} \int_{\mathcal{M}^c} H_m^2(\nu) d\nu &\leq \rho, \\ \frac{1}{|\mathcal{W}^c|} \int_{\mathcal{W}^c} H_w^2(\nu) d\nu &\leq \rho. \end{aligned}$$

It is interesting to note that according to (7.3) the above integrals can all be efficiently expressed in terms of sums of products of pairs of filter coefficients in which the constants involve sinc functions. Such expressions are nonlinear. The fact that these functions are convex is only revealed by noting their equality with the expression in (7.2). Finally, the constraint that the reconstructed sum of the three signals deviates as little as possible from the a uniform response can be written as

$$\int_{-\frac{1}{2}}^{\frac{1}{2}} (H_t(\nu) + H_m(\nu) + H_w(\nu) - 1)^2 d\nu \leq \epsilon.$$

At this juncture, there are several ways to formulate an optimization problem. We could fix ϵ to some small positive value and then minimize ρ, or we could fix ρ to some small positive value and minimize ϵ, or we could specify to proportional relation, such as equality, between ρ and ϵ and minimize both simultaneously. To be specific, for this tutorial, we choose the third approach.

In this paper (and in life), we formulate our optimization problems in AMPL, which is a small programming language designed for the efficient expression of optimization problems [7]. The AMPL model for this problem is shown in Figure 7.2. The three filters and their spectral response curves are shown in Figure 7.3.

For more information on FIR filter design, see for example [22, 5, 12, 6].

7.3 Shape Optimization (Telescope Design)

Until recently the search for extraterrestrial life has been the subject of science fiction stories—science itself was incapable of providing much help. Of course, there is so far one exception—the SETI project (SETI stands for *search for*

```
function sinc;
param n := 23;
param pi := 4*atan(1);
param aw := 0.05;
param am := 0.04;
param bm := 0.25;
param bt := 0.2;
var rho >= 0;
var hw {0..n};
var hm {0..n};
var ht {0..n};

minimize power_bnd: rho;

subject to passband:
    ((hw[0]+hm[0]+ht[0]-1)^2 + 2*sum {k in 1..n} (hw[k]+hm[k]+ht[k])^2)
    <= rho;

subject to wooferband:
    sum {k in -n..n} hw[abs(k)]^2
    -
    sum {k in -n..n, kk in -n..n} 2*aw*hw[abs(k)]*hw[abs(kk)] * sinc(2*pi*(k-kk)*aw)
    <= (1-2*aw)*rho;

subject to midrangeband:
    sum {k in -n..n} hm[abs(k)]^2
    -
    sum {k in -n..n, kk in -n..n} 2*bm*hm[abs(k)]*hm[abs(kk)] * sinc(2*pi*(k-kk)*bm)
    +
    sum {k in -n..n, kk in -n..n}
       2*am*hm[abs(k)]*hm[abs(kk)] * sinc(2*pi*(k-kk)*am)
    <= (1-2*(bm-am))*rho;

subject to tweeterband:
    sum {k in -n..n, kk in -n..n} 2*bt*ht[abs(k)]*ht[abs(kk)] * sinc(2*pi*(k-kk)*bt)
    <= 2*bt*rho;

solve;
printf {k in 0..n}: "%10.6f \n", hw[k] > hw;
printf {k in 0..n}: "%10.6f \n", hm[k] > hm;
printf {k in 0..n}: "%10.6f \n", ht[k] > ht;

printf {nu in 0..0.5 by 1/1000}: "%7.4f %10.3e \n",
    nu, 10*log10((hw[0] + 2* sum {k in 1..n} (hw[k]*cos(-2*pi*k*nu)))^2) > w.out;

printf {nu in 0..0.5 by 1/1000}: "%7.4f %10.3e \n",
    nu, 10*log10((hm[0] + 2* sum {k in 1..n} (hm[k]*cos(-2*pi*k*nu)))^2) > m.out;

printf {nu in 0..0.5 by 1/1000}: "%7.4f %10.3e \n",
    nu, 10*log10((ht[0] + 2* sum {k in 1..n} (ht[k]*cos(-2*pi*k*nu)))^2) > t.out;
```

Figure 7.2. A sample AMPL program for FIR filter design of coordinated woofer, midrange, and tweeter system.

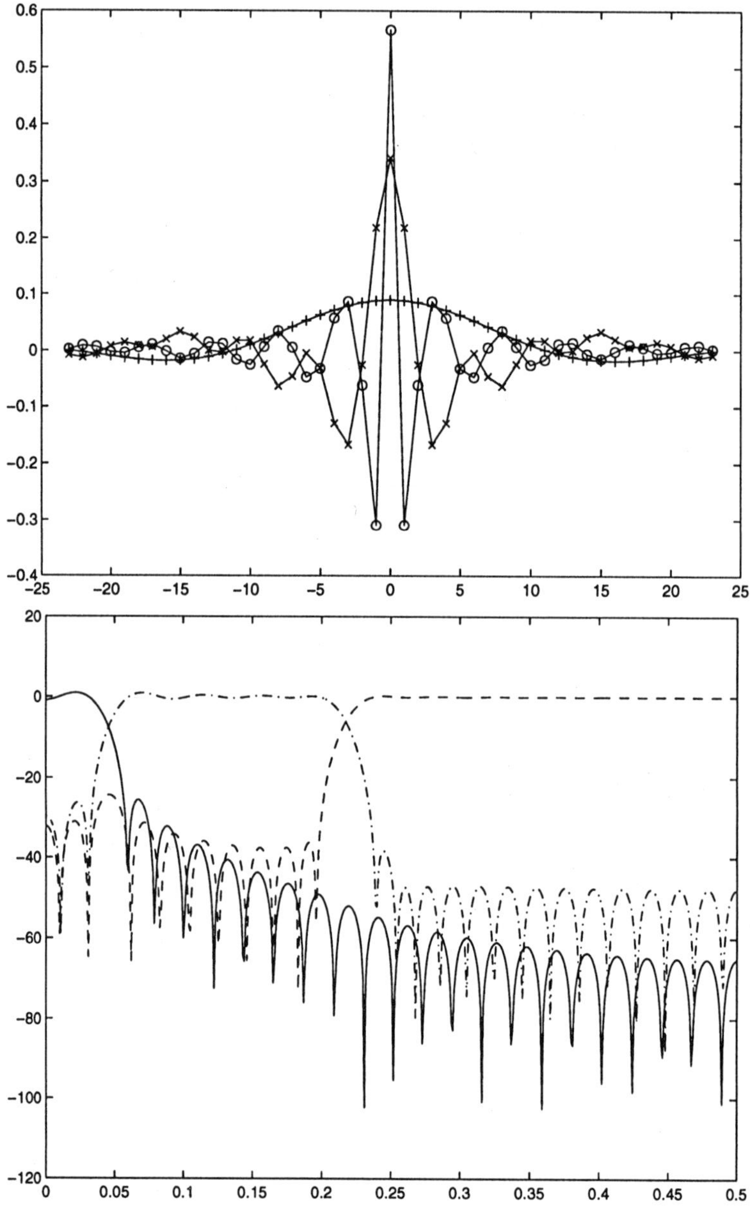

Figure 7.3. *Top.* The optimal filter coefficients. *Bottom.* The corresponding spectral response curves. In the top graph, the o's correspond to the tweeter filter, the $\times$'s correspond to the midrange filter, and the $+$'s correspond to the woofer filter. The spectral response curve is a plot of ten times log base ten of power as a function of frequency.

extraterrestrial intelligence), which has been operating for several years now. The idea here is to use radio telescopes to listen for radio transmissions from advanced civilizations. This project was started with the support of Carl Sagan and his book *Contact* was made into a Hollywood movie starring Jodie Foster. But, the universe is big and the odds that there is an advanced civilization in our neck of the woods is small so this project seems like a long shot. And, every year that goes by without hearing anything proves more and more what a long shot it is. Even if advanced civilizations are rare, there is every expectation that most stars have planets around them and even Earth-like planets are probably fairly common. It would be interesting if we could search for, catalog, and survey such planets. In fact, astrophysicists, with the support of NASA and JPL, are now embarking on this goal—imaging Earth-like planets around nearby Sun-like stars. We have already detected indirectly more than 100 Jupiter-sized planets around other stars and we will soon be able to take pictures of some of these planets. Once we can take pictures, we can start answering questions like: is there water? is there chlorophyl? is there carbon dioxide in the atmosphere? etc. But Jupiter-sized planets are not very Earth-like. They are mostly gas and very massive. There's not much place for an ET to get a foothold and if one could the gravity would be crushing. A more interesting but much more difficult problem is to survey Earth-like planets. NASA has made such a search and survey one of its key science projects for the coming decades. The idea is to build a large space telescope, called the *Terrestrial Planet Finder (TPF)* that is capable of imaging these planets. But just making it large and putting it into space is not enough. The planet, which will appear very close to its star, will still be invisible in the glare of its much brighter star.

To put the problem into perspective, here are a few numbers. Consider a star that is say 30 light years from our solar system. A planet that is as far from this star as we are from our Sun will appear to us here on Earth at an angular separation of about 0.1 arcseconds from its star. And the star will be 10^{10} times brighter than the planet. There will be an enormous amount of starlight that will "spill" onto that part of the image where the planet is supposed to be. This spillage is not one of engineering/construction imprecisions. Rather, it is a consequence of fundamental physics. Light is a wave, an electro-magnetic wave, and because of this it is impossible to focus the light from a point source (such as a star) to a perfect point in the image. Instead, for a telescope with a circular opening letting in the light, you get a small blob of light called an *Airy disk* surrounded by *diffraction rings*—see Figure 7.4. The planet is about as bright as the light in the 100th diffraction ring. Earlier rings are much brighter. The spacing of the rings is inversely proportional to the size of the telescope. To make it so that the rings are tight enough that we can image a planet like the one just described would require a telescope with a mirror having a 250 meter diameter. That is more than 10 times the diameter of the mirror in the

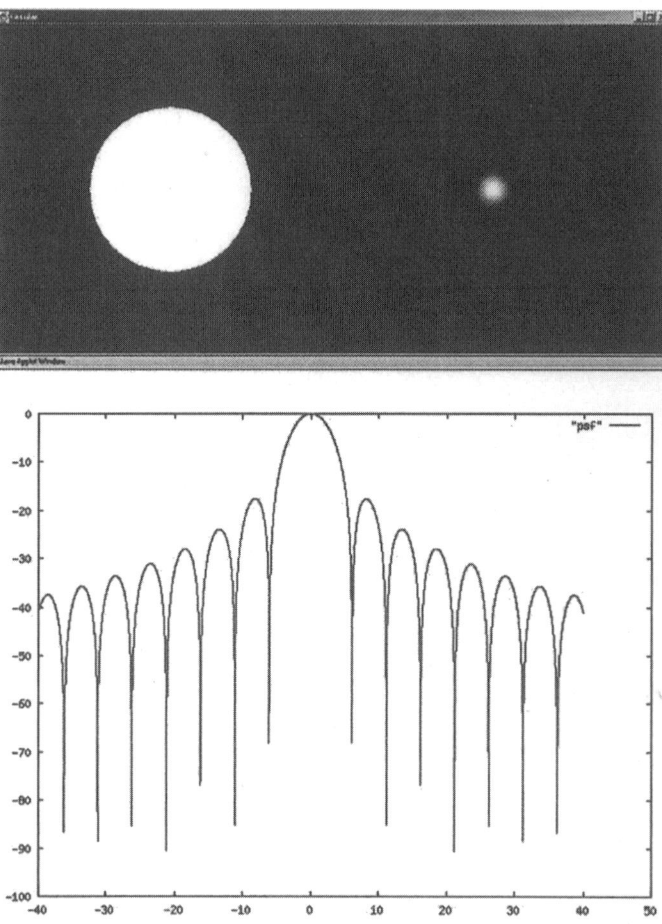

Figure 7.4. The top left shows a circular opening at the front of a telescope. The top right shows the corresponding Airy disk and a few diffraction rings. The plot on the bottom shows the cross-sectional intensity on a log scale. The desired level of 10^{-10} corresponds to an intensity level of -100 on this log plot. It is way off to the side. It occurs out somewhere around the 100th diffraction ring.

Hubble space telescope. There are not any rockets in existence, or on the drawing boards, that will be capable of lifting such a large monolithic object into space at any time in the foreseeable future. For this reason some clever ideas are required. A few have been proposed. Perhaps the most promising one exploits the idea that the ring pattern is a consequence of the circular shape of the telescope. Different shapes provide different patterns—perhaps some of them provide a very dark zone very close to the Airy disk. An even broader generalization is to consider a telescope that has a filter over its opening that has light transmission properties that vary over the surface of the filter. If the transmission is everywhere either zero or one then the filter acts to create a different shaped opening. Such filters are called *apodizations*. The problem is to find an apodization that provides a very dark area very close to the Airy disk.

Okay, enough with the words already—we need to give a mathematical formulation of the problem. The diffraction pattern produced by the star in the image is the square of the electric field at the image plane and the electric field at the image plane turns out to be just the Fourier transform of the apodization function A defining the transmissivity of the apodized pupil:

$$E(\xi,\zeta) = \iint_S e^{-2\pi i(x\xi+y\zeta)} A\,(x,y)\,dxdy,$$

where

$$S = \{(x,y):\ 0 \le r(x,y) \le 1/2,\ \theta(x,y) \in [0,2\pi]\},$$

and $r(x,y)$ and $\theta(x,y)$ denote the polar coordinates associated with point (x,y). Here, and throughout this section, x and y denote coordinates on the filter measured in units of the mirror diameter D and ξ and ζ denote angular (radian) deviation from on-axis measured in units of wavelength λ over mirror-diameter (λ/D) or, equivalently, physical distance in the image plane measured in units of focal-length times wavelength over mirror-diameter ($f\lambda/D$).

For circularly-symmetric apodizations, it is convenient to work in polar coordinates. To this end, let r and θ denote polar coordinates in the filter plane and let ρ and ϕ denote the image plane coordinates:

$$\begin{aligned} x &= r\cos\theta & \qquad \xi &= \rho\cos\phi \\ y &= r\sin\theta & \qquad \zeta &= \rho\sin\phi. \end{aligned}$$

Hence,

$$\begin{aligned} x\xi + y\zeta &= r\rho(\cos\theta\cos\phi + \sin\theta\sin\phi) \\ &= r\rho\cos(\theta-\phi). \end{aligned}$$

The electric field in polar coordinates depends only on ρ and is given by

$$E(\rho) = \int_0^{1/2} \int_0^{2\pi} e^{-2\pi i r \rho \cos(\theta - \phi)} A(r) r d\theta dr, \quad (7.4)$$

$$= 2\pi \int_0^{1/2} J_0(2\pi r \rho) A(r) r dr, \quad (7.5)$$

where J_0 denotes the 0-th order Bessel function of the first kind. Note that the mapping from apodization function A to electric field E is linear. Furthermore, the electric field in the image plane is real-valued (because of symmetry) and its value at $\rho = 0$ is the *throughput* of the apodization:

$$E(0) = 2\pi \int_0^{1/2} A(r) r dr.$$

As mentioned already, the diffraction pattern, which is called the *point spread function* (psf), is the square of the electric field. The contrast requirement is that the psf in the dark region be 10^{-10} of what it is at the center of the Airy disk. Because the electric field is real-valued, it is convenient to express the contrast requirement in terms of it rather than the psf, resulting in a field requirement of $\pm 10^{-5}$.

The apodization that maximizes throughput subject to contrast constraints can be formulated as an infinite dimensional linear programming problem:

$$\begin{array}{lll} \text{maximize} & E(0) & \\ \text{subject to} & -10^{-5} E(0) \leq E(\rho) \leq 10^{-5} E(0), & \rho_{\text{iwa}} \leq \rho \leq \rho_{\text{owa}}, \\ & 0 \leq A(r) \leq 1, & 0 \leq r \leq 1/2, \end{array}$$

where ρ_{iwa} denotes a fixed *inner working angle* and ρ_{owa} a fixed *outer working angle*. Discretizing the sets of r's and ρ's and replacing the integrals with their Riemann sums, the problem is approximated by a finite dimensional linear programming problem that can be solved to a high level of precision.

The solution obtained for $\rho_{\text{iwa}} = 4$ and $\rho_{\text{owa}} = 40$ is shown in Figure 7.5. Note that the solution is of a bang-bang type. That is, the apodization function is mostly 0 or 1 valued. This suggests looking for a mask that is about as good as this apodization. Such a mask can be found by solving the following nonlinear optimization problem. A mask consists of a set of concentric opaque rings,

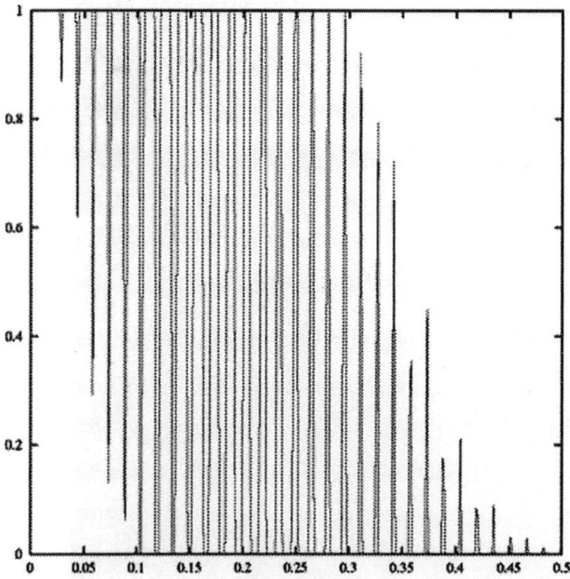

Figure 7.5. The optimal apodization function turns out to be of bang-bang type.

formulated in terms of the inner and outer radii of the openings between the rings:

$$\begin{array}{ll} [r_0, r_1] & \text{first opening} \\ [r_2, r_3] & \text{second opening} \\ [r_4, r_5] & \text{third opening} \\ & \vdots \\ [r_{2m-2}, r_{2m-1}] & m\text{-th opening} \end{array}$$

With this notation, the formula for $E(\rho)$ given in (7.5) can be rewritten as a sum of integrals over these openings:

$$\begin{aligned} E(\rho) &= 2\pi \sum_{k=0}^{m-1} \int_{r_{2k}}^{r_{2k+1}} J_0(2\pi r\rho) r dr, \\ &= \frac{1}{\rho} \sum_{k=0}^{m-1} \left(r_{2k+1} J_1(2\pi r_{2k+1}\rho) - r_{2k} J_1(2\pi r_{2k}\rho) \right) \end{aligned}$$

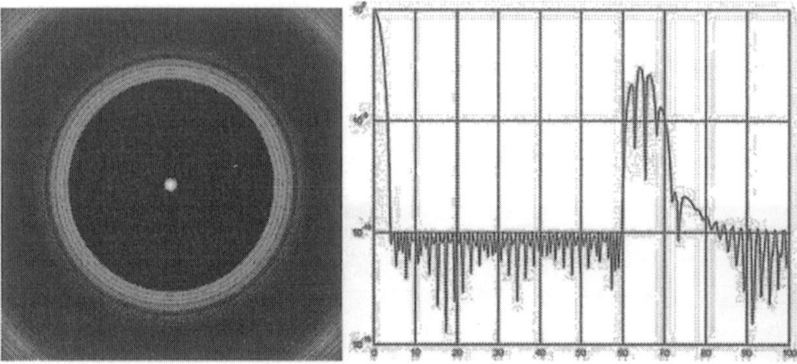

Figure 7.6. At the top is shown the concentric ring apodization. The second row shows the psf as a 2-D image and in cross-section. Note that it achieves the required 10^{-10} level of darkness at a distance of only 4 from the center of the Airy disk.

Treating the r_k's as variables and using this new expression for the electric field, the mask design problem becomes:

$$\begin{array}{lll} \text{maximize} & \pi \displaystyle\sum_{k=0}^{m-1} \left(r_{2k+1}^2 - r_{2k}^2\right) & \\ \text{subject to} & -10^{-5}E(0) \leq E(\rho) \leq 10^{-5}E(0), & \rho_{\text{iwa}} \leq \rho \leq \rho_{\text{owa}}, \\ & 0 \leq r_0 \leq r_1 \leq \cdots \leq r_{2m-1} \leq 1/2. & \end{array}$$

This problem is a nonconvex nonlinear optimization problem and hence the best hope for solving it in a reasonable amount of cpu time is to use a "local-search" method starting the search from a solution that is already close to optimal. The bang-bang solution from the linear programming problem can be used to generate a starting solution. Indeed, the discrete solution to the linear programming problem can be used to find the inflection points of A which can be used as initial guesses for the r_k's. LOQO was used to perform this local optimization. Figure 7.6 shows an optimal concentric-ring mask computed using an inner working angle of 4 and an outer working angle of 60. Using this mask over a 10 meter primary mirror makes it possible to image the Earth-like planet 30 light-years away from us. Even a telescope with a 10 meter primary mirror is larger than anything we have launched into space to date but it is a size that fits into the realm of possibility. And, if a 10 circular mirror is too large, we could fall back on elliptical designs say using a 4×10 mirror. Such a mirror could be put into space using currently available Delta rockets. The mask designs presented here and many others can be found in the following references: [15, 11, 10, 18, 17].

7.4 Minimum weight truss design

As we saw in the previous section, designing a space telescope with an apodization, or mask, over the mirror makes it possible to image Earth-like planets around nearby stars. However, it is still just on the edge of tractability. The desire remains to launch a much larger telescope. But, given constraints on mirror manufacture and launch capabilities, the prospect of using a huge telescope remains well out of reach for the foreseeable future. A compromise idea it to launch several, say four, individual telescopes, attach them to a common structure so that they are configured along a straight line, and then combine the light from the four telescopes to make one image. In some sense, this is equivalent to making a huge circular-mirror telescope and masking out everything except for four circular areas spread out along the mirror. As explained in the previous section, this masking changes the diffraction pattern but a basic principle holds which says that the further apart you can get the telescopes, the tighter the diffraction pattern will be. One of the main challenges with this design concept is to design and build a truss-like structure that is light enough to be launchable into space yet stiff enough that it can hold four massive telescopes in position with a precision that is small relative to the wavelength of light.

Such truss optimization problems have a long history starting, I believe, with Jim Ho's Ph.D. thesis at Stanford which was published in [8]. In more recent times, Ronnie Ben-Tal has collaborated with Bendsøe and Zowe on just this family of problems. They wrote many papers including the seminal paper [1]. Anyway, in the following subsection, I will outline the basic optimization problem and its reduction to a linear programming problem. Then in the following subsection, I will describe how it is being applied to the truss-design problem mentioned above.

7.4.1 Mathematical Formulation

We assume we are given a design space, which consists of a set $\mathcal{N}$ of nodes (aka joints) at fixed locations and a set $\mathcal{A}$ of undirected arcs (aka members) connecting various pairs of nodes—see Figure 7.7. The unknowns in the problem are the tensions x_{ij} in each member. We assume that there are many more members than are needed to make a rigid truss, so it follows that the system is underdetermined and there is lots of freedom as to how the forces "flow" through the structure. The tensions x_{ij} are allowed to go negative. Negative tensions are simply compressions. The constraints in our optimization problem are that force be balanced at each node. For example, if we look at node 2, the force balance equations are:

$$x_{12}\begin{bmatrix} -1 \\ 0 \end{bmatrix} + x_{23}\begin{bmatrix} -0.6 \\ 0.8 \end{bmatrix} + x_{24}\begin{bmatrix} 0 \\ 1 \end{bmatrix} = -\begin{bmatrix} b_2^1 \\ b_2^2 \end{bmatrix}.$$

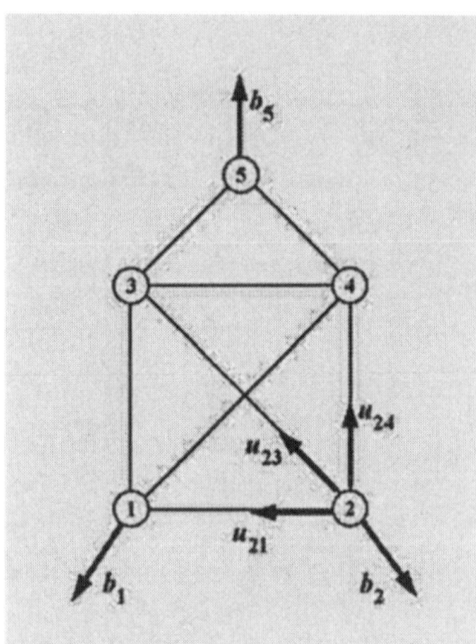

Figure 7.7. A design space showing 5 nodes and 8 arcs. Also shown are three externally applied forces and the tensile forces they induce in node 2.

To write down the equations in more generality, we need to introduce some notation:

$$\begin{aligned} p_i &= \text{position vector for joint } i \\ u_{ij} &= \frac{p_j - p_i}{\|p_j - p_i\|} \end{aligned}$$

Note that $u_{ji} = -u_{ij}$. With these notations, the general force balance constraints can be written as

$$\sum_{\substack{j: \\ \{i,j\} \in \mathcal{A}}} u_{ij} x_{ij} = -b_i, \quad i = 1, \ldots, m$$

It is instructive to write these equations in matrix form as $Ax = -b$, where

$$x^T = [\; x_{12} \quad x_{13} \quad x_{14} \quad x_{23} \quad x_{24} \quad x_{34} \quad x_{35} \quad x_{45} \;]$$

$$A = \begin{array}{c} 1 \\ \\ 2 \\ \\ 3 \\ \\ 4 \\ \\ 5 \\ \\ \end{array} \left[\begin{array}{cccccccc} 1 & 0 & .6 & & & & & \\ 0 & 1 & .8 & & & & & \\ -1 & & & -.6 & 0 & & & \\ 0 & & & .8 & 1 & & & \\ & 0 & & .6 & & 1 & .6 & \\ & -1 & & -.8 & & 0 & .8 & \\ & & -.6 & & 0 & -1 & & -.6 \\ & & -.8 & & -1 & 0 & & .8 \\ & & & & & & -.6 & .6 \\ & & & & & & -.8 & -.8 \end{array} \right], \quad b = \begin{bmatrix} b_1^1 \\ b_1^2 \\ b_2^1 \\ b_2^2 \\ b_3^1 \\ b_3^2 \\ b_4^1 \\ b_4^2 \\ b_5^1 \\ b_5^2 \end{bmatrix}.$$

Note that $\|u_{ij}\| = \|u_{ji}\| = 1$ and $u_{ij} = -u_{ji}$. Also, each column contains a u_{ij}, a u_{ji}, and the rest are zero. If the problem were one dimensional, this would be exactly a node-arc incidence matrix. In fact, much of the theory that has been developed for minimum-cost network flow problems has an immediate analogue in these truss design problems. These connections are described at length in [21].

So far we have only written down the force balance constraints. The optimization problem is to minimize the weight of the final structure. We assume that weight is related to tension/compression in the members by assuming that the cross-sectional area of a member must be proportional to the tension/compression that must be carried by that member (the constants of proportionality can be different for tension and for compression but in what follows we assume that they are equal). Hence, the minimum weight structural design problem can be formulated like this:

$$\begin{array}{lll} \text{minimize} & \displaystyle\sum_{\{i,j\}\in\mathcal{A}} l_{ij}|x_{ij}| & \\ \text{subject to} & \displaystyle\sum_{\substack{j:\\ \{i,j\}\in\mathcal{A}}} u_{ij}x_{ij} = -b_i & i = 1, 2, \ldots, m. \end{array}$$

This is not quite a linear programming problem. But it is easy to convert it into one using a common trick of splitting every variable into the difference between its positive and negative parts:

$$\begin{array}{rcll} x_{ij} & = & x_{ij}^+ - x_{ij}^-, & x_{ij}^+, x_{ij}^- \geq 0, \qquad x_{ij}^+ x_{ij}^- = 0 \\ |x_{ij}| & = & x_{ij}^+ + x_{ij}^- & \end{array}$$

In terms of these new variables, the problem can be written as follows:

$$\begin{array}{llll} \text{minimize} & \displaystyle\sum_{\{i,j\}\in\mathcal{A}} (l_{ij}x_{ij}^+ + l_{ij}x_{ij}^-) & & \\ \text{subject to} & \displaystyle\sum_{\substack{j:\\ \{i,j\}\in\mathcal{A}}} (u_{ij}x_{ij}^+ - u_{ij}x_{ij}^-) = -b_i & & i = 1, 2, \ldots, m \\ & \qquad\qquad x_{ij}^+ x_{ij}^- = 0 & \{i,j\} \in \mathcal{A}, & \\ & \qquad\qquad x_{ij}^+,\ x_{ij}^- \geq 0 & \{i,j\} \in \mathcal{A}. & \end{array}$$

It is easy to argue that one can drop the complementarity type constraints, $x_{ij}^+ x_{ij}^- = 0$, $\{i,j\} \in \mathcal{A}$, since these constraints will automatically be satisfied at optimality. With these constraints gone, the problem is a linear programming problem that can be solved very efficiently.

It was shown in [1] that this minimum weight structural design problem is dual to a maximum stiffness structural design problem and therefore that the structure found according to this linear programming methodology is in fact maximally stiff.

7.4.2 Telescope Truss Design

As mentioned at the beginning of this section, it is a very real problem of current interest to design a maximally stiff truss-like structure to support four telescopes. In order to apply the ideas described in the previous subsection, we need to add one extra twist to the model, which is that the design must be stiff relative to a collection of different loading scenarios. In particular, we assume that the structure must be stiff with respect to accelerations in each of the three main coordinate directions and also that it is stiff relative to torques about the three principle axes of rotation. Torques are modeled as pairs of forces that are equal and opposite but applied at points that are not colinear with the direction of the force. So, our basic model has six load scenarios. Forces must be balanced in each scenario. Of course, the tensions/compressions are scenario dependent but the beam cross-sections must be chosen independently of the scenario (since one physical structure must be stiff under each scenario).

The underlying design space consists of 26 truss-like cells, 1.6 m wide with triangular cross-section 2 m on a side. The middle two cells attach to the satellite. Forces are applied by thrusters on the satellite. Of course, in the real satellite, forces are applied to perform various motions but in our model we don't want, or need, to model moving objects. Instead we assume that countervaling forces are applied exactly where the massive objects are—i.e., at the point of attachment of the four telescopes. Figure 7.9 shows the underlying design space together with the optimal design. The force vectors corresponding to the six load scenarios are shown as elongated, shaded cones.

7.5 New orbits for the n-body problem

Since the time of Lagrange and Euler precious few solutions to the n-body problem have been discovered. Lagrange proved that two bodies being mutually attracted to the other by gravity will execute elliptical orbits where each of the two ellipses has a focus at the center of mass of the two-body system. This can be proved mathematically. Not only does this solution to Newton's equations of motion exist, but it is also stable. Euler pointed out that a third body can be placed stationarily at the center of mass and this makes a solution to the three-body problem. However, this 3-body system is unstable—if the third body is perturbed ever so slightly the whole system will fall apart. A few other simple solutions have been known for hundreds of years. For example, you can distribute n equal-mass bodies uniformly around a circle and start each one off with a velocity perpendicular to the line through the center and this system will behave much like the 2-body system. But, for three bodies or more, this system

```
param m default 26;  param n default 39;

set X := {0..n}; set Y := {0..m};
set NODES := X cross Y;      # A lattice of Nodes
set ANCHORS within NODES
    := { x in X, y in Y : x == 0 && y >= floor(m/3) && y <= m-floor(m/3) };

param xload {(x,y) in NODES: (x,y) not in ANCHORS} default 0;
param yload {(x,y) in NODES: (x,y) not in ANCHORS} default 0;
param gcd {x in -n..n, y in -n..n} :=
    (if x <  0 then gcd[-x,y] else
    (if x == 0 then y else
    (if y < x  then gcd[y,x] else
        (gcd[y mod x, x])  )));

set ARCS := { (xi,yi) in NODES, (xj,yj) in NODES:
    abs(xj-xi) <= 3 && abs(yj-yi) <=3 &&
    abs(gcd[ xj-xi, yj-yi ]) == 1 &&
    ( xi > xj || (xi == xj && yi > yj) ) };

param length {(xi,yi,xj,yj) in ARCS} := sqrt( (xj-xi)^2 + (yj-yi)^2 );

var comp {ARCS} >= 0;
var tens {ARCS} >= 0;
minimize volume:
    sum {(xi,yi,xj,yj) in ARCS}
        length[xi,yi,xj,yj] * (comp[xi,yi,xj,yj] + tens[xi,yi,xj,yj]);

subject to Xbalance {(xi,yi) in NODES: (xi,yi) not in ANCHORS}:
    sum { (xi,yi,xj,yj) in ARCS }
        ((xj-xi)/length[xi,yi,xj,yj]) * (comp[xi,yi,xj,yj]-tens[xi,yi,xj,yj])
    +
    sum { (xk,yk,xi,yi) in ARCS }
        ((xi-xk)/length[xk,yk,xi,yi]) * (tens[xk,yk,xi,yi]-comp[xk,yk,xi,yi])
    = xload[xi,yi];

subject to Ybalance {(xi,yi) in NODES: (xi,yi) not in ANCHORS}:
    sum { (xi,yi,xj,yj) in ARCS }
        ((yj-yi)/length[xi,yi,xj,yj]) * (comp[xi,yi,xj,yj]-tens[xi,yi,xj,yj])
    +
    sum { (xk,yk,xi,yi) in ARCS }
        ((yi-yk)/length[xk,yk,xi,yi]) * (tens[xk,yk,xi,yi]-comp[xk,yk,xi,yi])
    = yload[xi,yi];

let yload[n,m/2] := -1;

solve;
```

Figure 7.8. A sample AMPL program for minimum weight truss-like structures designed to accommodate a single loading scenario.

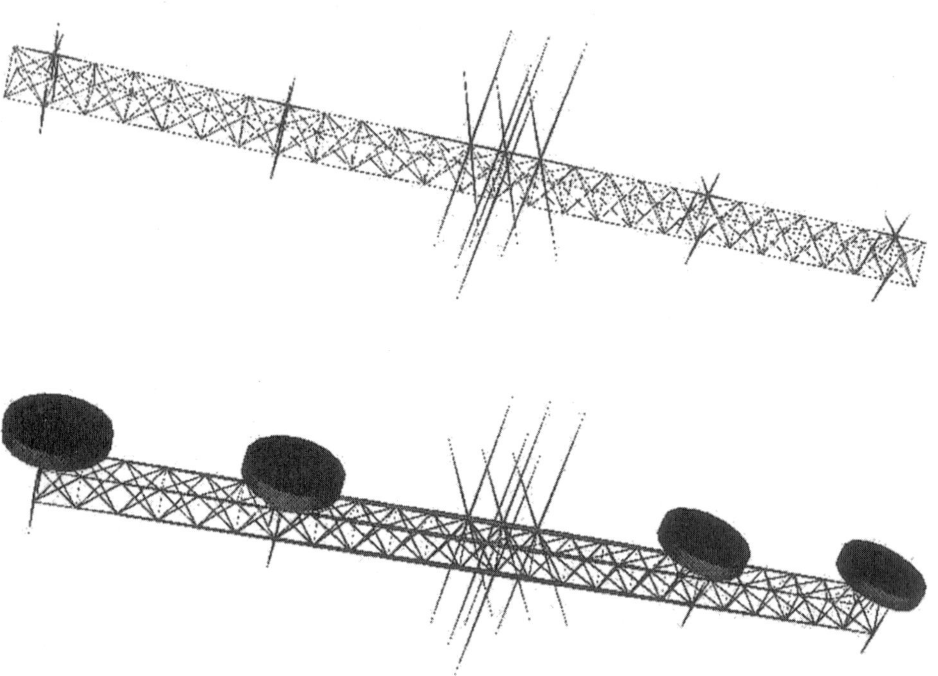

Figure 7.9. *Top.* The design space. *Bottom.* The optimal design.

is again unstable. So, it was a tremendous shock a few years ago when Cris Moore at the Sante Fe Institute discovered a new, stable solution to the equal-mass 3-body problem. This discovery has created a tremendous level of interest in the celestial mechanics community. Not only was the solution he discovered both new and stable, it is also aesthetically beautiful because each of the three bodies follow the exact same path. At any given moment they are at different parts of this path. Such orbital systems have been called *choreographies*. Many celestial mechanics have been working hard to discover new choreographies. The interesting thing for us is that the main tool is to minimize the so-called action functional.

In this section, we will describe how it is that minimizing the action functional provides solutions to the n-body problem and we will illustrate several new solutions that we have found.

7.5.1 Least Action Principle

Given n bodies, let m_j denote the mass and $z_j(t)$ denote the position in $\mathbb{R}^2 = \mathbb{C}$ of body j at time t. The *action functional* is a mapping from the space of all trajectories, $z_1(t), z_2(t), \ldots, z_n(t), 0 \le t \le 2\pi$, into the reals. It is defined as the integral over one period of the kinetic minus the potential energy:

$$A = \int_0^{2\pi} \left(\sum_j \frac{m_j}{2} \|\dot{z}_j\|^2 + \sum_{j,k:k<j} \frac{m_j m_k}{\|z_j - z_k\|} \right) dt.$$

Stationary points of the action function are trajectories that satisfy the equations of motions, i.e., Newton's law gravity. To see this, we compute the first variation of the action functional,

$$\begin{aligned} \delta A &= \int_0^{2\pi} \sum_\alpha \left(\sum_j m_j \dot{z}_j^\alpha \delta \dot{z}_j^\alpha - \sum_{j,k:k<j} m_j m_k \frac{(z_j^\alpha - z_k^\alpha)(\delta z_j^\alpha - \delta z_k^\alpha)}{\|z_j - z_k\|^3} \right) dt \\ &= -\int_0^{2\pi} \sum_j \sum_\alpha \left(m_j \ddot{z}_j^\alpha + \sum_{k:k\neq j} m_j m_k \frac{z_j^\alpha - z_k^\alpha}{\|z_j - z_k\|^3} \right) \delta z_j^\alpha dt, \end{aligned}$$

and set it to zero. We get that

$$m_j \ddot{z}_j^\alpha = -\sum_{k:k\neq j} m_j m_k \frac{z_j^\alpha - z_k^\alpha}{\|z_j - z_k\|^3}, \qquad j = 1, 2, \ldots, n, \quad \alpha = 1, 2 \quad (7.6)$$

Note that if $m_j = 0$ for some j, then the first order optimality condition reduces to $0 = 0$, which is *not* the equation of motion for a massless body. Hence, we must assume that all bodies have strictly positive mass.

7.5.2 Periodic Solutions

Our goal is to use numerical optimization to minimize the action functional and thereby find periodic solutions to the n-body problem. Since we are interested only in periodic solutions, we express all trajectories in terms of their Fourier series:

$$z_j(t) = \sum_{k=-\infty}^{\infty} \gamma_k e^{ikt}, \qquad \gamma_k \in \mathbb{C}.$$

Abandoning the efficiency of complex-variable notation, we can write the trajectories with components $z_j(t) = (x_j(t), y_j(t))$ and $\gamma_k = (\alpha_k, \beta_k)$. So doing, we get

$$\begin{aligned} x(t) &= a_0 + \sum_{k=1}^{\infty} (a_k^c \cos(kt) + a_k^s \sin(kt)) \\ y(t) &= b_0 + \sum_{k=1}^{\infty} (b_k^c \cos(kt) + b_k^s \sin(kt)) \end{aligned}$$

where

$$\begin{array}{llll} a_0 = \alpha_0, & a_k^c = \alpha_k + \alpha_{-k}, & a_k^s = \beta_{-k} - \beta_k, \\ b_0 = \beta_0, & b_k^c = \beta_k + \beta_{-k}, & b_k^s = \alpha_k - \alpha_{-k}. \end{array}$$

Since we plan to optimize over the space of trajectories, the parameters a_0, a_k^c, a_k^s, b_0, b_k^c, and b_k^s are the decision variables in our optimization model. The objective is to minimize the action functional.

Figure 7.10 shows the AMPL program for minimizing the action functional.

Note that the action functional is a nonconvex nonlinear functional. Hence, it is expected to have many local extrema and saddle points. We use the author's local optimization software called LOQO (see [19], [16]) to find local minima in a neighborhood of an arbitrary given starting trajectory. One can provide either specific initial trajectories or one can give random initial trajectories. The four lines just before the call to `solve` in Figure 7.10 show how to specify a random initial trajectory. Of course, AMPL provides capabilities of printing answers in any format either on the standard output device or to a file. For the sake of brevity and clarity, the print statements are not shown in Figure 7.10. AMPL also provides the capability to loop over sections of code. This is also not shown but the program we used has a loop around the four initialization statements, the call to solve the problem, and the associated print statements. In this way, the program can be run once to solve for a large number of periodic solutions.

Choreographies. Recently, [4] introduced a new family of solutions to the n-body problem called choreographies. A *choreography* is defined as a

```
param N := 3;  # number of masses
param n := 15;  # number of terms in Fourier series representation
param m := 100;  # number of terms in numerical approx to integral

set Bodies := {0..N-1};
set Times  := {0..m-1} circular; # "circular" means that next(m-1) = 0

param theta {t in Times} := t*2*pi/m;
param dt := 2*pi/m;

param a0 {i in Bodies} default 0;      param b0 {i in Bodies} default 0;
var as {i in Bodies, k in 1..n} := 0;  var bs {i in Bodies, k in 1..n} := 0;
var ac {i in Bodies, k in 1..n} := 0;  var bc {i in Bodies, k in 1..n} := 0;

var x {i in Bodies, t in Times}
  = a0[i]+sum {k in 1..n} ( as[i,k]*sin(k*theta[t]) + ac[i,k]*cos(k*theta[t]) );
var y {i in Bodies, t in Times}
  = b0[i]+sum {k in 1..n} ( bs[i,k]*sin(k*theta[t]) + bc[i,k]*cos(k*theta[t]) );

var xdot {i in Bodies, t in Times} = (x[i,next(t)]-x[i,t])/dt;
var ydot {i in Bodies, t in Times} = (y[i,next(t)]-y[i,t])/dt;

var K {t in Times} = 0.5*sum {i in Bodies} (xdot[i,t]^2 + ydot[i,t]^2);
var P {t in Times}
  = - sum {i in Bodies, ii in Bodies: ii>i}
            1/sqrt((x[i,t]-x[ii,t])^2 + (y[i,t]-y[ii,t])^2);

minimize A: sum {t in Times} (K[t] - P[t])*dt;

let {i in Bodies, k in 1..n} as[i,k] := 1*(Uniform01()-0.5);
let {i in Bodies, k in 1..n} ac[i,k] := 1*(Uniform01()-0.5);
let {i in Bodies, k in n..n} bs[i,k] := 0.01*(Uniform01()-0.5);
let {i in Bodies, k in n..n} bc[i,k] := 0.01*(Uniform01()-0.5);

solve;
```

Figure 7.10. AMPL program for finding trajectories that minimize the action functional.

```
param N := 3;  # number of masses
param n := 15;  # number of terms in Fourier series representation
param m := 99;  # terms in num approx to integral.  must be a multiple of N

param lagTime := m/N;

set Bodies := {0..N-1};
set Times  := {0..m-1} circular; # "circular" means that next(m-1) = 0

param theta {t in Times} := t*2*pi/m;
param dt := 2*pi/m;

param a0 default 0;            param b0 default 0;
var as {k in 1..n} := 0;       var bs {k in 1..n} := 0;
var ac {k in 1..n} := 0;       var bc {k in 1..n} := 0;

var x {i in Bodies, t in Times}
  = a0+sum {k in 1..n} ( as[k]*sin(k*theta[(t+i*lagTime) mod m])
                       + ac[k]*cos(k*theta[(t+i*lagTime) mod m]) );
var y {i in Bodies, t in Times}
  = b0+sum {k in 1..n} ( bs[k]*sin(k*theta[(t+i*lagTime) mod m])
                       + bc[k]*cos(k*theta[(t+i*lagTime) mod m]) );

var xdot {i in Bodies, t in Times} = (x[i,next(t)]-x[i,t])/dt;
var ydot {i in Bodies, t in Times} = (y[i,next(t)]-y[i,t])/dt;

var K {t in Times} = 0.5*sum {i in Bodies} (xdot[i,t]^2 + ydot[i,t]^2);
var P {t in Times}
  = - sum {i in Bodies, ii in Bodies: ii>i}
          1/sqrt((x[i,t]-x[ii,t])^2 + (y[i,t]-y[ii,t])^2);

minimize A: sum {t in Times} (K[t] - P[t])*dt;

let {k in 1..n} as[k]  := 1*(Uniform01()-0.5);
let {k in 1..n} ac[k]  := 1*(Uniform01()-0.5);
let {k in n..n} bs[k]  := 0.01*(Uniform01()-0.5);
let {k in n..n} bc[k]  := 0.01*(Uniform01()-0.5);

solve;
```

Figure 7.11. AMPL program for finding choreographies by minimizing the action functional.

solution to the n-body problem in which all of the bodies share a common orbit and are uniformly spread out around this orbit. Such trajectories are even easier to find using the action principle. Rather than having a Fourier series for each orbit, it is only necessary to have one master Fourier series and to write the action functional in terms of it. Figure 7.11 shows the AMPL model for finding choreographies.

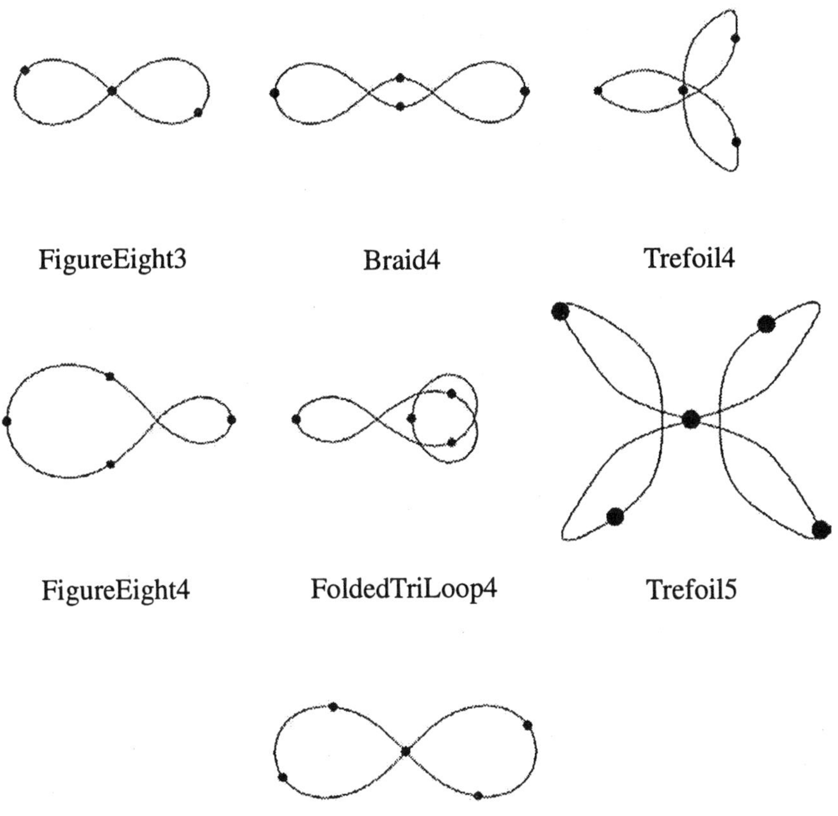

FigureEight5

Figure 7.12. Periodic Orbits — Choreographies.

7.5.3 Stable vs. Unstable Solutions

Figure 7.12 shows some simple choreographies found by minimizing the action functional using the AMPL model in Figure 7.11. The famous 3-body figure eight, first discovered by [14] and later analyzed by [4], is the first one shown—labeled FigureEight3. It is easy to find choreographies of arbitrary complexity. In fact, it is not hard to rediscover most of the choreographies given in [3], and more, simply by putting a loop in the AMPL model and finding various local minima by using different starting points.

However, as we discuss in a later section, simulation makes it apparent that, with the sole exception of FigureEight3, all of the choreographies we found are unstable. And, the more intricate the choreography, the more unstable it is. Since the only choreographies that have a chance to occur in the real world are stable ones, many cpu hours were devoted to searching for other stable choreographies. So far, none have been found. The choreographies shown in Figure 7.12 represent the ones closest to being stable.

Given the difficulty of finding stable choreographies, it seems interesting to search for stable nonchoreographic solutions using, for example, the AMPL model from Figure 7.10. The most interesting such solutions are shown in Figure 7.13. The one labeled Ducati3 is stable as are Hill3_15 and the three DoubleDouble solutions. However, the more exotic solutions (OrthQuasiEllipse4, Rosette4, PlateSaucer4, and BorderCollie4) are all unstable.

For the interested reader, a JAVA applet can be found at [20] that allows one to watch the dynamics of each of the systems presented in this paper (and others). This applet actually integrates the equations of motion. If the orbit is unstable it becomes very obvious as the bodies deviate from their predicted paths.

7.5.4 Ducati3 and its Relatives

The Ducati3 orbit first appeared in [14] and has been independently rediscovered by this author, Broucke [2], and perhaps others. Simulation reveals it to be a stable system. The JAVA applet at [20] allows one to rotate the reference frame as desired. By setting the rotation to counter the outer body in Ducati3, one discovers that the other two bodies are orbiting each other in nearly circular orbits. In other words, the first body in Ducati3 is executing approximately a circular orbit, $z_1(t) = -e^{it}$, the second body is oscillating back and forth roughly along the x-axis, $z_2(t) = \cos(t)$, and the third body is oscillating up and down the y-axis, $z_3(t) = i\sin(t)$. Rotating so as to fix the first body means multiplying by e^{-it}:

$$
\begin{aligned}
\bar{z}_1(t) &= e^{-it}(-e^{it}) = -1 \\
\bar{z}_2(t) &= e^{-it}\cos(t) = (1 + e^{-2it})/2 \\
\bar{z}_2(t) &= e^{-it} i\sin(t) = (1 - e^{-2it})/2.
\end{aligned}
$$

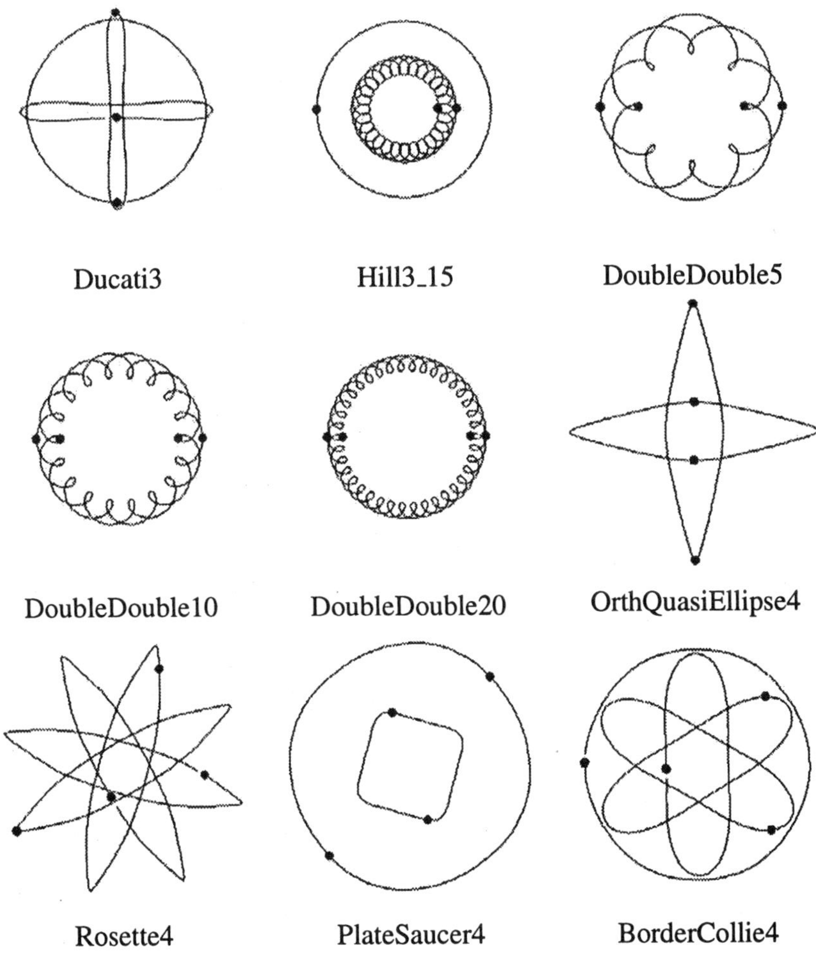

Figure 7.13. Periodic Orbits — Non-Choreographies.

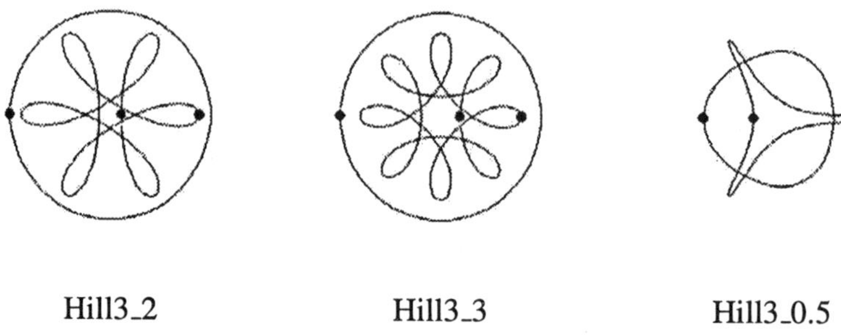

Figure 7.14. Periodic Orbits — Hill-type with equal masses.

Now it is clear that bodies 2 and 3 are orbiting each other at half the distance of body 1. So, this system can be described as a Sun, Earth, Moon system in which all three bodies have equal mass and in which one (sidereal) month equals one year. The synodic month is shorter—half a year.

This analysis of Ducati3 suggests looking for other stable solutions of the same type but with different resonances between the length of a month and a year. Hill3_15 is one of many such examples we found. In Hill3_15, there are 15 sidereal months per year. Let Hill3_n denote the system in which there are n months in a year. All of these orbits are easy to calculate and they all appear to be stable. This success suggests going in the other direction. Let Hill3_$\frac{1}{n}$ denote the system in which there are n years per month. We computed Hill3_$\frac{1}{2}$ and found it to be unstable. It is shown in Figure 7.14.

In the preceding discussion, we decomposed these Hill-type systems into two 2-body problems: the Earth and Moon orbit each other while their center of mass orbits the Sun. This suggests that we can find stable orbits for the 4-body problem by splitting the Sun into a binary star. This works. The orbits labeled DoubleDouble$n$ are of this type. As already mentioned, these orbits are stable.

Given the existence and stability of FigureEight3, one often is asked if there is any chance to observe such a system among the stars. The answer is that it is very unlikely since its existence depends crucially on the masses being equal. The Ducati and Hill type orbits, however, are not constrained to have their masses be equal. Figure 7.15 shows several Ducati-type orbits in which the masses are not all equal. All of these orbits are stable. This suggests that stability is common for Ducati and Hill type orbits. Perhaps such orbits can be observed.

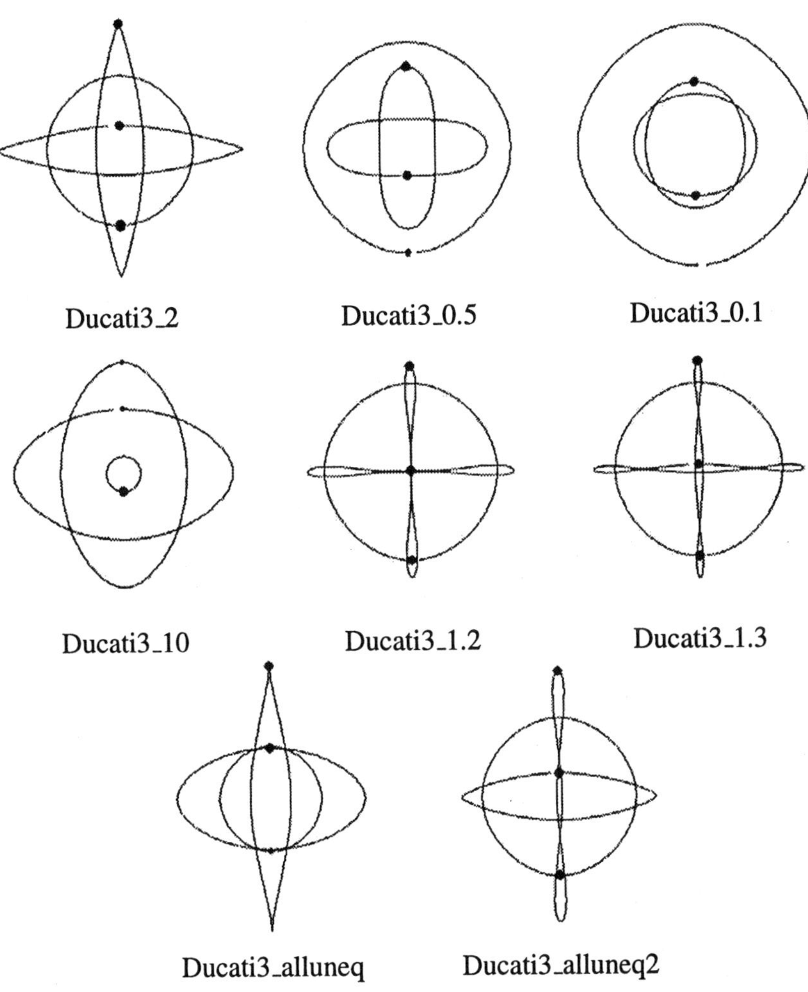

Figure 7.15. Periodic Orbits — Ducati's with unequal masses.

7.5.5 Limitations of the Model

The are certain limitations to the approach articulated above. First, the Fourier series is an infinite sum that gets truncated to a finite sum in the computer model. Hence, the trajectory space from which solutions are found is finite dimensional.

Second, the integration is replaced with a Riemann sum. If the discretization is too coarse, the solution found might not correspond to a real solution to the n-body problem. The only way to be sure is to run a simulator.

Third, as mentioned before, all masses must be positive. If there is a zero mass, then the stationary points for the action function, which satisfy (7.6), don't necessarily satisfy the equations of motion given by Newton's law.

Lastly, the model, as given in Figure 7.10, can't solve 2-body problems with eccentricity. We address this issue in the next section.

7.5.6 Elliptic Solutions

An ellipse with semimajor axis a, semiminor axis b, and having its left focus at the origin of the coordinate system is given parametrically by:

$$x(t) = f + a\cos t, \qquad y(t) = b\sin t,$$

where $f = \sqrt{a^2 - b^2}$ is the distance from the focus to the center of the ellipse.

However, this is *not* the trajectory of a mass in the 2-body problem. Such a mass will travel faster around one focus than around the other. To accommodate this, we need to introduce a time-change function $\theta(t)$:

$$x(t) = f + a\cos\theta(t), \qquad y(t) = b\sin\theta(t).$$

This function θ must be increasing and must satisfy $\theta(0) = 0$ and $\theta(2\pi) = 2\pi$.

The optimization model can be used to find (a discretization of) $\theta(t)$ automatically by changing `param theta` to `var theta` and adding appropriate monotonicity and boundary constraints. In this manner, more realistic orbits can be found that could be useful in real space missions.

In particular, using an eccentricity $e = f/a = 0.0167$ and appropriate Sun and Earth masses, we can find a periodic Hill-Type satellite trajectory in which the satellite orbits the Earth once per year.

References

[1] M.P. Bendsøe, A. Ben-Tal, and J. Zowe. Optimization methods for truss geometry and topology design. *Structural Optimization*, 7:141–159, 1994.

[2] R. Broucke. New orbits for the n-body problem. In *Proceedings of Conference on New Trends in Astrodynamics and Applications*, 2003.

[3] A. Chenciner, J. Gerver, R. Montgomery, and C. Simó. Simple choreographic motions on n bodies: a preliminary study. In *Geometry, Mechanics and Dynamics*, 2001.

[4] A. Chenciner and R. Montgomery. A remarkable periodic solution of the three-body problem in the case of equal masses. *Annals of Math*, 152:881–901, 2000.

[5] J.O. Coleman and D.P. Scholnik. Design of Nonlinear-Phase FIR Filters with Second-Order Cone Programming. In *Proceedings of 1999 Midwest Symposium on Circuits and Systems*, 1999.

[6] J.O. Coleman. Systematic mapping of quadratic constraints on embedded fir filters to linear matrix inequalities. In *Proceedings of 1998 Conference on Information Sciences and Systems*, 1998.

[7] R. Fourer, D.M. Gay, and B.W. Kernighan. *AMPL: A Modeling Language for Mathematical Programming*. Scientific Press, 1993.

[8] J.K. Ho. Optimal design of multi-stage structures: a nested decomposition approach. *Computers and Structures*, 5:249–255, 1975.

[9] N.K. Karmarkar. A new polynomial time algorithm for linear programming. *Combinatorica*, 4:373–395, 1984.

[10] N.J. Kasdin, R.J. Vanderbei, D.N. Spergel, and M.G. Littman. Extrasolar Planet Finding via Optimal Apodized and Shaped Pupil Coronagraphs. *Astrophysical Journal*, 582:1147–1161, 2003.

[11] N. J. Kasdin, D. N. Spergel, and M. G. Littman. An optimal shaped pupil coronagraph for high contrast imaging, planet finding, and spectroscopy. *submitted to Applied Optics*, 2002.

[12] M.S. Lobo, L. Vandenberghe, S. Boyd, and H. Lebret. Applications of second-order cone programming. Technical report, Electrical Engineering Department, Stanford University, Stanford, CA 94305, 1998. To appear in *Linear Algebra and Applications* special issue on linear algebra in control, signals and imaging.

[13] I.J. Lustig, R.E. Marsten, and D.F. Shanno. Interior point methods for linear programming: computational state of the art. *Operations Research Society of America Journal on Computing*, 6:1–14, 1994.

[14] C. Moore. Braids in classical gravity. *Physical Review Letters*, 70:3675–3679, 1993.

[15] D. N. Spergel. A new pupil for detecting extrasolar planets. *astro-ph/0101142*, 2000.

[16] R.J. Vanderbei and D.F. Shanno. An interior-point algorithm for nonconvex nonlinear programming. *Computational Optimization and Applications*, 13:231–252, 1999.

[17] R.J. Vanderbei, D.N. Spergel, and N.J. Kasdin. Circularly Symmetric Apodization via Starshaped Masks. *Astrophysical Journal*, 599:686–694, 2003.

[18] R.J. Vanderbei, D.N. Spergel, and N.J. Kasdin. Spiderweb Masks for High Contrast Imaging. *Astrophysical Journal*, 590:593–603, 2003.

[19] R.J. Vanderbei. LOQO user's manual—version 3.10. *Optimization Methods and Software*, 12:485–514, 1999.

[20] R.J. Vanderbei. http://www.princeton.edu/~rvdb/JAVA/astro/galaxy/Galaxy.html, 2001.

.

[21] R.J. Vanderbei. *Linear Programming: Foundations and Extensions*. Kluwer Academic Publishers, 2nd edition, 2001.

[22] S.-P. Wu, S. Boyd, and L. Vandenberghe. Magnitude filter design via spectral factorization and convex optimization. *Applied and Computational Control, Signals and Circuits*, 1997. To appear.

Chapter 8

CONNECTING MRP, MRP II AND ERP — SUPPLY CHAIN PRODUCTION PLANNING VIA OPTIMIZATION MODELS

Stefan Voß

Institute of Information Systems, University of Hamburg, Germany

stefan.voss@uni-hamburg.de

David L. Woodruff

Graduate School of Management, University of California, Davis

dlwoodruff@ucdavis.edu

Abstract Supply chain management rose to prominence as a major management issue in the last decade. While the focus of managing supply chains has undergone a drastic change as a result of improved information technology, production planning remains a critical issue. We begin with optimization models that map to mrp and MRP II and end up as a basis for a planning model. In order to produce a useful model, we add extensions and ultimately arrive at models that bear little resemblance to mrp and certainly solutions for the optimization problems cannot be obtained using mrp or MRP II processing or logic. Connections with hierarchical systems for planning and detailed scheduling are outlined.

Keywords: Supply Chain Planning, Production Planning, Optimization Models

Introduction

The word "optimization" gets a lot of air time in the supply chain arena these days. Often, it simply refers to processes that are the best possible. But sometimes it is used in the mathematical sense employed by Operations Research professionals, which is the sense of our work here.

Long before the words "Supply Chain" were popular, *materials requirements planning* was seen as the best possible way to do production planning. The model is often referred to as *mrp* with lower case letters (or sometimes "little-mrp") to make clear the distinction between mrp and *MRP II* (*manufacturing resources planning*). In this chapter, following [39, 40], we show how to create an optimization model that matches the behavior of mrp. We then show how the model can be extended to create useful planning models. The first step along the way is creation of a model that is very similar to MRP II.

Enterprise resource planning (ERP) may be referred to as an integrated software application designed to support a wealth of business functions. After the early focus on mrp switched over to MRP II, later it was extended to cover enterprise-wide business functions from finance, human resources and the like. The basic functions incorporated into ERP, however, still heavily rely on mrp and MRP II. And even if so-called advanced planning systems (APS), incorporating powerful planning procedures and methodologies together with concepts for dealing with exceptions and variability, are being developed this has not yet changed (for popular books on ERP and APS see, e.g., [20, 34]).

In order to discuss planning hierarchies, it will be useful to make a careful distinction between *scheduling* and *planning*. It is often said that schedules are for people who do the work and plans are for people who talk about work. In the planning activity we include aggregate planning and assignment of gross production quantities to plants and/or lines. We also include routine capacity decisions such as assignment of work to a qualified subcontractor or activation of production lines. But we do not include sequencing decisions. Typical time frames for this type of activity are weekly time buckets extending to, e.g., six or eighteen months. However, some organizations use monthly or daily planning time buckets.

Many of the planning and scheduling models proposed in the research literature are labelled as lot sizing. An excellent example of this literature is a paper by Trigeiro, Thomas, and McClain [36], which addresses a variant of the capacitated lot sizing problem (CLSP). Their assumptions are consistent with the CLSP in that they ignore sequencing. They assume that setup requirements production of all parts from a family in a period is done in one batch that requires a setup and that the setup characteristics do not vary with the sequence. The objective function considers holding costs (due to early completion), temporal variations in production costs, and setup costs. Their model is very similar to the first model given here except that they provide the modeling details necessary to allow production of a stock keeping unit to span two time buckets. They also provide details of a special purpose solution technique based on Lagrangian relaxation that proves to be very effective for this problem.

The lot sizing literature is quite large; see, e.g., [10, 11, 19, 35]. Each of these models makes different assumptions that result in a different model. All of them are intended to determine production quantities. The lot sizing models typically assume a fixed lead time of one period and do not consider alternative routings. These models are similar to our basic model (SCPc) given below in that they are useful for production scheduling within a factory, but not appropriate for assigning production to factories within a supply chain. Alternatively, our work can also be seen as extending the category denoted as "Multiple-stage production planning with limited resources" by Erenguc and Simpson [30].

There is also a considerable literature on setting inventory control policies for an entire bill of materials simultaneously. The multiechelon inventory and related literature [9, 15, 28] provides methods for setting policies to control inventory levels for a complete production/distribution system under a variety of conditions. We seek to produce models that are appropriate for plans based on the best information available at the time of the planning process as opposed to policies that are parameters for making decisions.

In the next two sections we provide a mathematical representation of mrp and MRP II. Together with a discussion of some extended modeling considerations this is then carried over into a simple supply chain model. The last three sections provide some additional extensions, an embedding of the models into hierarchical planning approaches as well as a short summary of the chapter.

Advanced mathematicians as well as practitioners should not be misled by the exposition of our models. While mrp and MRP II may be regarded as "philosophy," it is important to bridge the gap between such philosophy and advanced mathematical programming models in production planning and supply chain management. This chapter aims to bridge this gap. While much of the material is taken from or at least closely related to [40], the interested reader is referred to this source for further reading as well as an implementation of some of our models in popular modeling languages.

8.1 MRP

The ideas behind mrp come from a data processing perspective rather than an optimization perspective, but we can create an optimization model that corresponds to it. This will constitute a useful starting point for further modeling. One does not need an optimization model for mrp, but we will use our model to further understand the limitations of mrp and we will use it as a basis for more sophisticated models.

Orlicky is widely credited with having invented mrp, or at least with popularizing it. The second edition of his seminal work on mrp is [26]. This book explains the methods and associated record keeping needed for mrp. Bear in mind that mrp was a tremendous improvement over older management systems that were better suited to a make-to-stock environment. Shorter product life cycles and make-to-order environments require a planning system that anticipates the need for varying mixes of components.

The mrp model makes use of data about items and components. The term *Stock Keeping Unit* (SKU) is used to refer to items that are sold as well as components and components of components, etc. For each SKU we need to know

- the *lead time*, which is an estimate of the time between the release of an order to the shop floor or to a supplier and the receipt of the goods;
- if there is a minimum production quantity (referred to as a smallest possible *lot size* for items that are manufactured in-house) or if there is a minimum order quantity for purchased items;
- the current inventory levels;
- components needed, which is often referred to as a bill of materials (BOM).

The data needed for an mrp optimization model is summarized in Table 8.1. The ubiquitous big M is needed to force the computer to make some-or-none decisions that are needed to enforce the minimum lot sizes. Theoretically, this can be any number, provided it is larger than any possible production quantity.

An optimization model is not needed to use mrp, but we can create one and then extend it. In other words, our goal is to create an optimization model that matches mrp not for its own sake but to get started with models that match classic planning systems. Using this model as a starting point, we can go on to more sophisticated models.

To mimic mrp, we really need only one decision variable, $x_{i,t}$, which is the quantity of SKU i to start or order in period t. In order to enforce lot sizing rules, we make use of a variable that indicates production of an SKU in period t and call it $\delta_{i,t}$.

Table 8.1. Data for the mrp Formulation

P	Number of SKUs
T	Number of time buckets (i.e., the planning horizon)
$LT(i)$	Lead time for SKU i
$R(i,j)$	Number i's needed to make one j
$D(i,t)$	External demand for i in period t
$I(i,0)$	Beginning inventory of SKU i
$LS(i)$	Lot size for SKU i
M	A large number (e.g., 1 + largest $D(i,t)$ times largest $R(i,j)$)

The objective in mrp is to make things as late as possible but no later. So one possible objective is to minimize

$$\sum_{i=1}^{P}\sum_{t=1}^{T}(T-t)x_{i,t}$$

that will result in things being made "as late as possible" and we count on the constraints to enforce the "but no later" part of the mrp model. We can think of better objectives than this, and later we will describe some. However, for now our goal is to match the decisions that would be made by mrp. Stated in classic optimization form, the mrp problem is to

$$\text{minimize: } \sum_{i=1}^{P}\sum_{t=1}^{T}(T-t)x_{i,t} \qquad \text{(mrp)}$$

subject to:

$$\begin{array}{rll}
\displaystyle\sum_{\tau=1}^{t-LT(i)} x_{i,\tau} + I(i,0) & & \\
\displaystyle -\sum_{\tau=1}^{t}\left(D(i,\tau)+\sum_{j=1}^{P} R(i,j)x_{j,\tau}\right) & \geq 0 & i=1,\ldots,P,\ t=1,\ldots,T \\
x_{i,t}-\delta_{i,t}LS(i) & \geq 0 & i=1,\ldots,P,\ t=1,\ldots,T \\
\delta_{i,t}-\dfrac{x_{i,t}}{M} & \geq 0 & i=1,\ldots,P,\ t=1,\ldots,T \\
\delta_{i,t} & \in\{0,1\} & i=1,\ldots,P,\ t=1,\ldots,T \\
x_{i,t} & \geq 0 & i=1,\ldots,P,\ t=1,\ldots,T
\end{array}$$

The first constraint requires that the sum of initial inventory and production up to each period has to be at least equal to the total of external demand and demand for assemblies that make use of the SKU. The summation is to $t-LT(i)$ for each period (there will be one constraint for each value of t) because work must be started LT periods before it can be used to satisfy demand. The product $R(i,j)x_{j,\tau}$ anticipates the demand for SKU i that results when it is a component of SKU j.

The demand and materials constraint contains the term

$$\sum_{\tau=1}^{t-LT(i)} x_{i,\tau}$$

that captures the production that will be completed up to time t, while the term

$$\sum_{\tau=1}^{t}\left(D(i,\tau)+\sum_{j=1}^{P} R(i,j)x_{j,\tau}\right)$$

is the total demand that will have occurred up to the same time period.

The lot size constraint specifies that if there is any production of a SKU during a period it must be at least as much as the minimum lot size. The modeling constraint for the production indicator forces it to take a value greater than zero if there is production for the SKU in the period, while the integer constraint forces it to be either zero or one.

8.2 Extending to an MRP II Model

MRP II was inspired by shortcomings in mrp, but the data processing orientation is preserved in MRP II. As was the case with mrp, we first explain the concepts behind MRP II, then we develop an optimization model to mimic and improve its behavior. After we have this model in hand, we extend it to produce a model that can give us production plans that trade off alternative capacity uses, holding inventory and tardiness in an optimized way.

8.2.1 MRP II Mechanics

To make better use of mrp, deeper understanding of the methods and particularly the relationships between inventory and lead time was needed. Early work by Wight (such as [42]) helped make mrp successful. In fact, Wight is often credited with inventing MRP II as a way to make mrp logic work correctly.

The actual practice of MRP II was, and is, invented and reinvented by the software firms, consultants, planners and schedulers who make it work. A number of books and a large number of articles provide practical tips for implementing and using MRP II [24, 41]. In such works, MRP II is referred to as a *closed loop* production planning system because the capacity check is followed by adjustments to the data followed by another execution of mrp and so forth.

A classic text by Vollmann, Berry, and Whybark [38] places mrp, MRP II and the associated planning tools in a broader perspective of planning and scheduling tools popularized in the 1970's and 80's. At the time of their second edition in 1988, mathematical programming approaches to production planning were considered to be an advanced concept. Simultaneous consideration of an objective function along with the materials requirements constraint and the capacity constraint was treated as part of the human aided processing of MRP II (a popular advertising called it planning "at your fingertips"). The book presents a number of sophisticated and practical methods for planning as well as scheduling and forecasting.

At about the same time, some of the shortcomings of the overall philosophy of MRP II were beginning to be discussed (see, e.g., [17, 33]). This process is on-going (see, e.g., [32]) in the academic literature. Meanwhile, a literature surrounding new technologies from ERP and SCM software vendors is beginning to appear (see, e.g., [12]).

There are a number of well-known deficiencies in the model that underlies mrp. Potentially the most severe is the fact that it ignores capacity. To discuss this issue it is useful to remember that we are making a distinction between planning and scheduling. Although we have introduced it as a planning tool, mrp is also often used as a scheduling tool as well. A severe problem is that there is no guarantee that there will be enough capacity to actually carry out the plan produced by mrp. In fact, for capacity constrained production systems, it is seldom possible to implement an mrp plan as a schedule. This is debilitating when mrp is used as a scheduling tool, but is also bad for mrp as a planning tool because the absence of capacity considerations can make the plans so unrealistic that they are not useful.

The data processing professionals who were developing and selling mrp software in its early years recognized this deficiency and MRP II was developed in response to it. The database for mrp is extended to include routing and capacity information. Each production resource is entered into the database along with its maximum production during a time bucket. We will refer to the maximum production by a resource during a time bucket as its *capacity.* The list of resources used to produce a particular SKU is known as the *routing* for the SKU.

With this information the data processing specified by MRP II can be carried out. The processing begins by executing mrp to determine a production plan. Then, for each time bucket each SKU is "followed along its routing" and the utilization of each resource is updated. In the end, those resources whose capacity would be exceeded by the mrp plan are identified. The user can be given a list of the resources and time buckets along with the SKUs that would use the resources in those time buckets and perhaps the list of the end-items that would use these "offending" SKUs and the time buckets in which the end-items would be produced.

The information concerning capacity infeasibilities can be used by the planner and/or some software to attempt to change the input data so that a feasible plan results. The most common method is to "repair" the master production schedule (i.e., change the timing of the external demands). Such repairs are difficult, so one of our goals will be to use optimization models to produce feasible plans in the first place.

Before we can proceed, we need to deal with the issue of units of measurement. In a specific production facility capacity may be measured in hours, or tons, or pieces or something else. Since we want to create abstract models, we will designate the capacity of every resource during every time bucket to be one. This will allow us to state resource utilizations as fractions in our models. We will represent the fraction of the capacity of resource k in one time bucket used by production of one SKU i as $U(i, k)$.

Table 8.2. Data for MRP II Formulation

P	Number of SKUs
T	Number of time buckets
K	Number of resources
$I(i,0)$	Beginning inventory of SKU i
$LT(i)$	Lead time for SKU i
$R(i,j)$	Number of i's needed to make one j
$D(i,t)$	External demand for SKU i in period t
$U(i,k)$	Fraction of resource k needed to make one unit of SKU i
M	A big number; e.g., 1 + 1/(smallest U)

8.2.2 MRP II Data and Constraints

As was the case with mrp, MRP II did not begin as an optimization model. However, we can create an optimization model that will mimic its behavior and do more. To be specific, we can mimic its intended behavior. That is, we can schedule things as late as possible without violating capacity constraints. The objective function for mrp is retained, but additional data is needed for constraints. The data needed to mimic MRP II is given in Table 8.2.

The data requirements are nearly the same as for mrp except that we have dropped the lot sizing information and added information about utilization. We use the same variables as for mrp, namely $x_{i,t}$, which is the quantity of SKU i to start or order in period t. We will not need $\delta_{i,t}$ unless we need to add information about changeovers. The major change from the mrp model is the addition of a capacity constraint. The MRP II constraints are as follows:

Stated in classic optimization form, the MRP II problem is to

$$\text{minimize: } \sum_{i=1}^{P}\sum_{t=1}^{T}(T-t)x_{i,t} \qquad \text{(MRP II)}$$

subject to:

$$\sum_{\tau=1}^{t-LT(i)} x_{i,\tau} + I(i,0) - \sum_{\tau=1}^{t}\left(D(i,\tau)+\sum_{j=1}^{P}R(i,j)x_{j,\tau}\right) \geq 0 \quad i=1,\ldots,P,\ t=1,\ldots,T$$

$$\sum_{i=1}^{P}U(i,k)x_{i,t} \leq 1 \quad t=1,\ldots,T,\ k=1,\ldots,K$$

$$x_{i,t} \geq 0 \quad i=1,\ldots,P,\ t=1,\ldots,T$$

For the purpose of discussion it is useful to isolate the two most important constraints and refer to them by name. We will refer to the *requirements constraint,* which is

$$\sum_{\tau=1}^{t-LT(i)} x_{i,\tau} + I(i,0) - \sum_{\tau=1}^{t} \left[D(i,\tau) + \sum_{j=1}^{P} R(i,j) x_{j,\tau} \right] \geq 0$$

and we will call the constraint

$$\sum_{i=1}^{P} U(i,k) x_{i,t} \leq 1$$

the *capacity constraint.*

8.2.3 Discussion of MRP II

Using classic software, problem (MRP II) would not be solved directly. Instead, problem (mrp) would be solved and then the capacity constraint for (MRP II) would be checked. To be more specific, suppose the solution to the problem (mrp) was given as $X(i,t)$ for the production of SKU i to plan to start in time t. Then those SKUs for which

$$\sum_{i=1}^{P} U(i,k) X(i,t) > 1$$

would violate the capacity constraints. SKUs for which the capacity constraint was violated would be subject of reports and the production planner would change the data in an attempt to find a solution to mrp that was also feasible for (MRP II). By "change the data" we mean that due dates, lot sizes, and capacity utilizations would be changed. Due dates for end items are typically adjusted to be later. This data is then given to the software that uses the new data to compute a new solution to (mrp) and then checks the constraints for (MRP II).

This process is very hard work for the planner. Even though MRP II software provides numerous reports, it is often still not possible for the planner to produce a capacity feasible schedule given the fact that they often have only a few hours to produce it. With a few dozen time buckets and a few thousand SKUs, the problem is just too big to be solved easily by a person, even if assisted by software that can check the constraints and solve problem (mrp). An important thing to note is that the classic iterative MRP II solution process tends to use a "meandering" and an undefined objective function because the solution is obtained by changing the data rather than finding the best solution given the best estimate of the data.

In spite of the severe difficulties, at the time of this writing the ideas behind MRP II are included in many modern ERP systems. Some academics are of the opinion that MRP II should be done away with. If they mean that we should dump the solution methods, then we agree, but if they mean dump it completely, then they are hopelessly misguided. Regardless of how you articulate the methods, one should include the requirements constraint and the capacity constraint. They represent physical reality. The processing for MRP II is not normally explained by giving the constraints as we have shown them, but the constraints are there. They have to be. They have to be there in any other system as well. If the software does not include them explicitly or implicitly, then the planners will have to enforce them or else produce a plan that is very unrealistic and therefore not very valuable. Ultimately, the constraints are present in the production process.

Direct solution of the optimization model is a much better idea and this is basis of much of the newest planning software that is sold as modules or add-ins for ERP systems. In practice, the problem is bigger and harder to solve than the simple (MRP II) model that we have presented. However, (MRP II) provides us with a good jumping off point for more sophisticated models because it mimics a widely used planning tool.

We can and will embed these constraints in a model that captures costs and constraints which are important to the manufacturing organization or the supply chain. By solving the optimization problem directly, we can include in the objective function guidance for solutions. We want to find solutions that correspond to the goals of the organization in addition to merely satisfying the two constraints.

8.3 Changeover Modeling Considerations

The simple (MRP II) model will be the basis for many additional features. However, we might also want to remove features. For example, not all resources need to be modeled. Often, it is easy to see that some resources do not pose a problem. Such resources should simply be omitted from the model.

One feature that has been dropped from the (mrp) model in creating the (MRP II) model is lot sizes. Usually, the valid reason for minimum lot sizes is that a significant effort or cost is required to changeover for production, hence small lots might not be cost effective. Setting a fixed lot size is a crude response to this problem, so we will try to build models that take changeovers explicitly into account.

However, there are cases where minimum lot sizes are needed. For example, some chemical processes must be done in batches with a minimum size. If needed, the δ variables can be put in for the SKUs that require lot sizes along with the lot sizing constraints for those SKUs.

In many production environments, the proper modeling of capacity requires modeling of changeovers. By the word "changeover" we mean the effort re-

Table 8.3. Data for Changeover Constraints

$S(i,k)$	Fraction of resource k used to changeover to SKU i
$W(i,j)$	Waste of SKU i to setup (changeover to) SKU j

quired to switch a resource from the production of one SKU to the production of another. In fact, it is changeover avoidance that results in the need for lots that are larger than what would be needed to satisfy immediate customer demands. Changeover modeling can be quite involved.

The first thing that we will need is $\delta_{i,t}$, which will be one if any of SKU i will be started in period t. This, in turn, requires that we include the constraints introduced for (mrp) to enforce the meaning of the δ variables.

- Modeling constraint for production indicator for all SKUs i and times t:

$$\delta_{i,t} \geq \frac{x_{i,t}}{M}$$

- Binary constraint for production indicator for all SKUs i and times t:

$$\delta_{i,t} \in \{0,1\}$$

8.3.1 A Straightforward Modification

In a simple model of changeovers, we use the data given in Table 8.3. As usual we do not insist that this data be given for every resource or every SKU. In particular, we would expect that W values will be essentially zero for most or all SKUs. If they are all zero, then of course there is no need to add them to the model.

The following replacements for the requirements and capacity constraints are needed:

- Demand and materials requirement for all times t and all SKUs i:

$$\begin{array}{ll} \sum_{\tau=1}^{t-LT(i)} x_{i,\tau} + I(i,0) & \\ -\sum_{\tau=1}^{t} \left[D(i,\tau) + \sum_{j=1}^{P} \left(R(i,j)x_{j,\tau} + W(i,j)\delta_{j,\tau}\right)\right] & \geq 0 \end{array}$$

- Capacity constraint for all resources k and times t:

$$\sum_{i=1}^{P} \left(U(i,k)x_{i,t} + S(i,k)\delta_{i,t}\right) \leq 1$$

Table 8.4. Basic Cost Data for an Improved MRP II Objective Function

$H(i)$	Per period holding cost for SKU i
$C(i)$	Total (out of pocket) changeover cost for SKU i

8.4 From a Cost Based Objective Function to a Useful Supply Chain Planning Model

In order to create a good or optimal decision using OR techniques, we need an objective. The first issue that must be resolved is a very important one: are we trying to minimize costs or maximize profits? For now, we are trying to develop an improved MRP II model, so it seems best to try to minimize costs. This is consistent with the mrp philosophy that demands are given to production planners and their job is to meet them.

8.4.1 Costs

When constructing the objective function, we make use of marginal costs. A minimal set of cost data is described in Table 8.4. The cost of holding an inventory item for one period is a discount rate times the cost of the item. The out-of-pocket costs for a changeover is also fairly straightforward as soon as we get past the difference between a *setup* and a *changeover.* If a machine or workgroup must spend some extra time for each batch of some SKU i, then we call this a setup and we can add the capacity utilization into our calculations of U. Any material that is wasted for each batch should be rolled into R.

Armed with these cost estimates, we can formulate a cost minimizing objective function. Before doing this we will find it useful to introduce the idea of *macros* or *notational variables*. These will not really be variables, they will be simple functions of variables that we will create to make it easier to read the models. We have used notation like $I(i,0)$ for data and notation like $x_{i,t}$ for variables. So data has the list indexes in parenthesis and variables have them as subscripts. For macros, we will show the list indexes as subscripts and the variables on which the macro depends will be shown in parenthesis.

We need to emphasize that these macros behave like typical computer macros, which is to say that they rely on the computer to replace the macro literally with the corresponding string. They are not new variables or data. We show the macro definition using $\equiv$. The first macro we will use denotes the inventory of SKU i in period t:

$$\begin{aligned} I_{i,t}(x,\delta) \equiv & \sum_{\tau=1}^{t-LT(i)} x_{i,\tau} + I(i,0) \\ & - \sum_{\tau=1}^{t} \left(D(i,\tau) + \sum_{j=1}^{P} (R(i,j)x_{j,\tau} + W(i,j)\delta_{j,\tau}) \right) \end{aligned}$$

We then write the demand and materials requirement constraint for all times t and all SKUs i as

$$I_{i,t}(x, \delta) \geq 0$$

with the understanding that this will be interpreted as

$$\begin{array}{ll} \sum_{\tau=1}^{t-LT(i)} x_{i,\tau} + I(i,0) & \\ -\sum_{\tau=1}^{t} \left[D(i,\tau) + \sum_{j=1}^{P} \left(R(i,j)x_{j,\tau} + W(i,j)\delta_{j,\tau} \right) \right] & \geq \ 0. \end{array}$$

In other words, we use the macro $I_{i,t}(x, \delta)$ to be able to write exactly the same constraint that appears in model (MRP II) in a much more condensed form. Combining the macro expression with $I_{i,t}(x, \delta) \geq 0$ results in the same constraint.

Note that I always represents inventory, but $I(i, 0)$ is given as data, while $I_{i,t}(x, \delta)$ is a macro. We use the macro to give us a shorthand to write it, but we should bear in mind that $I_{i,t}(x, \delta)$ represents a complicated expression for the planned inventory of SKU i at time t.

We can use the macro to write our objective to be the minimization of

$$\sum_{t=1}^{T} \sum_{i=1}^{P} \left[H(i) I_{i,t}(x, \delta) + C(i)\delta_{i,t} \right]$$

subject to the MRP II constraints. Inspection of this objective function indicates that the changeover cost, C, should include the cost of wasted material. The data element W allows us to take into account the fact that the wasted material will have to be produced but does not "charge" us for the extra money spent. However, by adding the term $W(i, j)\delta_{j,\tau}$ to the inventory expression, we are "charged" for the capacity used.

8.4.2 Overtime and Extra Capacity

It is often overly simplistic to impose a single, hard capacity constraint for each resource in each period. In most cases capacity can be added, particularly with some advance notice as would be the case if the production plan called for it. Classic examples are the use of overtime or an extra shift. In other cases, it may be possible to bring in extra resources on relatively short notice. To keep our language simple, we will refer to all such short term capacity additions as overtime. We now extend the MRP II constraints to capture this notion in a number of different ways. The data needed is described in Table 8.5. In order to produce a model that can be used easily in a variety of settings, we continue with the convention of working with capacity fractions rather than absolute measures of capacity.

Table 8.5. Data for Short Term Capacity Expansion

$F(k,t)$	Maximum fraction of resource k that can be added in t
$O(k,t)$	Marginal cost per fraction of resource k added in t

We must introduce a new variable, $y_{k,t}$, to represent the overtime fraction for resource k in period t. This allows us to add the term

$$\sum_{k=1}^{K} O(k,t) y_{k,t}$$

to the objective function to capture the cost of overtime. We must add constraints to establish the overtime fraction for all resources k and times t and then constrain overtime and capacity:

$$\sum_{i=1}^{P} U(i,k) x_{i,t} \leq 1 + y_{k,t}$$

$$y_{k,t} \geq 0$$

$$y_{k,t} \leq F(k,t)$$

The capacity constraint is shown here without changeovers, which can be added if needed.

8.4.3 Allowing Tardiness

For the purposes of mathematical modeling, it is useful to make a clear distinction between *deadlines* and *due dates*. A deadline is modeled using constraints, while a due date appears only in the objective function. The clear implication is that deadlines *must* be respected while due dates are more of a target.

The word "deadline" sounds a bit extreme, since it is extremely unusual for a customer to actually kill someone if their order is late (although not at all unusual for them to threaten it). For end items, deadlines are typically the correct model only for items that are of the highest priority. Other items can be allowed to be tardy and a penalty function can be used to model the fact that tardiness is more of a problem for some SKUs and/or for some customers.

Deadlines are the appropriate way to model intermediate SKUs. Since we are creating a plan, it makes no sense to plan for the production of an end item to begin before the completion of its components. Consequently, the completion times for items that will be used in assemblies must be enforced in the constraints in order for the model to make sense. This statement may not always be exactly correct (e.g., if there are end items that require some components initially and others a few time buckets later). This can often be addressed by splitting the item into two SKUs one of which is made entirely using the other. Our concern now is to make a simple model that allows end items to be tardy but requires components to be on time.

8.4.4 A Simple Model

We introduce as data $A(i)$, which is the per period tardiness cost for external demand for SKU i. To make use of it, we need to make another significant change to the materials requirements and demand constraint. To allow tardiness, instead of using

$$I_{i,t}(x, \delta) \geq 0,$$

use

$$I_{i,t}(x, \delta) \geq -\sum_{\tau=1}^{t} D(i, \tau).$$

In other words, we allow a negative inventory position. Items with negative inventory positions are often referred to as *backordered*. The idea is that we still need to require production of items that are needed as components or else the entire plan is invalid. However, we want to allow the inventory position to be negative for an SKU to the extent that we have external demand for that SKU. This can be seen by expanding the macro $I_{i,t}(x, \delta)$ and then simplifying it to eliminate the summation over D. The result after expanding the macro and rearranging terms is:

$$\sum_{\tau=1}^{t-LT(i)} x_{i,\tau} + I(i,0) - \sum_{\tau=1}^{t} \sum_{j=1}^{P} (R(i,j)x_{j,\tau} + W(i,k)\delta_{j,\tau}) \geq 0$$

In order to construct an objective function that takes into account backorder quantities, we have to distinguish between negative and positive inventory positions. We let I^- be $-I$ if $I < 0$ and zero otherwise and let I^+ be I if $I > 0$ and zero otherwise. To get I^+ and I^-, add the following constraints for all i and t: $I^+_{i,t} - I^-_{i,t} = I_{i,t}(x, \delta)$, both $I^+_{i,t} \geq 0$, and $I^-_{i,t} \geq 0$. These constraints would allow I^+ and I^- to become arbitrarily large, but they will not do so because there will be positive costs associated with them in the objective function.

We can add terms like the following to the objective function:

$$\sum_{t=1}^{T} \sum_{i=1}^{P} \left(A(i)I^-_{i,t} + H(i)I^+_{i,t} \right)$$

8.4.5 Summarizing the Model

We now give a base model. In actual practice, many of the data elements will be zero and some of the constraints and variables can be omitted altogether. The

Table 8.6. Data for the (SCPc) Model

P	Number of SKUs
T	Number of time Buckets
K	Number of resources
$LT(i)$	Lead time for SKU i
$I(i,0)$	Beginning inventory of SKU i
$D(i,t)$	External demand for SKU i in period t
$R(i,j)$	Number of i's needed to make one j
$U(i,k)$	Fraction of resource k needed by one unit of SKU i
$F(k,t)$	Max fraction of resources k that can be added in t
$S(i,k)$	Fraction of resource k used to changeover to SKU i
$W(i,j)$	Waste of SKU i to setup (changeover to) SKU j
$H(i)$	Per period holding cost for SKU i
$C(i)$	Total (out of pocket) changeover cost for SKU i
$O(k,t)$	Marginal cost per fraction added
$A(i)$	Per period tardiness cost for external demand for SKU i.
M	A big number; e.g., 1 + 1/(smallest U)

Table 8.7. Variables Set by the (SCPc) Model

$x_{i,t}$	Order release quantity for SKU i in time t
$y_{k,t}$	"Overtime" Fraction of resource k in time t
$\delta_{i,t}$	Binary indicator of production of SKU i in time t
$I^+_{i,t}$	Inventory of SKU i to carry in time t
$I^-_{i,t}$	Quantity of SKU i backordered in time t

complete picture is given here to summarize the things in one concise statement. We refer to the model as (SCPc) to emphasize that it is a cost minimization model for supply chain planning. The data is given in Table 8.6 and the model is shown as Figure 8.1 making use of the macro shown in Figure 8.2. The variables that are to be set by the model are shown in Table 8.7. In some cases, the constraints have been rearranged to put a zero on the right hand side. The modifications listed as complications are not shown here.

The use of the symbol I requires a little explaining. We use the symbol consistently to represent inventory position, but that requires it to appear with the data, variables and the macros. The initial inventory, $I(i,0)$, is given as data. The inventory expression depends on existing variables and data, so it is given as a macro. Backorder and positive inventory quantities have to be constrained, so they must be given as variables.

Minimize:

$$\sum_{t=1}^{T}\left[\sum_{i=1}^{P}\left(A(i)I^-_{i,t} + H(i)I^+_{i,t} + C(i)\delta_{i,t}\right) + \sum_{k=1}^{K} O(k,t)y_{k,t}\right] \qquad \text{(SCPc)}$$

subject to:

$$\begin{array}{rll}
I_{i,t}(x,\delta) + \sum_{\tau=1}^{t} D(i,\tau) & \geq 0 & i=1,\dots,P,\ t=1,\dots,T \\
\sum_{i=1}^{P} [U(i,k)x_{i,t} + S(i,k)\delta_{i,t}] & \leq 1 + y_{k,t} & t=1,\dots,T,\ k=1,\dots,K \\
y_{k,t} & \leq F(k,t) & t=1,\dots,T,\ k=1,\dots,K \\
\delta_{i,t} & \geq \frac{x_{i,t}}{M} & i=1,\dots,P,\ t=1,\dots,T \\
\delta_{i,t} & \in \{0,1\} & i=1,\dots,P,\ t=1,\dots,T \\
x_{i,t} & \geq 0 & i=1,\dots,P,\ t=1,\dots,T \\
y_{k,t} & \geq 0 & t=1,\dots,T,\ k=1,\dots,K \\
I^+_{i,t} & \geq 0 & i=1,\dots,P,\ t=1,\dots,T \\
I^-_{i,t} & \geq 0 & i=1,\dots,P,\ t=1,\dots,T \\
I^+_{i,t} - I^-_{i,t} & = I_{i,t}(x,\delta) & i=1,\dots,P,\ t=1,\dots,T
\end{array}$$

Figure 8.1. (SCPc) Model

$$\begin{aligned}
I_{i,t}(x,\delta) \equiv & \sum_{\tau=1}^{t-LT(i)} x_{i,\tau} + I(i,0) \\
& - \sum_{\tau=1}^{t}\left(D(i,\tau) + \sum_{j=1}^{P}(R(i,j)x_{j,\tau} + W(i,j)\delta_{j,\tau})\right)
\end{aligned}$$

Figure 8.2. (SCPc) Macro

While interorganizational and intraorganizational boundaries are certainly an important concern for a distinction between supply chain planning and production planning this need not be the case when we use a certain level of abstraction as it is often done when developing a model. In that sense many problems in supply chain planning including multiple companies and production planning within just one company may be represented by the same model. Based on our earlier remark about the distinction between planning and scheduling this can also be related to the term supply chain production planning that we used in the title of this chapter.

8.5 More Extensions to the Model

We have now extended our models well beyond the data processing concepts embodied by MRP II to make use of the capabilities of optimization software and modelling languages. In this section, we continue by describing a number of important extensions to the model.

8.5.1 Substitutes, Multiple Routings and/or Subcontractors

There are two issues to consider with substitutes, multiple routings and/or subcontractors: the substantive issue of planning when there are multiple routings or substitutes available and the modelling issues or ways of representing these mechanisms. The academic literature has, of course, been primarily concerned with the former.

Chandra and Tombak [8] look at ways to evaluate the flexibility that is provided by alternate routings. This work is useful during the design of supply chains. A paper by Kamien and Li [16] examines subcontracting at the aggregate level and discusses economic effects and the structure of the subcontracting relationship. They show that, under certain conditions, subcontracting reduces the variability in production and inventory. This paper, like [37], provides insight into subcontracting policies but does not prescribe methods for production planning in the presence of multiple routing opportunities.

A paper more directly related to our model for multiple routings is one by Logendran and Ramakrishna [23] who create a mathematical programming model for the problem of scheduling work in manufacturing cells that have duplicate machines available for bottleneck operations and/or subcontractors who can perform the bottleneck operations. They also give details necessary to use a general-purpose heuristic search solution method.

Solution methodologies for single products with subcontracting are provided in [1]. This work also considers the interaction between the operational decision to subcontract and tactical capacity decisions. The paper provides algorithms and insights for both aspects of the problem.

Multiple routings are very common since modeling alternate routing is also reasonable for the use of multiple vendors (with differing characteristics) or subcontractors for parts that are also produced in-house.

This feature is very important if the model is to be used in a supply chain. There are choices to be made between suppliers, or between different factories or between different production lines in the same factory. In this situation, we would say that multiple routings are available. We would like to create a model that picks between the alternative in an optimal way while simultaneously considering the effect on all other decisions regarding production and capacity utilization.

Furthermore, a single supplier might have multiple offerings that can be used to create the needed SKU. These multiple offerings might differ in minor physical ways or they might be exactly the same part but with different lead times and different costs. For example, sometimes a larger size can be substituted for a smaller one with no loss in functionality or two SKUs can serve the same purpose, but one can make use of material of a higher quality than necessary. In other cases, suppliers offer expedited delivery which can result in a shorter lead time at higher price. If we think of the lead time as one of the important attributes of the SKU, then it is natural to consider an SKU shipped sooner to be a substitute for the SKU with the standard lead time.

Another possibility is that the SKU can be made either "in-house" or by an outside subcontractor. Generally, both the cost and the lead time will be different for subcontractors. With just a bit of abstraction, we can see that the choice between subcontractors or the decision regarding whether to use a subcontractor is the same as choosing between alternative routings or selecting one SKU from among a list of substitutes.

With a little effort, all three of these possibilities can be modeled using the same basic mechanisms. There needs to be a separate SKU for each substitute, alternate routing and/or for each subcontractor. So if an SKU, say AJ8172, could go along two routings, there would be SKUs AJ8172a and AJ8172b. This results in the need for a new data structure, the list of substitutes sets, $\mathcal{L}$. This list has P elements in it. Each element of $\mathcal{L}$, $\mathcal{L}(i)$ is a set of substitute SKU indexes for the SKU i. For those SKUs that do not have substitutes the substitutes list is empty, which we write as $\mathcal{L}(i) = \emptyset$.

We assume that regardless of the substitute used, the substitute parts are essentially indistinguishable upon completion for the purpose of computing holding and tardiness costs. The completed parts are planned for using a *master* SKU, which is constrained to have zero production (i.e. for master SKU i, $x_{i,t} = 0$ for all t), and which is created for each group of substitutes. To use mathematical notation, master SKUs are those SKUs i for which $\mathcal{L}(i) \neq \emptyset$. To make it convenient to distinguish SKUs that are in an alternates list for at least

Table 8.8. Additional Data Required for Alternate Routings

$\mathcal{L}(i)$	Alternates (or substitutes) SKU list for SKU i
$V(i)$	Marginal cost of using alternate SKU i

one master SKU, we use the unusual notation $j \in \mathcal{L}$ to signify that SKU j is in the list $\mathcal{L}(i)$ for some master SKU i.

Note that the value of P includes the substitute SKUs. Our expressions are cleaner if we simply treat the master SKUs and substitutes the same as SKUs that neither have, nor are, substitutes. We rely on data such as the substitutes list to provide the distinctions.

We insist that all requirements be for the master SKU. In other words, all external demands will be for the master SKU and any product that can use one of the substitutes as component will have an $R(i,j)$ value for the master only. This means that the inventory macro for master SKUs must be changed to

$$I_{i,t}(x,\delta) \equiv \sum_{\ell \in \mathcal{L}(i)} \sum_{\tau=1}^{t-LT(\ell)} x_{\ell,\tau} + I(i,0) - \sum_{\tau=1}^{t} D(i,\tau) - \sum_{j=1}^{P} (R(i,j)x_{j,\tau}) .$$

So the production is done for one or more of the substitutes to meet the demand and requirements for the master. Since there is no production of master SKUs, the term $W(i,j)\delta_{j,\tau}$ is not needed in the macro for a master SKU. The original inventory macro can be used for the SKUs for the substitutes themselves as well as for SKUs that do not have substitutes. However, the inventory position variables for the substitutes, I^+ and I^-, must be fixed at zero. This will comply with the idea that inventory is held only as the master SKU. In the unusual situation where substitutes have significantly different holding costs, then additional binary variables must be used to indicate which of the substitutes is in inventory. This will not be developed here.

If costs differ for the substitutes, then terms must be added to the objective function. Once again, we are reminded of the fact that we are basing our models on marginal costs so the lowest cost substitute will have a cost of zero and the others will have cost that is the difference between their cost and the lowest cost substitute. The idea is that as soon as we accept the order a cost at least as high as the lowest cost substitute is sunk. The marginal costs of substitutes should include any significant difference that arise as a result of holding cost differences caused by widely varying lead times.

The additional data required is shown in Table 8.8. The full formulation for alternative routings as an extension to problem (SCPc) is given as (SCPA).

(SCPA) Minimize:

$$\sum_{t=1}^{T}\left[\sum_{i=1}^{P}\left(A(i)I_{i,t}^{-}+H(i)I_{i,t}^{+}+C(i)\delta_{i,t}+V(i)x_{i,t}\right)+\sum_{k=1}^{K}O(k,t)y_{k,t}\right]$$

subject to:

$$\begin{array}{rll}
I_{i,t}(x,\delta)+\sum_{\tau=1}^{t}D(i,\tau) & \geq 0 & i=1,\ldots,P,\ t=1,\ldots,T \\
\sum_{i=1}^{P}[U(i,k)x_{i,t}+S(i,k)\delta_{i,t}] & \leq 1+y_{k,t} & t=1,\ldots,T,\ k=1,\ldots,K \\
y_{k,t} & \leq F(k,t) & t=1,\ldots,T,\ k=1,\ldots,K \\
\delta_{i,t} & \geq \frac{x_{i,t}}{M} & i=1,\ldots,P,\ t=1,\ldots,T \\
\delta_{i,t} & \in \{0,1\} & i=1,\ldots,P,\ t=1,\ldots,T \\
x_{i,t} & \geq 0 & i=1,\ldots,P,\ t=1,\ldots,T \\
y_{k,t} & \geq 0 & t=1,\ldots,T,\ k=1,\ldots,K \\
I_{\ell,t}^{+} & =0 & \forall\, t,\ell\in\bigcup_{i=1,\ldots,P}\mathcal{L}(i) \\
I_{\ell,t}^{-} & =0 & \forall\, t,\ell\in\bigcup_{i=1,\ldots,P}\mathcal{L}(i) \\
I_{\ell,t}^{+} & \geq 0 & \forall\, t,\ell\notin\bigcup_{i=1,\ldots,P}\mathcal{L}(i) \\
I_{\ell,t}^{-} & \geq 0 & \forall\, t,\ell\notin\bigcup_{i=1,\ldots,P}\mathcal{L}(i) \\
I_{i,t}^{+}-I_{i,t}^{-} & =I_{i,t}(x,\delta) & i=1,\ldots,P,\ t=1,\ldots,T
\end{array}$$

where the shorthand notation $\forall$ and $\in$ is abused in the interest of brevity to mean that we want indexes for all appropriate values of the indexes listed. For example, $t,\ell\notin\bigcup_{i=1,\ldots,P}\mathcal{L}(i)$ means the constraints are repeated for $t=1,\ldots,T$ and for values of ℓ in the set formed by removing from $1,\ldots,P$ the indexes that appear in the list of substitute sets $\mathcal{L}$.

8.6 Hierarchical Planning Issues

The idea of dividing planning up into a hierarchy with longer range issues at the top, and more detail at the bottom is a well-established paradigm. Various detailed proposals have appeared in the production planning literature. For example, a hierarchical planning system was proposed by [31] for a specific type of control system known as CONWIP (see [33]), where CONWIP stands for "constant work in process." These systems proposed the use of a hierarchical system as a way of dividing the problem along sensible lines to improve the ability to solve the resulting problems, which is our goal as well. A good reference for hierarchical planning is [29]. A somewhat extended view of aggregation can also be found in [22].

8.6.1 Aggregation and Consolidation

The models described to this point will often be far too large to solve in a reasonable amount of time. The level of detail must be reduced in order to find a solution. The resolution can then be increased.

We will work with the concept of *aggregation* and *disaggregation.* Products and resources are grouped together and considered as single entities. Once a plan is found in terms of aggregate entities, a good plan with more detail can be created that corresponds to it. As with all modeling, decisions about how to group parts and resources is part art and part science.

Consolidating Resources. An important concept for consolidating resources is the notion of a *bottleneck.* The word is so overused that its meaning has become somewhat blurred, but typically serves to signify a limiting factor. Consider an extreme example. Suppose that a supply chain consists entirely of three resources and every SKU is routed through them in order. If for all SKUs, the first and third servers are always much faster and more reliable than the second, then the second server would clearly be the bottleneck. For many planning purposes there would be no need to include resources one and three in the model.

What effect do non-bottleneck resources have on the model? For non-bottleneck resources, k, constraints such as

$$\sum_{i=1}^{P} U(i,k)x_{i,t} \leq 1 + y_{k,t}$$
$$y_{k,t} \geq 0$$
$$y_{k,t} \leq F(k,t),$$

will not affect the solutions obtained. The value of $\sum_{i=1}^{P} U(i,k)x_{i,t}$ will always be less than one. The non-bottleneck resources make it harder to solve because there are variables and constraints associated with each resource. But the y variables are always zero and the constraints are never *binding.*

This constraint is said to bind when the left hand side is equal to the right hand side. One can see that a constraint cannot be affecting the solution unless it binds. Consider simple capacity constraints without overtime:

$$\sum_{i=1}^{P} U(i,k)x_{i,t} \leq 1$$

If $\sum_{i=1}^{P} U(i,k)x_{i,t}$ is strictly less than one for some particular resource, k, for all SKUs i and all times t, then some other constraint(s) and/or the objective function are determining the values of $x_{i,t}$ for all i and t. To put it another way: the solution will be the same with or without the constraints

$$\sum_{i=1}^{P} U(i,k)x_{i,t} \leq 1$$

for this particular k.

We will refer to the difference between the left hand side and the right hand side as the *slack*. For some purposes, it is useful to collect all of the variables on the left hand side before using the word, but for our purposes this is not helpful. We can define a bottleneck as a resource with no slack in some or all of its capacity constraints. A non-bottleneck is a resource with slack in all of its capacity constraints when written in either form:

$$\sum_{i=1}^{P} U(i,k)x_{i,t} \leq 1 + y_{k,t}$$

or

$$\sum_{i=1}^{P} U(i,k)x_{i,t} \leq 1$$

Resources that cannot conceivably become bottlenecks can safely be removed from the models that we have developed and the solutions will not be affected. However, the time required to find a solution will be reduced. The only trick is to know which resources would have no slack if they were in the model.

One possibility is to use intuition, engineering estimates and historical data to eliminate non-bottlenecks. Typically, the resources for which capacity constraints are binding (or overtime is needed) do change over time as the demand mix evolves. Obviously, the installation of new capacity or removal of resources can also change the location of the bottleneck. One must update the model to reflect changing conditions.

Another method of looking for capacity slack is to execute the model. This can also be used to supplement estimates and intuition. Of course, if the reason to reduce the number of resources in the model is that the model is too big, this can be problematic. Modern modelling languages provide features that can be helpful for trying the model with some resources included and then again with them excluded. In this way one can test suspicions of capacity limitations without the need to solve a model that is excessively large. More resources can be checked at the same time if the model size can be reduced via other means. The model size can also be reduced by aggregating SKUs which we now consider.

Aggregating Parts. Part aggregation is based on the idea of part *families* that share similar production characteristics. We use the symbol $\mathcal{F}(\ell)$ to refer to part family number ℓ. The set $\mathcal{F}(\ell)$ contains the list of part indexes that are in the family. We define the production quantity for an entire family, ℓ in period t to be $\hat{x}_{\ell,t}$. We could use any of the models that we have developed so far, but the x variables would be replaced by $\hat{x}$ with appropriate adjustments to the data.

If the variables x are used for the disaggregated plan, it is easy to see that:

$$\hat{x}_{\ell,t} = \sum_{i \in \mathcal{F}(\ell)} x_{i,t}$$

But usually, we will have solved the aggregated model to get values for $\hat{x}$ variables for the aggregate parts and we will want to estimate the corresponding x values for the individual SKUs. For this purpose we will use the data $Pr(i, \ell)$, which gives the fraction (or average fraction, or estimated fraction) of family ℓ production that is SKU i production. We can think of the letters Pr as representing either "proportion" or "probability" depending on the context. Once we have this data, and we have obtained values for $\hat{x}$, then we can compute the corresponding values for x using

$$x_{i,t} = Pr(i, \ell)\hat{x}_{\ell,t}.$$

By using data about past production quantities, we can estimate good values for Pr. We use the symbol $Avg(x_i)$ to refer to the average per period production of SKU i and $Avg(\hat{x}_\ell)$ for the average production quantities for the entire family of which SKU i is a member. A good estimate for Pr is then given by

$$Pr(i, \ell) = \frac{Avg(x_i)}{Avg(\hat{x}_\ell)}$$

if $i \in \mathcal{F}(\ell)$, and zero if $i \notin \mathcal{F}(\ell)$. These values can be adjusted if there are reasons to believe that the mixture within a family will change. If there is no historical production data for an SKU, the values of Pr must be based on an educated guess or marketing estimates.

We can use the values of Pr to switch back and forth between aggregated and disaggregated data and construct the requirements data between families. Consider families ℓ and ℓ'. We can write:

$$\begin{aligned} \hat{U}(\ell, k) &= \sum_{i \in \mathcal{F}_\ell} Pr(i, \ell)U(i, k) \\ \hat{R}(\ell, \ell') &= \sum_{i \in \mathcal{F}_\ell} \sum_{j \in \mathcal{F}(\ell')} Pr(j, \ell)R(i, j) \end{aligned}$$

and so forth for the other data values, except changeover data. This is intertwined with issues related to part family formation, so it is discussed in the next subsection.

Obviously, the quality of the solution found using aggregate SKUs will depend in large measure on the quality of the families. In an organization with thousands of SKUs assignments of parts to families can be a big undertaking. It is non-trivial even with hundreds of SKUs. Many organizations have grouped their parts into families for purposes other than using optimization models for production planning, but these families may work quite well for our purposes. Industrial and manufacturing engineers often create part families to facilitate process planning and computer aided (re)design of parts.

Some of the earliest work on systematic formation of part families was done by Soviet engineers in the later 1950's to support application of cellular manufacturing. Collection of a group of machines into a cell to process a family of similar parts remains an important topic. There are a number of algorithms available for dividing SKUs into groups to support manufacture by different cells [25]. Another important application of part families is computer assisted process planning (see, e.g., [21]). This results in a hierarchy of families driven by a part classification scheme. The classifications are done so that it is possible to reuse process plans for similar parts when a new part is designed. Such classification schemes therefore are based on manufacturing characteristics and can be very useful for family formation for the purposes.

Our application of part family formation was for the problem of aggregate planning. This has also been the subject of a significant research literature. In the 1970's researchers at the MIT collectively developed a planning system that they referred to as hierarchical production planning (HPP). The HPP system is designed to translate aggregate forecasts into part production requirements. The overall HPP approach is described and justified in a paper by Hax and Meal [14].

The hierarchy is based on scheduling parts in increasing levels of disaggregation. At the lowest level, parts are scheduled. Before that, families of parts are scheduled (intra-family changes do not require a setup, but inter-family changes do). At the highest level are *types* which are part families grouped according to similarities in production and demand. The scheduling of types is referred to as aggregate planning and the decomposition into family and part schedules is referred to as disaggregation. Bitran and Hax [6] suggested optimization sub-problems for the various levels and a rigorous means of linking them. We will now discuss simplified versions of the optimization sub-problems to make their methodology more concrete.

Aggregate planning is a cost minimization with respect to temporal variations for production to meet demand forecasts. There are constraints to assure that demand is met, periods are correctly linked via inventory, and capacity is not exceeded. Disaggregation to families is done with the objective of minimizing the total setup *cost*. A survey of disaggregation procedures is contained in [4]. The disaggregation to parts is done with the objective of minimizing setup cost subject to capacity feasibility, producing the quantities specified in the family disaggregation and keeping inventories within safety stock and overstock limits. Bitran, Haas, and Hax [5] have proposed a two stage version of the model which also schedules production of components.

Although this approach is demonstrably better than mrp in some circumstances (see [13]), it is not universally applicable. It does not consider due dates which can be an important factor as flow times drop and competitive pressures increase. Their work is perhaps most appropriate when production of components can and must be begun in response to forecasted demand. A more severe problem is that the minimization of setup "costs" is not generally appropriate. In some cases there are actual costs associated with a setup such as materials and supplies. But more generally, the costs are due to lost capacity and in fact depend on the schedule and part mix [18]. Thus, the actual costs depend on the detailed schedule, which is our next topic.

8.6.2 Detailed Scheduling

If we think of a model such as SCPc as being in the middle of the hierarchy, then aggregate planning is above it. Below it, in some sense, is detailed scheduling. Textbooks on the topic of scheduling include [3, 7, 27]. A survey of research concerning MIP modeling of changeovers in production planning and scheduling is provided in [43]. A survey of research on sequencing with earliness and tardiness penalties is provided by Baker and Scudder [2].

To create a concrete example, consider a situation where there are no out-of-pocket costs for a changeover and no wasted material so C and W are zero. However, suppose further that a significant amount of time is needed to change over some bottleneck resource, k', and this time depends on both the current job and the next job. We will reuse the symbol S for these changeover data. For simplicity, assume that each job is one piece. In such a situation, we might make use of a direct sequence representation as a basis for heuristic search.

In order to write the model in a concise way, we will make use of the fact that we can go back and forth between the familiar quantity representation, x_{it}, and a sequence representation where σ_j gives the job number of the jth job. We could modify the model as follows: For resource k' for each time t replace the constraints

$$\begin{aligned}\sum_{i=1}^{P}[U(i,k)x_{i,t}+S(i,k)\delta_{i,t}] &\leq 1+y_{k,t}\\ y_{k,t} &\leq F(k,t)\end{aligned}$$

$$y_{k,t} \geq 0$$

with the embedded problem

$$\text{minimize:} \quad y_{k',t}$$

subject to:

$$\begin{aligned}
\sum_{j=1}^{n} U(\sigma_j, k') + S(\sigma_{j-1}, \sigma_j, k') &\leq 1 + y_{k',t} \\
y_{k',t} &\leq F(k',t) \\
y_{k',t} &\geq 0
\end{aligned}$$

The data $S(\sigma_{j-1}, \sigma_j, k')$ could be an array giving the changeover times or a complicated function. For this to make sense, the value of σ_0 must be given as data or an assumed value must be used.

The main thing to notice is that constraints have been replaced with an optimization problem. In doing so, we have embedded a higher level of scheduling detail in the planning model. This modification will result in a model that will check for the existence of a feasible sequence that corresponds to the plan while at the same time seeking a good sequence. For a general changeover matrix or function, the sequencing problems are too hard to solve to proven optimality.

For such embedded problems, heuristic search methods are appropriate. These methods offer a great deal of flexibility, e.g., the changeover time can be a function rather than data at the expense of not finding a solution that is provably optimal. The flexibility also makes it possible to solve models that have non-zero C and W cost data, but the statement of the model is a little more complicated.

8.7 Conclusions

Enterprise resource planning is still in dire need of realistic planning functionality. In this chapter we have seen that we can begin with MRP II and end up with a useful basis for a planning model. In some sense, we must begin with MRP II because the requirements constraints and capacity constraints are a representation of physical reality. In order to produce a useful model, we ultimately arrive at models that bear little resemblance to mrp and certainly solutions for the optimization problems cannot be obtained using mrp or MRP II processing or logic. In that sense, we agree with those who suggest that we should abandon mrp completely. However, we have shown that it can be the basis for developing and understanding models that are much more useful, which provides a connection between ERP and OR roots.

References

[1] A. Atamtürk and D.S. Hochbaum. Capacity acquisition, subcontracting, and lot sizing. *Management Science*, 47:1081–1100, 2001.

[2] K.R. Baker and G.D. Scudder. Sequencing with earliness and tardiness penalties: A review. *Operations Research*, 38:22–36, 1990.

[3] K.R. Baker. *Introduction to Sequencing and Scheduling*. Wiley, New York, 1974.

[4] G.R. Bitran, E.A. Haas, and A.C. Hax. Hierarchical production planning: A single stage system. *Operations Research*, 29:717–743, 1981.

[5] G.R. Bitran, E.A. Haas, and A.C. Hax. Hierarchical production planning: A two stage system. *Operations Research*, 30:232–251, 1982.

[6] G.R. Bitran and A.C. Hax. On the design of hierarchical production planning systems. *Decision Sciences*, 8:28–54, 1977.

[7] J. Blazewicz, K.H. Ecker, E. Pesch, G. Schmidt, and J. Weglarz. *Scheduling Computer and Manufacturing Processes*. Springer, Berlin, 2 edition, 2001.

[8] P. Chandra and M.M. Tombak. Models for the evaluation of routing and machine flexibility. *European Journal of Operational Research*, 60:156–165, 1992.

[9] F. Chen. Stationary policies in multiechelon inventory systems with deterministic demand and backlogging. *Operations Research*, 46:26–34, 1998.

[10] A.R. Clark and V.A. Armentano. A heuristic for a resource-capacitated multi-stage lot-sizing problem with lead times. *Journal of the Operational Research Society*, 46:1208–1222, 1995.

[11] A. Drexl and A. Kimms. Lot sizing and scheduling – Survey and extensions. *European Journal of Operational Research*, 99:221–235, 1997.

[12] R. Gumaer. Beyond ERP and MRP II – Optimized planning and synchronized manufacturing. *IIE Solutions*, 28(9):32–35, 1996.

[13] A.C. Hax and D. Candea. *Production and Inventory Management*. Prentice-Hall, Englewood Cliffs, NJ, 1984.

[14] A.C. Hax and H.C. Meal. Hierarchical integration of production planning and scheduling. In M.A. Geisler, editor, *TIMS Studies in Management Science*, volume 1: Logistics, pages 53–69, New York, 1975. North Holland/American Elsevier.

[15] J. Hwang and M.R. Singh. Optimal production policies for multi-stage systems with setup costs and uncertain capacities. *Management Science*, 44:1279–1294, 1998.

[16] M.I. Kamien and L. Li. Subcontracting, coordination, flexibility, and production smoothing in aggregate planning. *Management Science*, 36:1352–1363, 1990.

[17] J.J. Kanet. Mrp 96: Time to rethink manufacturing logistics. *Production and Inventory Management Journal*, 29(2):57–61, 1988.

[18] U.S. Karmarkar. Lot sizes, lead times and in-process inventories. *Management Science*, 33:409–418, 1987.

[19] E. Katok, H.S. Lewis, and T.P. Harrison. Lot sizing in general assembly systems with setup costs, setup times, and multiple constrained resources. *Management Science*, 44:859–877, 1998.

[20] G. Knolmayer, P. Mertens, and A. Zeier. *Supply Chain Management based on SAP Systems*. Springer, Berlin, 2002.

[21] D.T. Koenig. *Manufacturing Engineering: Principles for Optimization.* Taylor & Francis, Washington, 2 edition, 1994.

[22] R. Leisten. An LP-aggregation view on aggregation in multi-level production planning. *Annals of Operations Research*, 82:413–434, 1998.

[23] R. Logendran and P. Ramakrishna. A methodology for simultaneously dealing with machine duplication and part subcontracting in cellular manufacturing systems. *Computers & Operations Research*, 24:97–116, 1997.

[24] M. Luscombe. *MRP II: Integrating the Business.* Butterworth-Heinemann, Oxford, 1993.

[25] J. Miltenburg and W. Zhang. A comparative evaluation of nine well-known algorithms for solving the cell formation problem in group technology. *Journal of Operations Management*, 10(1):44–72, 1991.

[26] J. Orlicky. *Material Requirements Planning: The New Way of Life in Production and Inventory Management.* McGraw-Hill, New York, 1975.

[27] M. Pinedo. *Scheduling: Theory, algorithms and systems.* Prentice Hall, Englewood Cliffs, 1995.

[28] R. Roundy. A 98%-effective lot-sizing rule for a multi-product, multi-stage production/inventory system. *Mathematics of Operations Research*, 11:699–727, 1986.

[29] C.A. Schneeweiss. *Hierarchies in Distributed Decision Making.* Springer, New York, 1999.

[30] N.C. Simpson and S.S. Erenguc. Multiple-stage production planning research: History and opportunities. *International Journal of Operations and Production Management*, 16(6):25–40, 1996.

[31] M.L. Spearman, W.J. Hopp, and D.L. Woodruff. A hierarchical control architecture for CONWIP production systems. *Journal of Manufacturing and Operations Management*, 2:147–171, 1989.

[32] M.L. Spearman and W.J. Hopp. Teaching operations management from a science of manufacturing. *Production and Operations Management*, 7(2):132–145, 1998.

[33] M.L. Spearman, D.L. Woodruff, and W.J. Hopp. CONWIP: A pull alternative to Kanban. *International Journal of Production Research*, 28:879–894, 1990.

[34] H. Stadtler and C. Kilger, editors. *Supply Chain Management and Advanced Planning*, Berlin, 2002. Springer.

[35] H. Tempelmeier and M. Derstroff. A Lagrangean-based heuristic for dynamic multilevel multiitem constrained lotsizing with setup times. *Management Science*, 42:738–757, 1996.

[36] W.W. Trigeiro, L.J. Thomas, and J.O. McClain. Capacitated lot sizing with setup times. *Management Science*, 35:353–366, 1989.

[37] J.A. van Mieghem. Coordinating investment, production, and subcontracting. *Management Science*, 45:954–971, 1999.

[38] T.E. Vollmann, W.L. Berry, and D.C. Whybark. *Manufacturing Planning and Control Systems.* Dow Jones-Irwin, Homewood, 2 edition, 1988.

[39] S. Voß and D.L. Woodruff. Supply chain planning: Is mrp a good starting point? In H. Wildemann, editor, *Supply Chain Management*, pages 177–203, München, 2000. TCW.

[40] S. Voß and D.L. Woodruff. *Introduction to Computational Optimization Models for Production Planning in a Supply Chain.* Springer, Berlin, 2003.

[41] T.F. Wallace. *MRP II: Making it Happen*. Oliver Wight Limited Publications, Essex Junction, VT, 1990.

[42] O.W. Wight. *Production and Inventory Management in the Computer Age*. Cahners, Boston, MA, 1974.

[43] L.A. Wolsey. MIP modelling of changeovers in production planning and scheduling problems. *European Journal of Operational Research*, 99:154–165, 1997.

Early Titles in the
INTERNATIONAL SERIES IN OPERATIONS RESEARCH & MANAGEMENT SCIENCE
Frederick S. Hillier, Series Editor, *Stanford University*

Saigal/ *A MODERN APPROACH TO LINEAR PROGRAMMING*
Nagurney/ *PROJECTED DYNAMICAL SYSTEMS & VARIATIONAL INEQUALITIES WITH APPLICATIONS*
Padberg & Rijal/ *LOCATION, SCHEDULING, DESIGN AND INTEGER PROGRAMMING*
Vanderbei/ *LINEAR PROGRAMMING*
Jaiswal/ *MILITARY OPERATIONS RESEARCH*
Gal & Greenberg/ *ADVANCES IN SENSITIVITY ANALYSIS & PARAMETRIC PROGRAMMING*
Prabhu/ *FOUNDATIONS OF QUEUEING THEORY*
Fang, Rajasekera & Tsao/ *ENTROPY OPTIMIZATION & MATHEMATICAL PROGRAMMING*
Yu/ *OR IN THE AIRLINE INDUSTRY*
Ho & Tang/ *PRODUCT VARIETY MANAGEMENT*
El-Taha & Stidham/ *SAMPLE-PATH ANALYSIS OF QUEUEING SYSTEMS*
Miettinen/ *NONLINEAR MULTIOBJECTIVE OPTIMIZATION*
Chao & Huntington/ *DESIGNING COMPETITIVE ELECTRICITY MARKETS*
Weglarz/ *PROJECT SCHEDULING: RECENT TRENDS & RESULTS*
Sahin & Polatoglu/ *QUALITY, WARRANTY AND PREVENTIVE MAINTENANCE*
Tavares/ *ADVANCES MODELS FOR PROJECT MANAGEMENT*
Tayur, Ganeshan & Magazine/ *QUANTITATIVE MODELS FOR SUPPLY CHAIN MANAGEMENT*
Weyant, J./ *ENERGY AND ENVIRONMENTAL POLICY MODELING*
Shanthikumar, J.G. & Sumita, U./ *APPLIED PROBABILITY AND STOCHASTIC PROCESSES*
Liu, B. & Esogbue, A.O./ *DECISION CRITERIA AND OPTIMAL INVENTORY PROCESSES*
Gal, T., Stewart, T.J., Hanne, T. / *MULTICRITERIA DECISION MAKING: Advances in MCDM Models, Algorithms, Theory, and Applications*
Fox, B.L. / *STRATEGIES FOR QUASI-MONTE CARLO*
Hall, R.W. / *HANDBOOK OF TRANSPORTATION SCIENCE*
Grassman, W.K. / *COMPUTATIONAL PROBABILITY*
Pomerol, J-C. & Barba-Romero, S. / *MULTICRITERION DECISION IN MANAGEMENT*
Axsäter, S. / *INVENTORY CONTROL*
Wolkowicz, H., Saigal, R., & Vandenberghe, L. / *HANDBOOK OF SEMI-DEFINITE PROGRAMMING: Theory, Algorithms, and Applications*
Hobbs, B.F. & Meier, P. / *ENERGY DECISIONS AND THE ENVIRONMENT: A Guide to the Use of Multicriteria Methods*
Dar-El, E. / *HUMAN LEARNING: From Learning Curves to Learning Organizations*
Armstrong, J.S. / *PRINCIPLES OF FORECASTING: A Handbook for Researchers and Practitioners*
Balsamo, S., Personé, V., & Onvural, R./ *ANALYSIS OF QUEUEING NETWORKS WITH BLOCKING*
Bouyssou, D. et al. / *EVALUATION AND DECISION MODELS: A Critical Perspective*
Hanne, T. / *INTELLIGENT STRATEGIES FOR META MULTIPLE CRITERIA DECISION MAKING*
Saaty, T. & Vargas, L. / *MODELS, METHODS, CONCEPTS and APPLICATIONS OF THE ANALYTIC HIERARCHY PROCESS*
Chatterjee, K. & Samuelson, W. / *GAME THEORY AND BUSINESS APPLICATIONS*
Hobbs, B. et al. / *THE NEXT GENERATION OF ELECTRIC POWER UNIT COMMITMENT MODELS*
Vanderbei, R.J. / *LINEAR PROGRAMMING: Foundations and Extensions, 2nd Ed.*
Kimms, A. / *MATHEMATICAL PROGRAMMING AND FINANCIAL OBJECTIVES FOR SCHEDULING PROJECTS*
Baptiste, P., Le Pape, C. & Nuijten, W. / *CONSTRAINT-BASED SCHEDULING*
Feinberg, E. & Shwartz, A. / *HANDBOOK OF MARKOV DECISION PROCESSES: Methods and Applications*
Ramík, J. & Vlach, M. / *GENERALIZED CONCAVITY IN FUZZY OPTIMIZATION AND DECISION ANALYSIS*
Song, J. & Yao, D. / *SUPPLY CHAIN STRUCTURES: Coordination, Information and Optimization*
Kozan, E. & Ohuchi, A. / *OPERATIONS RESEARCH/ MANAGEMENT SCIENCE AT WORK*
Bouyssou et al. / *AIDING DECISIONS WITH MULTIPLE CRITERIA: Essays in Honor of Bernard Roy*

Early Titles in the
INTERNATIONAL SERIES IN
OPERATIONS RESEARCH & MANAGEMENT SCIENCE
(Continued)

Cox, Louis Anthony, Jr. / *RISK ANALYSIS: Foundations, Models and Methods*
Dror, M., L'Ecuyer, P. & Szidarovszky, F. / *MODELING UNCERTAINTY: An Examination of Stochastic Theory, Methods, and Applications*
Dokuchaev, N. / *DYNAMIC PORTFOLIO STRATEGIES: Quantitative Methods and Empirical Rules for Incomplete Information*
Sarker, R., Mohammadian, M. & Yao, X. / *EVOLUTIONARY OPTIMIZATION*
Demeulemeester, R. & Herroelen, W. / *PROJECT SCHEDULING: A Research Handbook*
Gazis, D.C. / *TRAFFIC THEORY*

*** A list of the more recent publications in the series is at the front of the book ***